181 Topics in Current Chemistry

Springer-Verlag Berlin Heidelberg GmbH

Density Functional Theory II

Relativistic and Time Dependent Extensions

Volume Editor: R. F. Nalewajski

With contributions by
F.A. Buot, R.M. Dreizler, J.F. Dobson,
E. Engel, E.K.U. Gross,
M. Petersilka, A.K. Rajagopal

With 18 Figures and 12 Tables

Springer

This series presents critical reviews of the present position and future trends in modern chemical research. It is addressed to all research and industrial chemists who wish to keep abreast of advances in the topics covered.

As a rule, contributions are specially commissioned. The editors and publishers will, however, always be pleased to receive suggestions and supplementary information. Papers are accepted for "Topics in Current Chemistry" in English.

In references Topics in Current Chemistry is abbreviated Top.Curr.Chem. and is cited as a journal.

Springer WWW home page: http://www.springer.de

ISSN 0340-1022
ISBN 978-3-642-08249-8 ISBN 978-3-540-49946-6 (eBook)
DOI 10.1007/978-3-540-49946-6

Library of Congress Catalog Card Number 74-644622

Originally published by Springer-Verlag Berlin Heidelberg New York in 1996
Softcover reprint of the hardcover 1st edition 1996

Typesetting: Macmillan India Ltd., Bangalore-25
SPIN: 10509056 66/3020 – 5 4 3 2 1 0 – Printed on acid-free paper

Volume Editor

Prof. R. F. Nalewajski
Jagiellonian University
Faculty of Chemistry
ul. Ingardena 3
30-060 Krakow, Poland

Editorial Board

Foreword

Density functional theory (DFT) is an entrancing subject. It is entrancing to chemists and physicists alike, and it is entrancing for those who like to work on mathematical physical aspects of problems, for those who relish computing observable properties from theory and for those who most enjoy developing correct qualitative descriptions of phenomena in the service of the broader scientific community.

DFT brings all these people together, and DFT needs all of these people, because it is an immature subject, with much research yet to be done. And yet, it has already proved itself to be highly useful both for the calculation of molecular electronic ground states and for the qualitative description of molecular behavior. It is already competitive with the best conventional methods, and it is particularly promising in the applications of quantum chemistry to problems in molecular biology which are just now beginning. This is in spite of the lack of complete development of DFT itself. In the basic researches in DFT that must go on, there are a multitude of problems to be solved, and several different points of view to find full expression.

Thousands of papers on DFT have been published, but most of them will become out of date in the future. Even collections of works such as those in the present volumes, presentations by masters, will soon be of mainly historic interest. Such collections are all the more important, however, when a subject is changing so fast as DFT is. Ative workers need the discipline imposed on them by being exposed to the works of each other. New workers can lean heavily on these sources to learn the different viewpoints and the new discoveries. They help allay the difficulties associated with the fact that the literature is in both physics journals and chemistry journals. [For the first two-thirds of my own scientific career, for example, I felt confident that I would miss nothing important if I very closely followed the Journal of Chemical Physics. Most physicists, I would guess, never felt the need to consult JCP. What inorganic or organic chemist in the old days took the time to browse in the physics journals?] The literature of DFT is half-divided, and DFT applications are ramping into chemical and physical journals, pure and applied. Watch JCP, Physical Review A and Physical Review B, and watch even Physical Review Letters, if you are a chemist interested in applying DFT. Or ponder the edited volumes, including the present two. Then you will not be surprised by the next round of improvements in DFT methods. Improvements are coming.

The applications of quantum mechanics to molecular electronic structure may be regarded as beginning with Pauling's Nature of the Chemical Bond, simple molecular orbital ideas, and the Huckel and Extended Huckel Methods. The molecular orbital method then was systematically quantified in the Hartree-Fock SCF Method; at about the same time, its appropriateness for chemical description reached its most elegant manifestation in the analysis by Charles Coulson of the Huckel method. Chemists interested in structure learned and taught the nature of the Hartree-Fock orbital description and the importance of electron correlation in it. The Hartree-Fock single determinant is only an approximation. Configurations must be mixed to achieve high accuracy. Finally, sophisticated computational programs were developed by the professional theoreticians that enabled one to compute anything. Some good methods involve empirical elements, some do not, but the road ahead to higher and higher accuracy seemed clear: Hartree-Fock plus correction for electron correlation. Simple concepts in the everyday language of non theoretical chemists can be analyzed (and of course have been much analyzed) in this context.

Then, however, something new came along, density functional theory. This is, of course, what the present volumes are about. DFT involves a profound change in the theory. We do not have merely a new computational gimmick that improves accuracy of calculation. We have rather a big shift of emphasis. The basic variable is the electron density, not the many-body wavefunction. The single determinant of interest is the single determinant that is the exact wavefuntion for a noninteracting (electron-electron repulsion-less) system corresponding to our particular system of interest, and has the same electron density as our system of interest. This single determinant, called the Kohn-Sham single determinant, replaces the Hartree-Fock determinant as the wavefunction of paramount interest, with electron correlation now playing a lesser role than before. It affects the potential which occurs in the equation which determines the Kohn-Sahm orbitals, bus once that potential is determined, there is no configurational mixing or the like required to determine the accurate electron density and the accurate total electronic energy. Hartree-Fock orbitals and Kohn-Sham orbitals are quantitatively very similar, it has turned out. Of the two determinants, the one of Kohn-Sham orbitals is mathematically more simple than the one of Hartree-Fock orbitals. Thus, each KS orbital has its own characteristic asymptotic decay; HF orbitals all share in the same asymptotic decay. The highest KS eigenvalue is the exact first ionization potential; the highest HF eigenvalue is an approximation to the first ionization potential. The KS effective potential is a local multiplicative potential; the HF potential is nonlocal and nonmultiplicative. And so on. When at the Krakow meeting I mentioned to a physicist that I thought that chemists and physicists all should be urged to adopt the KS determinant as the basic descriptor for electronic structure, he quickly replied that the physicists had already done so. So, I now offer that suggestion to the chemistry community.

On the conceptual side, the powers of DFT have been shown to be considerable. Without going into detail, I mention only that the Coulson work referred to above anticipated in large part the formal manner in which DFT describes molecular changes, and that the ideas of electronegativity and hardness fall into place, as do Ralph Pearson's HSAB and Maximum Hardness Principles.

It was Mel Levy, I think who first called density functional theory a charming subject. Charming it certainly is to me. Charming it should be revealed to you as you read the diverse papers in these volumes.

Chapel Hill, 1996

Robert G. Parr

Foreword

Thirty years after Hohenberg and myself realized the simple but important fact that the theory of electronic structure of matter can be rigorously based on the electronic density distribution *n(r)* a most lively conference was convened by Professor R. Nalewajski and his colleagues at the Jagiellonian University in Poland's historic capital city, Krakow. The present series of volumes is an outgrowth of this conference.

Significantly, attendees were about equally divided between theoretical physicists and chemists. Ten years earlier such a meeting would not have had much response from the chemical community, most of whom, I believe, deep down still felt that density functional theory (DFT) was a kind of mirage. Firmly rooted in a tradition based on Hartree Fock wavefunctions and their refinements, many regarded the notion that the many electron function, $\Psi(r_1 \dots r_N)$ could, so to speak, be traded in for the density *n(r)*, as some kind of not very serious slight-of-hand. However, by the time of this meeting, an attitudinal transformation had taken place and both chemists and physicists, while clearly reflecting their different upbringings, had picked up DFT as both a fruitful viewpoint and a practical method of calculation, and had done all kinds of wonderful things with it.

When I was a young man, Eugene Wigner once said to me that understanding in science requires understanding from *several different points of view.* DFT brings such a new point of view to the table, to wit that, in the ground state of a chemical or physical system, the electrons may be regarded as a *fluid* which is fully characterized by its density distribution, *n(r)*. I would like to think that this viewpoint has enriched the theory of electronic structure, including (via potential energy surfaces) molecular structure; the chemical bond; nuclear vibrations; and chemical reactions.

The original emphasis on electronic ground states of non-magnetic systems has evolved in many different directions, such as thermal ensembles, magnetic systems, time-dependent phenomena, excited states, and superconductivity. While the abstract underpinning is exact, implementation is necessarily approximate. As this conference clearly demonstrated, the field is vigorously evolving in many directions: rigorous sum rules and scaling laws; better understanding and description of correlation effects; better understanding of chemical principles and phenomena in terms of *n(r)*; application to systems consisting of thousands of atoms; long range polarization energies; excited states.

Here is my personal wish list for the next decade: (1) An improvement of the accurary of the exchange-correlation energy $E_{xc}[n(r)]$ by a factor of 3-5. (2) A practical, systematic scheme which, starting from the popular local density approach, can – with sufficient effort – yield electronic energies with any specified accuracy. (3) A sound DFT of excited states with an accuracy and practicality comparable to present DFT for ground states. (4) A practical scheme for calculating electronic properties of systems of 10^3 - 10^5 atoms with "chemical accuracy". The great progress of the last several years made by many individuals, as mirrored in these volumes, makes me an optimist.

Santa Barbara, 1996

Walter Kohn

Preface

Density functional methods emerged in the early days of quantum mechanics; however, the foundations of the modern density functional theory (DFT) were established in the mid 1960 with the classical papers by Hohenberg and Kohn (1964) and Kohn and Sham (1965). Since then impressive progress in extending both the theory formalism and basic principles, as well as in developing the DFT computer software has been reported. At the same time, a substantial insight into the theory structure and a deeper understanding of reasons for its successes and limitations has been reached. The recent advances, including new approaches to the classical Kohn-Sham problem and constructions of more reliable functionals, have made the ground-state DFT investigations feasible also for very large molecular and solid-state systems (of the order of 10^3 atoms), for which conventional CI calculations of comparable accuracy are still prohibitively expensive. The DFT is not free from difficulties and controversies but these are typical in a case of a healthy, robust discipline, still in a stage of fast development. The growing number of monographs devoted to this novel treatment of the quantum mechanical many body problem is an additional measure of its vigor, good health and the growing interest it has attracted.

In addition to a traditional, solid-state domain of appplications, the density functional approach also has great appeal to chemists due to both computational and conceptual reasons. The theory has already become an important tool within quantum chemistry, with the modern density functionals allowing one to tackle problems involving large molecular systems of great interest to experimental chemists. This great computational potential of DFT is matched by its already demonstrated capacity to both rationalize and quantify basic classical ideas and rules of chemistry, e.g., the electronegativity and hardness/softness characteristics of the molecular electron distribution, bringing about a deeper understanding of the nature of the chemical bond and various reactivity preferences. The DFT description also effects progress in the theory of chemical reactivity and catalysis, by offering a "thermodynamic-like" perspective on the electron cloud reorganization due to the reactant/catalyst presence at various intermediate stages of a reaction, e.g. allowing one to examine the relative importance of the polarization and charge transfer components in the resultant reaction mechanism, to study the influence of the infinite surface reminder of cluster models of heterogeneous catalytic systems, etc.

The 30th anniversary of the modern DFT was celebrated in June 1994 in Cracow, where about two hundred scientists gathered at the ancient Jagiellonian University Robert G. Parr were the honorary chairmen of the conference. Most of the reviewers of these four volumes include the plenary lecturers of this symposium; other leading contributors to the field, physicists and chemists, were also invited to take part in this DFT survey. The fifteen chapters of this DFT series cover both the basic theory (Parts I, II, and the first article of Part III), applications to atoms, molecules and clusters (Part III), as well as the chemical reactions and the DFT rooted theory of chemical reactivity (Part IV). This arrangement has emerged as a compromise between the volume size limitations and the requirements of the maximum thematic unity of each part.

In these four DFT volumes of the *Topics in Current Chemistry* series, a real effort has been made to combine the authoritative reviews by both chemists and physicists, to keep in touch with a wider spectrum of current developments. The Editor deeply appreciates a fruitful collaboration with Dr. R. Stumpe, Dr. M. Hertel and Ms B. Kollmar-Thoni of the Springer-Verlag Heidelberg Office, and the very considerable labour of the Authors in preparing these interesting and informative monographic chapters.

Cracow, 1996 Roman F. Nalewajski

Table of Contents

Table of Contents of Volume 180

Density Functional Theory I: Functionals and Effective Potentials

Table of Contents of Volume 182

Density Functional Theory III: Interpretation, Atoms, Molecules and Clusters

Table of Contents of Volume 183

Density Functional Theory IV: Theory of Chemical Reactivity

Relativistic Density Functional Theory

E. Engel and R. M. Dreizler

Institut für Theoretische Physik, Universität Frankfurt, Robert Mayer Str. 8–10, 60054 Frankfurt/Main, Germany

Table of Contents

An overview of relativistic density functional theory covering its foundations, the construction of explicit functionals and applications to spherical atoms is given. After a brief summary of the relevant field theoretical background we discuss the Hohenberg-Kohn theorem for quantum electrodynamical systems as well as the corresponding Kohn-Sham equations, emphasising in particular the renormalisation of ground state energies and currents required. We then outline the transition from the full quantum electrodynamical Kohn-Sham equations to the more practical variants which are actually used in applications. As an extension of the Kohn-Sham equations we also summarise the relativistic optimised-potential-method (OPM) which, in addition to the kinetic energy, also treats the exchange energy on the basis of Kohn-Sham orbitals. As far as the construction of explicit functionals is concerned, we review the local density approximation (LDA) and the weighted density approximation (WDA) for the exchange-correlation energy as well as the gradient expansion of the kinetic energy, again addressing in detail questions of renormalisation. The relativistic corrections to the ground state, single particle and exchange energies as well as exchange potentials of atoms are then examined within the exchange-only limit of the no-sea approximation to the full relativistic Kohn-Sham equations, comparing the LDA and the WDA with the results obtained by the relativistic OPM. In addition, we investigate transverse exchange and correlation contributions within the LDA by comparison with quantum chemical data.

1 Introduction

Density functional theory (DFT) of nonrelativistic many-particle systems has progressed steadily over the last thirty years, so that it is now able to compete successfully with more traditional many body techniques, as e.g. configuration interaction (CI) methods, for the ab initio determination of *ground state* properties [1, 2, 3, 4]. The CI approach involves a rather straightforward concept, whose application is, however, limited by technical aspects (computer speed and memory) to systems with modest particle numbers. In the DF approach the first task is to find a sufficiently accurate representation of the ground state energy and other ground state observables as functionals of the ground state density. If such a representation is available (at least for a wider class of systems as e.g. finite Coulomb systems) further application is much simpler and can be carried through for larger systems.

For the application of DFT two avenues are possible. The first is the exploitation of a direct variational principle [5] with the density (or some suitable extension as spin-up and spin-down densities) as the basic variational variable. With this approach, usually referred to as extended Thomas-Fermi

(ETF) models, no results of high accuracy have been obtained. The second, the Kohn-Sham (KS) scheme [6], relies on a representation of the density in terms of single-particle orbitals. In this sense it is akin to the Hartree-Fock (HF) approach. The difference is, however, that the KS orbitals experience a local (i.e. multiplicative) single-particle potential, which also includes correlation effects (at least to some degree, depending on the state of art). In the last years rather refined approximations for this potential have been developed (as e.g. generalised gradient approximations – GGAs), which yielded among others results that compared favourably with conventional many-body methods in recent comparative studies for a variety of small molecules [2, 3].

These quantum chemical studies were restricted to rather light elements [2, 3], so that relativistic effects did not play any role. However, as soon as heavier atoms are involved (with Au being the prime candidate) they become essential for understanding the physical and chemical properties of atoms and molecules. Beyond the well known kinematic effects as the contraction or expansion of relativistic orbitals and the corresponding bond length modifications in molecules, also the nonadditivity of relativistic and exchange-correlation (xc) effects has recently been emphasised [7] (for a detailed account of the importance of relativity for chemical bonding we refer the reader to Refs. [8, 9]). Clearly, a DFT description of such high-Z systems also has to reflect their relativistic character, in particular, as even for atoms with moderate Z relativistic contributions to the xc-energy are larger than the differences between the most refined nonrelativistic xc-energy functionals [10, 11].

The development of a DF approach to relativistic many-particle systems has, however, been much slower than that of its nonrelativistic counterpart, although a precursor of modern relativistic DFT in the form of the relativistic Thomas-Fermi (TF) model [12, 13] was established with only a minor time delay with respect to the nonrelativistic TF approach [14, 15]. This and quite a number of subsequent attempts to incorporate relativity in explicit density functionals [16] were based on the first quantised Dirac equation, which is not an appropriate basis for the discussion of the full scale of relativistic effects in many-particle systems. If one wishes to address photon retardation effects and antiparticle contributions besides the relativistic kinematics of the electrons the correct basis for the discussion of many-electron systems is quantum electrodynamics (QED). A comparison of the general structure (and in particular of the diagrammatic formulation) of QED and of nonrelativistic many-body theory reveals definite similarities, but the elimination of infinite zero point energies, the renormalisation of ultraviolet (UV) divergencies as well as the spinor and tensor structure of the quantities involved introduces substantial complications in the former case. These additional features are reflected in relativistic DFT (RDFT), both on the more formal level and in the derivation of explicit functionals. An extension of the Hohenberg-Kohn (HK) theorem [5] to the relativistic domain has first been formulated by Rajagopal and Callaway [17], showing that in a fully relativistic DF formalism the four current plays the role of the density as the basic DF variable. The corresponding KS-equations have been given by Rajagopal [18] and independently by MacDonald and Vosko [19]. These authors mainly focussed on the Dirac structure of the relativistic KS (RKS) equations (including the four-component form of the xc-potential) and the retardation corrections to the Coulomb interaction, but essentially neglected the fundamental questions related to radiative corrections. UV-divergencies first showed up in the RDFT context in the derivation of gradient corrections to the kinetic energy [16, 20, 21], but only very recently an attempt has been made [22] to explicitly deal with the issue of UV-renormalisation in the context of the relativistic HK-theorem and the RKS-equations. As a consequence of the complicated structure of the RKS-equations applications beyond their most simple version, the Dirac-Fock-Slater (DFS) approach [23] or its extensions [24], in which all relativistic effects in the xc-energy (as well as all radiative corrections) are neglected, are rare (compare Section 5).

In this review we attempt to give a summary of the still restricted knowledge of RDFT for Coulomb systems (for an earlier review see [25]). Activity in the field of hadronic physics is not covered (we refer the interested reader to the relevant section in [22]). We begin with an outline of the background material in Section 2, but have chosen to relegate nearly all details into the Appendices. The foundations of RDFT, i.e. the HK-theorem, the KS-scheme, its extension in the form of the optimised potential method (OPM) and some remarks on the nonrelativistic limit, are addressed in Section 3. We emphasise, in particular, practical aspects concerning the application of the relativistic KS-scheme. Explicit energy functionals for relativistic Coulomb systems are discussed in Section 4, where we outline the local density approximation (LDA) for exchange and, as far as they are known, for correlation contributions. In addition, we indicate first attempts to apply the weighted density approximation (WDA) in the relativistic regime. Results for the solution of the relativistic KS-equations for atoms are analysed in Section 5. To our knowledge no corresponding results for molecules are yet available. Some concluding remarks indicating open problems and possible extensions (e.g. for thermal relativistic systems) are the contents of the final Section 6.

As indicated, we provide a more detailed summary of the background material for the discussion of RDFT in the Appendices. In particular, we address the question of perturbative renormalisation in vacuum quantum electrodynamics (Appendix A), discuss the properties of the relativistic homogeneous electron gas (RHEG) (Appendix B) in order to provide the input for the relativistic LDA (RLDA) and add some remarks on QED in the presence of external fields (Appendix C). In addition, Appendix D gives some information concerning the systematic derivation of inhomogeneity corrections to the RLDA, illustrating in particular the gradient expansion (GE) on the basis of linear response. Although we do not present relativistic ETF-results in this review, we briefly summarise (Appendix E) the derivation of relativistic kinetic energy functionals via a semiclassical expansion of the electron propagator as it demonstrates the occurrence and subsequent renormalisation of UV-divergencies in a transparent fashion.

We use $\hbar = c = 1$ throughout the manuscript (except for Section 3.4) and our metric is chosen as in Ref. [26], i.e. the space-time coordinates x are given by $x^\mu = (x^0 = t, \boldsymbol{x})$ with $\boldsymbol{x}$ (or $\boldsymbol{r}$) being the usual position in space. Greek indices run from 0 to 3, Latin indices from 1 to 3. The summation convention is used throughout. The abbreviations used for the various functionals in this work are based on the following scheme: All relativistic functionals are labelled beginning with R (like RLDA for relativistic LDA), while their nonrelativistic counterparts start with NR. The same notation is used for the resulting computational schemes, e.g. NRLDA abbreviates results obtained by utilizing the nonrelativistic LDA functional in the non-relativistic KS-equations. The only exception to this rule is the Dirac-Fock-Slater approach (with $\alpha = 2/3$) characterised by DFS, which corresponds to the use of the nonrelativistic LDA for the exchange energy in the relativistic KS-equations. Finally, all differences between relativistic and nonrelativistic results are referenced by the generic name of the approximation, e.g. LDA.

2 Field Theoretical Background

An appropriate basis for a complete discussion of relativistic effects in many-electron systems (atoms, molecules, clusters, solids) is quantum electrodynamics. In view of the large difference between the electron and the nuclear masses it is legitimate to treat the nuclei as fixed external sources[1]. This standard approximation relies on the assumption of a common rest frame for all nuclei and thus partially breaks the covariance of the resulting many body theory. A system of Dirac particles, which interact by the exchange of photons and move in a given static external electromagnetic field $V^\mu(\boldsymbol{x})$ (representing the fixed nuclei–and additional fields if present) is characterised by a Lagrangian density of the form [27]

$$\mathcal{L}(x) = \mathcal{L}_e(x) + \mathcal{L}_\gamma(x) + \mathcal{L}_{int}(x). \tag{2.1}$$

The three terms represent the free Dirac Lagrangian[2] of the fermions (electrons and positrons),

$$\mathcal{L}_e(x) = \tfrac{1}{4}\{[\hat{\bar{\psi}}(x), (i\overrightarrow{\not\partial} - m)\hat{\psi}(x)] + [\hat{\bar{\psi}}(x)(-i\overleftarrow{\not\partial} - m), \hat{\psi}(x)]\}, \tag{2.2}$$

the free photon Lagrangian,

$$\mathcal{L}_\gamma(x) = -\frac{1}{16\pi}\hat{F}_{\mu\nu}(x)\hat{F}^{\mu\nu}(x) - \frac{\lambda}{8\pi}(\partial_\nu \hat{A}^\nu(x))^2, \tag{2.3}$$

---

[1] In a truly covariant QED-approach to atoms and molecules both the nuclei and the electrons would have to be treated as dynamical degrees of freedom (at least on a classical level in the case of the nuclei).

[2] The vector bars on top of the partial derivatives indicate the direction in which the derivative has to be taken, i.e. in the second term of $\mathcal{L}_e$ the partial derivatives act on $\bar{\psi}(x)$.

and the interaction term

$$\mathscr{L}_{int}(x) = -e\hat{j}^{\nu}(x)(\hat{A}_{\nu}(x) + V_{\nu}(\boldsymbol{x})). \tag{2.4}$$

The operators $\hat{\psi}(x)$ and $\hat{A}_{\mu}(x)$ are the fermion and photon field operators,

$$\hat{F}_{\mu\nu}(x) = \partial_{\mu}\hat{A}_{\nu}(x) - \partial_{\nu}\hat{A}_{\mu}(x), \tag{2.5}$$

and

$$\hat{j}^{\nu}(x) = \frac{1}{2}[\hat{\bar{\psi}}(x), \gamma^{\nu}\hat{\bar{\psi}}(x)] \tag{2.6}$$

is the fermion four current operator. For the photon fields we have chosen to work in the covariant gauge so that we had to introduce the gauge fixing term [27, 26]

$$-\frac{\lambda}{8\pi}(\partial_{\nu}\hat{A}^{\nu}(x))^2$$

and to use the Gupta-Bleuler indefinite metric quantisation. For brevity we shall often restrict explicit formulae to some particular gauge, as e.g. the Feynman gauge $\lambda = 1$ or Landau gauge $\lambda = \infty$. For the external potential we use the gauge $\partial_i V^i(\boldsymbol{x}) = \nabla \cdot \boldsymbol{V}(\boldsymbol{x}) = 0$.

One of the most important properties of the Lagrangian (2.1) is its gauge invariance: A gauge transformation of the photon field,

$$\hat{A}_{\nu}(x) \rightarrow \hat{A}'_{\nu}(x) = \hat{A}_{\nu}(x) + \partial_{\nu}\Lambda(x); \quad \partial_{\mu}\partial^{\mu}\Lambda(x) = 0, \tag{2.7}$$

can be absorbed by an accompanying phase transformation of the fermion field operator

$$\hat{\psi}(x) \rightarrow \hat{\psi}'(x) = \exp[-ie\Lambda(x)]\hat{\psi}(x), \tag{2.8}$$

leaving the Lagrangian (2.1) invariant,

$$\mathscr{L}(\hat{\psi}', \hat{A}') = \mathscr{L}(\hat{\psi}, \hat{A}). \tag{2.9}$$

On the other hand, due to the choice of a particular Lorentz frame the gauge invariance of $\mathscr{L}$ with respect to gauge transformations of the external potential has been partially broken: Only static gauge transformations,

$$V'_{\nu}(\boldsymbol{x}) = V_{\nu}(\boldsymbol{x}) + \partial_{\nu}\Lambda(t, \boldsymbol{x}) \tag{2.10}$$

$$\hat{\psi}'(x) = \exp[-ie\Lambda(t, \boldsymbol{x})]\hat{\psi}(x) \tag{2.11}$$

$$\Lambda(t, \boldsymbol{x}) = Ct + \lambda(\boldsymbol{x}); \quad \Delta\lambda(\boldsymbol{x}) = 0, \tag{2.12}$$

are admitted within the common rest frame of the nuclei. In addition to the Lagrangian the four current $\hat{j}^{\mu}(x)$, Eq. (2.6), is also invariant under the transformations (2.7, 2.8) and (2.10–2.12).

Moreover, both the Lagrangian and the electronic four current have been written in a charge conjugation invariant form [27], i.e. under the charge conjugation $\hat{\mathscr{C}}$ (transforming electrons into positrons and vice versa) the four current (as a charge current rather than a probability current) changes its sign,

$$\hat{\mathscr{C}}\hat{j}^\nu(x)\hat{\mathscr{C}}^+ = -\hat{j}^\nu(x), \tag{2.13}$$

while in $\mathscr{L}$ the fermion charge manifests itself in the coupling to external sources,

$$\hat{\mathscr{C}}\mathscr{L}[V^\nu]\hat{\mathscr{C}}^+ = \mathscr{L}[-V^\nu], \tag{2.14}$$

i.e. an external potential which attracts electrons repels positrons.

It is useful for the following discussion to consider the symmetries of the Lagrangian (2.1) in order to analyse the conservation laws of a system characterised by (2.1) on the most general level, i.e. without further specifying V^μ, and their consequences for the structure of a density functional approach to (2.1). We first consider continuous symmetries which in the field theoretical context are usually discussed on the basis of Noether's theorem (see e.g. [26, 28]). The most obvious symmetry of the Lagrangian (2.1), its gauge invariance (2.9), directly reflects current conservation,

$$\partial_\nu \hat{j}^\nu(x) = 0, \tag{2.15}$$

and thus conservation of the total charge,

$$\hat{Q} = \int d^3x \hat{j}^0(x) = \frac{1}{2}\int d^3x [\hat{\bar{\psi}}^+(x), \hat{\psi}(x)]. \tag{2.16}$$

As a consequence any ground state resulting from (2.1) can be classified with respect to its charge (but not particle number).

Energy and momentum conservation can be directly deduced from the 'continuity' equation for the energy momentum tensor[3] [27, 28]. For the $\hat{T}^{\mu\nu}$ resulting from (2.1) one finds

$$\begin{aligned}\hat{T}^{\mu\nu}(x) = {} & \frac{i}{8}[\hat{\bar{\psi}}(x), (\gamma^\mu\vec{\partial}^\nu + \gamma^\nu\vec{\partial}^\mu - \gamma^\mu\overleftarrow{\partial}^\nu - \gamma^\nu\overleftarrow{\partial}^\mu)\hat{\psi}(x)] \\ & + \frac{1}{4\pi}\Big\{\hat{F}^{\mu\rho}(x)\hat{F}_\rho{}^\nu(x) + \frac{1}{4}g^{\mu\nu}\hat{F}(x)^2 - \frac{\lambda}{2}g^{\mu\nu}(\partial_\rho\hat{A}^\rho(x))^2 \\ & - \lambda(\partial_\rho\partial_\tau\hat{A}^\tau(x))(g^{\mu\nu}\hat{A}^\rho(x) - g^{\mu\rho}\hat{A}^\nu(x) - g^{\nu\rho}\hat{A}^\mu(x))\Big\} \\ & - \frac{e}{2}(\hat{j}^\mu(x)\hat{A}^\nu(x) + \hat{j}^\nu(x)\hat{A}^\mu(x)) \\ & + \frac{e}{2}(\hat{j}^\mu(x)V^\nu(\boldsymbol{x}) - \hat{j}^\nu(x)V^\mu(\boldsymbol{x})).\end{aligned} \tag{2.17}$$

[3] We base our considerations on the 'symmetric' energy momentum tensor $\hat{T}^{\mu\nu}$ rather than the canonical $\hat{\Theta}^{\mu\nu}$. Both versions of the energy momentum tensor, of course, satisfy identical 'continuity' equations, i.e. all physical results are independent of this choice. $\hat{T}^{\mu\nu}$ represents a covariant combination of the energy, momentum and stress densities of the system.

From the last line of Eq. (2.17) it is obvious that, as we are dealing with an open system, the source field breaks the symmetry of $\hat{T}^{\mu\nu}$. As an immediate consequence $\hat{T}^{\mu\nu}$ does not satisfy a homogeneous 'continuity' equation but rather the external potential acts as a source of momentum,

$$\partial_\mu \hat{T}^{\mu\nu}(x) = e\hat{j}_\mu(x)\partial^\nu V^\mu(\boldsymbol{x}). \tag{2.18}$$

Only the zeroth component of the total four momentum, i.e. the energy, is conserved for general time-independent external potentials,

$$\partial_\mu \hat{T}^{\mu 0}(x) = 0 \Rightarrow \int d^3x\, \hat{T}^{00}(x) \text{ is conserved.} \tag{2.19}$$

This implies that the system can be regarded as stationary in the common rest frame of the sources which allows an identification of the Hamiltonian,

$$\hat{H} \equiv \int d^3x\, \hat{T}^{00}(x) = \hat{H}_e + \hat{H}_\gamma + \hat{H}_{\text{int}} + \hat{H}_{ext} \tag{2.20}$$

$$\hat{H}_e = \frac{1}{2}\int d^3x [\hat{\bar{\psi}}(x), (-i\boldsymbol{\gamma}\cdot\boldsymbol{\nabla} + m)\hat{\psi}(x)] \tag{2.21}$$

$$\hat{H}_\gamma = -\frac{1}{8\pi}\int d^3x \{\partial^0 \hat{A}_\nu(x)\partial^0 \hat{A}^\nu(x) + \boldsymbol{\nabla}\hat{A}_\nu(x)\cdot\boldsymbol{\nabla}\hat{A}^\nu(x)\} \tag{2.22}$$

$$\hat{H}_{\text{int}} = e\int d^3x \hat{j}^\mu(x)\hat{A}_\mu(x) \tag{2.23}$$

$$\hat{H}_{ext} = e\int d^3x \hat{j}^\mu(x) V_\mu(\boldsymbol{x}), \tag{2.24}$$

where we have chosen Feynman gauge, $\lambda = 1$, for simplicity. If, in addition, the potential is independent of one spatial coordinate the corresponding momentum is also conserved.

The discussion of angular momentum conservation is based on the generalised angular momentum tensor[4] (compare [27, 28]),

$$\hat{J}^{\mu,\rho\tau} = \hat{J}_e^{\mu,\rho\tau} + \hat{J}_\gamma^{\mu,\rho\tau} \tag{2.25}$$

$$\hat{J}_e^{\mu,\rho\tau} = -\frac{1}{4}\left[\hat{\bar{\psi}}, \left(i\gamma^\mu(x^\rho\vec{\partial}^\tau - \overleftarrow{\partial}^\tau x^\rho - x^\tau\vec{\partial}^\rho + \overleftarrow{\partial}^\rho x^\tau) + \left\{\gamma^\mu, \frac{\sigma^{\rho\tau}}{2}\right\}\right)\hat{\psi}\right]$$

[4]Here we have chosen to define the generalised angular momentum tensor via the canonical energy-momentum tensor $\Theta^{\mu\nu}$,

$$J^{\mu,\rho\tau} = \Theta^{\mu\rho}x^\tau - \Theta^{\mu\tau}x^\rho - \sum_{r,s}\frac{\partial\mathscr{L}}{\partial(\partial_\mu\phi_r)} S_{rs}^{\rho\tau}\phi_s,$$

where the $S_{rs}^{\rho\tau}$ characterise the transformation properties of the fields ϕ_s under Lorentz transformations [28], rather than via $T^{\mu\nu}$ in the form $M^{\mu,\rho\tau} = T^{\mu\rho}x^\tau - T^{\mu\tau}x^\rho$. Both tensors, of course, lead to identical continuity equations and conserved angular momenta.

$$\hat{J}_{\gamma}^{\mu,\rho\tau} = \frac{1}{4\pi}(\hat{F}^{\mu\nu} + \lambda g^{\mu\nu}(\partial\cdot\hat{A}))(x^\rho\partial^\tau\hat{A}_\nu - x^\tau\partial^\rho\hat{A}_\nu + g^\rho_\nu\hat{A}^\tau - g^\tau_\nu\hat{A}^\rho)$$
$$+ (g^{\mu\tau}x^\rho - g^{\mu\rho}x^\tau)\mathscr{L}_\gamma,$$

which represents the angular momentum densities of both the fermions ($J_e^{\mu,\rho\tau}$) and the photons ($J_\gamma^{\mu,\rho\tau}$) in a covariant form. Eq. (2.25) explicitly demonstrates the coupling of spin and orbital angular momentum for both fermions and photons on the most general level. In analogy to Eq. (2.18) the external potential acts as a source of angular momentum in the 'continuity' equation for $J^{\mu,\rho\tau}$,

$$\partial_\mu\hat{J}^{\mu,\rho\tau}(x) = e\hat{j}_\mu(x)(x^\tau\partial^\rho V^\mu(\boldsymbol{x}) - x^\rho\partial^\tau V^\mu(\boldsymbol{x}) + g^{\tau\mu}V^\rho(\boldsymbol{x}) - g^{\rho\mu}V^\tau(\boldsymbol{x})). \tag{2.26}$$

As for the linear momentum, in general no component of the angular momentum is conserved: As a fixed nuclear rest frame has been chosen no 'boost' momentum $\int d^3x\hat{J}^{0,0j}(x)$ can be conserved and the conservation of a conventional angular momentum component $\int d^3x\hat{J}^{0,ij}(x)$ requires specific spatial symmetries. For instance, if all spatial components V^j vanish and V^0 only depends on $(x^1)^2 + (x^2)^2$, i.e. for axially symmetric electrostatic potentials, one finds as expected that the angular momentum with respect to the x^3-axis, $\int d^3x\hat{J}^{0,12}(x)$, is a conserved quantity.

As far as discrete symmetries are concerned three types are usually considered within QED [26, 29]:

- For parity to be a good quantum number some reflection symmetry of the potential is required (the same holds, of course, for more complex discrete spatial symmetries).
- As for the Lagrangian, charge conjugation is no symmetry of the Hamiltonian,

  $$\hat{\mathscr{C}}\hat{H}[V^\nu]\hat{\mathscr{C}}^+ = \hat{H}[-V^\nu],$$

  as long as the external potential does not vanish.
- Finally, time reversal symmetry leads to a twofold degeneracy for purely electrostatic potentials $V^\mu = (V^0, \mathbf{0})$. In this special situation not only the total charge is conserved, but time reversal introduces an additional conserved quantum number.

As a consequence one finds that in the general case, in which $V^\mu(\boldsymbol{x})$ does not exhibit some specific spatial symmetry and/or some of its components vanish, only the charge and the total energy of the system are conserved. Thus the ground state corresponding to (2.20) is nondegenerate in general. In view of this fact there seems to be no need to introduce a coupling of the electrons (fermions) to an additional external magnetic field $\boldsymbol{B}$ [19],

$$\hat{H}_{mag} = \frac{e}{4m}\int d^3x\hat{\bar{\psi}}(x)\sigma_{ij}\hat{\psi}(x)F^{ij}_{ext}(\boldsymbol{x}) = -\frac{e}{2m}\int d^3x\hat{\bar{\psi}}(x)\Sigma^k\hat{\psi}(x)B^k(\boldsymbol{x}),$$

in order to lift possible degeneracies as long as the initial vector potential V is nonzero. Moreover, the interaction of the magnetisation density

$$\hat{m}^k(x) = -\frac{e}{2m}\hat{\bar{\psi}}(x)\Sigma^k\hat{\psi}(x)$$

with $\boldsymbol{B}$ does not constitute a fully consistent contribution to the QED-Hamiltonian, as it does not take into account the intrinsic coupling of spin and orbital angular momentum. Thus, while $\hat{H}_{mag}$ may be quite useful under certain physical circumstances (e.g. for weakly relativistic problems), it does not seem to be appropriate as a basis for a RDFT even in the case $V^\mu = (V^0, \mathbf{0})$ [19, 30], in which one might want to split up the degeneracy originating from time reversal symmetry.

As a prerequisite for the discussion of the many-body problem implied by the Lagrangian (2.1) a second point needs to be addressed: The theory based on (2.1) is not well defined but rather requires renormalisation of the resulting Greens functions as well as the expressions for physical observables as ground state energies and currents (see Appendix A for further details). In the present context the renormalisation procedure consists of two steps. The first is the removal of the divergent vacuum (zero point) energy of noninteracting fermions and photons. This is most easily achieved by explicit subtraction of the vacuum expectation value of the Hamiltonian. For instance, if one considers noninteracting electrons not subject to any external potential, i.e. the noninteracting homogeneous electron gas characterised by $\hat{H}_e$, Eq. (2.21), the renormalised Hamiltonian

$$\hat{H}_R = \hat{H}_e - \langle 0|\hat{H}_e|0\rangle \tag{2.27}$$

leads to a finite ground state energy. The same procedure can be applied to noninteracting photons.

The second part of the renormalisation program, addressing the removal of the ultraviolet (UV) divergencies of QED, which result from the perturbative treatment of the interaction of the fermions with photons and the external field, is more involved. It is instructive to first consider noninteracting fermions in a given external potential,

$$\hat{H}_R = \hat{H}_e + \hat{H}_{ext} - \langle 0|\hat{H}_e|0\rangle, \tag{2.28}$$

where $|0\rangle$ represents the homogeneous vacuum as in (2.27), so that the energy calculated from $\hat{H}_\mathrm{R}$ with respect to the perturbed vacuum (often called Casimir energy [31]) is nonzero. While the Greens functions of this theory like the fermion propagator (for a precise definition of the Greens functions see Appendices A, B),

$$\begin{aligned} iG(x,y) = {} & \Theta(x^0 - y^0) \sum_{\varepsilon_n > \varepsilon_F} \varphi_n(\boldsymbol{x})\bar{\varphi}_n(\boldsymbol{y})\exp[-i\varepsilon_n(x^0 - y^0)] \\ & - \Theta(y^0 - x^0) \sum_{\varepsilon_F \geqslant \varepsilon_n} \varphi_n(\boldsymbol{x})\bar{\varphi}_n(\boldsymbol{y})\exp[-i\varepsilon_n(x^0 - y^0)], \end{aligned} \tag{2.29}$$

where the $\varphi_n(\boldsymbol{x})$ represent the single-particle solutions of the corresponding Dirac equation,

$$\gamma^0\{-i\boldsymbol{\gamma}\cdot\boldsymbol{\nabla}+m+eV(\boldsymbol{x})\}\,\varphi_n(\boldsymbol{x})=\varepsilon_n\varphi_n(\boldsymbol{x})\,, \tag{2.30}$$

are finite without additional modification, the ground state energy and the ground state four current resulting from (2.28) are not [32, 33]. This is most easily seen by rewriting these quantities in terms of $G(x, y)$. Taking the expectation value with respect to the N-electron ground state of the noninteracting system, $|\Phi_0>$, one finds

$$E_{tot}=\langle\Phi_0|\hat{H}_e+\hat{H}_{ext}|\Phi_0\rangle-\langle 0|\hat{H}_e|0\rangle \tag{2.31}$$

$$=-i\int d^3x\,\lim_{y\to x}{}_s\,tr[(-i\boldsymbol{\gamma}\cdot\boldsymbol{\nabla}+m+eV(\boldsymbol{x}))G(x,y)] \tag{2.32}$$

$$+i\int d^3x\,\lim_{y\to x}{}_s\,tr[(-i\boldsymbol{\gamma}\cdot\boldsymbol{\nabla}+m)G_V^0(x,y)]$$

$$j^\nu(\boldsymbol{x})=\langle\Phi_0|\hat{j}^\nu(x)|\Phi_0\rangle \tag{2.33}$$

$$=-i\lim_{y\to x}{}_s\,tr[G(x,y)\gamma^\nu]\,, \tag{2.34}$$

where the symmetric limit,

$$\lim_{y\to x}{}_s\equiv\frac{1}{2}\left(\lim_{y\to x,\,y^0>x^0}+\lim_{y\to x,\,y^0<x^0}\right)\Bigg|_{(x-y)^2\geq 0}\,, \tag{2.35}$$

is a consequence of the charge conjugation invariant forms (2.21, 2.6) and $G_V^0(x, y)$ represents the noninteracting vacuum fermion propagator (A.8). If one now utilises a perturbative expansion of $G(x, y)$ in powers of the external potential,

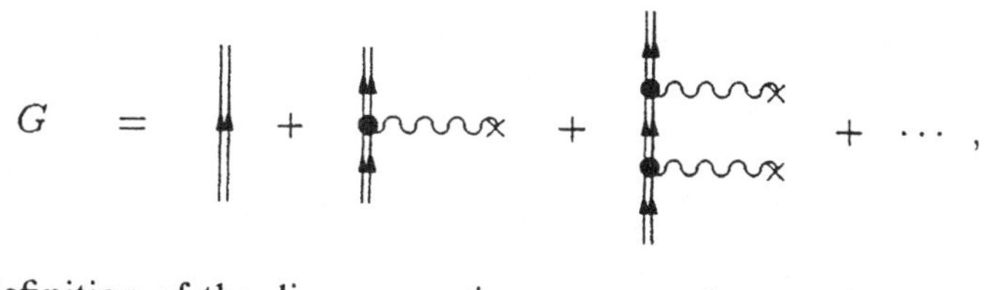

(2.36)

(for a definition of the diagrammatic representation see Appendices A–C) one realises that the evaluation of (2.32, 2.34) involves one further loop-integration induced by the symmetric limit, e.g.

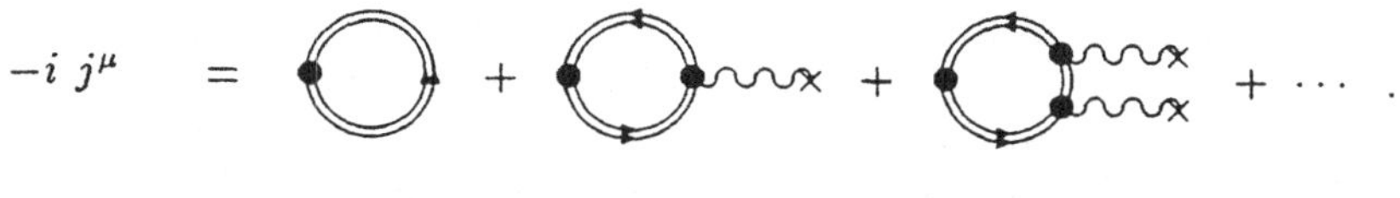

(2.37)

Thus the outermost loop integration, by which the quantities E_{tot} and $j^\nu(\boldsymbol{x})$ (which are of obvious interest in DFT) differ from the UV-finite Greens function G, introduces an UV-divergence completely analogous to the divergencies which arise in standard QED without external fields from the interaction of fermions and photons: Within a perturbative treatment it does not matter whether the external potential $V^{\nu}(x)$ or the quantised photon field creates virtual electron-positron pairs. As a consequence the renormalisation procedure for E_{tot} and $j^{\nu}(x)$ is completely determined by the renormalisation of the Greens functions of interacting vacuum QED without external potential. In particular, as discussed in detail in Appendices A–C only the second of the diagrams in (2.37) is UV-divergent. The corresponding counterterm $\Delta j^{(0),\nu}(x)$ (the superscript (0) indicates that $\Delta j^{(0),\nu}$ represents the lowest order of the complete counterterm Δj^{ν} with respect to the electron-electron coupling constant e^2) is explicitly given in Eq. (C.5). The same procedure has to be applied to the ground state energy leading to the counterterm $\Delta E_{tot}^{(0),inhom}$, Eq. (C.10). One thus has to define the renormalised E_{tot} and $j^{\nu}(x)$ by

$$E_{tot} = \langle \Phi_0 | \hat{H}_e + \hat{H}_{ext} | \Phi_0 \rangle - \langle 0 | \hat{H}_e | 0 \rangle + \Delta E_{tot}^{(0),inhom} \tag{2.38}$$

$$j^{\nu}(x) = \langle \Phi_0 | \hat{j}^{\nu}(x) | \Phi_0 \rangle + \Delta j^{(0),\nu}(x), \tag{2.39}$$

rather than via the initial relations (2.31, 2.33).

Finally both the external potential and the quantised photon field have to be considered together, i.e. we have to deal with the full Hamiltonian (2.20). In this case various approaches to the renormalisation procedure are possible. One could e.g. first utilise the standard renormalisation scheme for Greens functions (as summarised in Appendices A–C) to generate finite Greens and n-point functions. Expressing the interacting four current $j^{\nu}(x)$ then in terms of renormalised n-point functions as discussed in Appendix C one ends up with a finite $j^{\nu}(x)$. However, there still remains the basic problem of obtaining finite total energies as illustrated above: It is not possible to represent E_{tot} in terms of renormalised Greens functions in such a way that no further (outermost) loop integration is required. This outermost loop integration leads to additional UV-divergencies. Thus in the case of E_{tot} one is forced to renormalise each diagrammatic contribution to the perturbation expansion separately, following the standard rules for the renormalisation of vacuum (sub) graphs (often overlapping divergencies are involved). This procedure is explicitly demonstrated for the case of the exchange-correlation energy of the RHEG in Appendix B. Quite generally E_{tot} and $j^{\nu}(x)$ are then given by

$$E_{tot} = \langle \Phi_0 | \hat{H} | \Phi_0 \rangle - \langle 0 | \hat{H}_e + \hat{H}_{\gamma} + \hat{H}_{int} | 0 \rangle + \Delta E_{tot} \tag{2.40}$$

$$j^{\nu}(x) = \langle \Phi_0 | \hat{j}^{\nu}(x) | \Phi_0 > + \Delta j^{\nu}(x), \tag{2.41}$$

where the subtraction of the vacuum energy accounts for the fact that in the interacting case all energies can only be measured with respect to the energy of the interacting, homogeneous vacuum $|0>$. It is sometimes advantageous to decompose the total counterterm ΔE_{tot} into an electron gas part ΔE_{tot}^{RHEG}, which is independent of the external potential, and an inhomogeneity correction.

$$\Delta E_{tot} = \Delta E_{tot}^{RHEG} + \Delta E_{tot}^{inhom}, \tag{2.42}$$

where ΔE_{tot}^{inhom} also involves the full perturbation series with respect to the electron-electron interaction. Eqs. (2.40, 2.41) form a suitable starting point for RDFT, as one is now dealing with finite quantities only.

The preceding discussion has been completely based on the Heisenberg representation. The foundations of DFT, on the other hand, are usually formulated within the framework of the Schrödinger picture, so that one might ask in how far this field theoretical procedure can be useful. It is, however, possible to go over to an appropriately chosen Schrödinger representation as long as one does not try to eliminate the quantised photon fields (compare Sections 7d, 10g of Ref. [34]). The Hamiltonian then reads

$$\hat{H}_S = \frac{1}{2}\int d^3x \Big\{ [\hat{\bar{\psi}}(\boldsymbol{x}), (-i\boldsymbol{\gamma}\cdot\nabla + m + e\hat{\not{A}}(\boldsymbol{x}) + e\not{V}(\boldsymbol{x}))\hat{\psi}(\boldsymbol{x})] \qquad (2.43)$$

$$- \frac{1}{4\pi}(\hat{\Pi}_\nu(\boldsymbol{x})\hat{\Pi}^\nu(\boldsymbol{x}) + \nabla\hat{A}_\nu(\boldsymbol{x})\cdot\nabla\hat{A}^\nu(\boldsymbol{x}))\Big\},$$

where the field operators $\psi(\boldsymbol{x})$ and $A^\nu(\boldsymbol{x})$ are now in Schrödinger representation and

$$\hat{\Pi}^\nu(\boldsymbol{x}) = \exp(-i\hat{H}x^0)\partial_0\hat{A}_H^\nu(x)\exp(i\hat{H}x^0).$$

Moreover, all corresponding counterterms (being expectation values) are independent of the representation, so that the renormalisation scheme remains unchanged and it is just a matter of convenience which representation is used. While the Heisenberg representation (2.20) is more suitable for the derivation of explicit functionals, the Hamiltonian (2.43) (together with Eq. (2.40)) can be utilised for the proof of a relativistic Hohenberg-Kohn theorem.

3 Foundations

In this section we discuss the formal basis of relativistic DFT, that is the relativistic HK-theorem [17, 22], the resulting RKS-equations [18, 19] as well as the relativistic OPM (ROPM) [35, 36]. The discussion of the HK-theorem (in Section 3.1) is restricted to bare essentials, as details have been given in Ref. [22]. In the discussion of the RKS-equations (in Sections 3.2) we attempt to outline the transition from the complete equations, including all radiative effects, to practical variants involving various stages of approximations. In Section 3.3 we introduce the ROPM as an extension of the RKS-approach, which, in addition to the kinetic energy also treats the exchange energy within an orbital scheme[5]. The section is concluded (Section 3.4) by consideration of the weakly relativistic

[5] Extension of the OPM to parts of the correlation energy are possible, but far from being standard.

limit, which allows a connection with and comments on the problem of (non-relativistic) current density functional theory [37, 38], a topic that involves a number of conceptual questions if one approaches it from the point of view of nonrelativistic theory including magnetic fields.

3.1 Relativistic Hohenberg-Kohn Theorem

The extension of the HK-theorem to relativistic systems was first formulated by Ragagopal and Callaway [17] (see also MacDonald and Vosko [19]). The arguments of these authors are based on QED, but the question of possible UV divergencies was not addressed. As, however, the proof of the theorem on the basis of the celebrated *reductio ad absurdum* involves the comparison of energy values, one has to make sure that proper finite quantities for both the ground state energy as well as the ground state four current are used, i.e. the relativistic HK-theorem must be based on the renormalised quantities (2.40, 2.41). In particular, one has to make sure that the structure of the counterterms involved does not invalidate the proof. Here we shall not go through the *reductio ad absurdum* in full detail (for which the interested reader is referred to [22]), but rather summarise the essentials:

- In the first step one shows that there exists a unique map between the set of four potentials $V^\mu(\boldsymbol{x})$ (up to a global constant in V^0) and the set of ground states $|\Phi_0\rangle$ generated from these potentials. This part of the proof is based on the Hamiltonian (2.43). As only the lack of collinearity of ground states resulting from different external potentials is used at this point (but no energy values are compared) full renormalisation is not required, but rather a suitable regularisation is sufficient (one could e.g. modify the space-time dimension).
- In the second step one analyses the relation between the potential $V^\mu(\boldsymbol{x})$ and the resulting renormalised ground state four current $j^\nu(\boldsymbol{x})$, Eq. (2.41). As $j^\mu(\boldsymbol{x})$ is gauge invariant no one-to-one correspondence between the four current and $V^\mu(\boldsymbol{x})$ can exist: A unique map can only be established between classes of four potentials $V^\mu(\boldsymbol{x})$ differing by no more than a static gauge transformation (2.10–2.12) and $j^\mu(\boldsymbol{x})$. Moreover, the heart of this part of the proof is an inequality between ground state energies, so that the renormalised energies (2.40) must be used. The crucial observation in this context is the fact that the counterterms in (2.40) and (2.41) are completely determined by the external potential: Within a perturbation expansion with respect to $V^\mu(\boldsymbol{x})$ this is explicitly obvious from Eqs. (2.42, C.9), as the homogeneous counterterm ΔE_{tot}^{RHEG} is identical for all $V^\mu(\boldsymbol{x})$ and the inhomogeneity correction ΔE_{tot}^{inhom} can be written as a (gauge invariant) functional of $V^\mu(\boldsymbol{x})$.
- For the inequality between ground state energies, required for the second step, a minimum principle for the ground state energy (2.40) is used. However, while the Ritz variational principle is well established in the non-relativistic

context, we are not aware of any rigorous minimum principle for the renormalised ground state energies resulting from (2.1). There are nevertheless a number of arguments which can be given in favour of such a minimum principle. First of all, with increasing speed of light, i.e. in the nonrelativistic limit ($v/c \to 0$), the energies (2.40) continuously approach values which do satisfy the Ritz principle. There seems to be no reason to assume the minimum principle to be restricted to the isolated value $c = \infty$. Secondly, one can explicitly verify the minimum principle for a noninteracting inhomogeneous system characterised by the Hamiltonian (2.28), i.e. for the renormalised energies (2.38) within the Furry picture (compare [32, 34]). Finally, real atoms and molecules are stable (indicating that there exists a lower bound for energies) and QED has proven to be the most accurate theory available to date to describe these systems [39] (note further that, as a matter of principle, one need not rely on perturbation theory to deal with QED-systems so that the asymptotic character of this expansion does not contradict this argument).

In summary, one finds that there exists a one-to-one correspondence between the class of external potentials just differing by gauge transformations, the associated class of ground states and the ground state four current,

$$\{V_\nu | V_\nu + \partial_\nu \Lambda\} \Leftrightarrow \{|\Phi_0\rangle| \text{ with } |\Phi_0\rangle \quad \text{from } V_\nu + \partial_\nu \Lambda\} \Leftrightarrow j^\nu(x), \quad (3.1)$$

i.e. the class of physically equivalent ground states is uniquely determined by the ground state four current. Choosing some arbitrary representative of this class, i.e. fixing the gauge once and for all, one can understand this representative $|\Phi_0>$ as a functional of j^ν, $|\Phi_0[j^\nu]>$, and finally ends up with the statement that all ground state observables are unique functionals of the four current,

$$O[j^\nu] = \langle \Phi_0[j^\nu] | \hat{O} | \Phi_0[j^\nu] \rangle + \Delta O. \quad (3.2)$$

Of course, the functional $O[j^\nu]$ has to reflect an eventual gauge dependence of the operator $\hat{O}$ and may require renormalisation, indicated by the addition of a counterterm ΔO. Note that, by virtue of the unique correspondence between j^ν and V^ν also all counterterms become functionals of j^ν. For instance, the ground state energy itself, including all counterterms, would be

$$E_{tot}[j^\nu] = \langle \Phi_0[j^\nu] | \hat{H} | \Phi_0[j^\nu] \rangle - \langle 0 | \hat{H}_e + \hat{H}_\gamma + \hat{H}_{int} | 0 \rangle \quad (3.3)$$
$$+ \Delta E_{tot}^{RHEG} + \Delta E_{tot}^{inhom}.$$

This energy functional contains not only all relativistic kinetic effects for both electrons and photons but also all radiative (that is field-theoretical) effects.

With the Ritz principle and avoiding the question of interacting v-representability, one may then formulate the basic variational principle of RDFT as

$$\frac{\delta}{\delta j^\nu(\boldsymbol{r})} \{E_{tot}[j^\nu] - \mu \int d^3x j^0(\boldsymbol{x})\} = 0. \quad (3.4)$$

The subsidiary condition implies charge conservation and all quantities involved are supposed to be fully renormalised.

For the case of a purely electrostatic external potential, $V^{\nu} = (V^0, \mathbf{0})$, the complete proof of the relativistic HK-theorem can be repeated using just the zeroth component $j^0(\mathbf{x})$ of the four current (in the following often denoted by the more familiar $n(\mathbf{x})$), i.e. the structure of the external potential determines the minimum set of basic variables for a DFT approach. As a consequence the ground state and all observables, in this case, can be understood as unique functionals of the density n only. This does, however, not imply that the spatial components of the current vanish, but rather that $\mathbf{j}(\mathbf{x}) = \langle \Phi_0[n] | \hat{\mathbf{j}}(x) | \Phi_0[n] \rangle$ has to be interpreted as a functional of $n(\mathbf{x})$. Thus for standard electronic structure problems one can choose between a four current DFT description and a formulation solely in terms of $n(\mathbf{x})$, although one might expect the former approach to be more useful in applications to systems with $\mathbf{j}(\mathbf{x}) \neq \mathbf{0}$ as soon as approximations are involved. This situation is similar to the nonrelativistic case where for a spin-polarised system not subject to an external magnetic field $\mathbf{B}$ both the $\mathbf{B} \to \mathbf{0}$ limit of spin-density functional theory as well as the original pure density functional theory can be used. While the former leads in practice to more accurate results for actual spin-polarised systems (as one additional symmetry of the system is take into account explicitly), both approaches coincide for unpolarized systems.

In view of the two degrees of freedom resulting from time reversal symmetry, one might also set up a two-component formulation of the corresponding RDFT in the special case $V^{\nu} = (V^0, \mathbf{0})$, which would possibly allow a direct extension of the nonrelativistic spin-density functional formalism. Such an approach has not been investigated on the fully relativistic level. One may, however, interpret the suggestion to use the magnetisation density together with the charge density [19, 30] (and thus the spin-polarised relativistic homogeneous electron gas [30, 40, 41, 42]) as a basis for RDFT as an approximate realisation of such an approach (see also [43, 44]). In the following we shall not address this issue further.

We also mention that recently a density functional approach to excited states of relativistic systems has been formulated [45], using ensembles of unequally weighted states. This formalism is restricted to the electrostatic limit and the no-sea approximation (see Section 3.2). Moreover, it remains unclear how the spontaneous emission of photons, which is possible in QED in contrast to the standard nonrelativistic many-body theory, is handled for the excited states involved.

3.2 Relativistic Kohn-Sham Equations

The basic variational principle (3.4) is applied directly in relativistic (extended) Thomas-Fermi models [12, 21, 46] in which an approximate density functional representation for the complete $E_{tot}[j^{\nu}]$ is utilised. The mainstay of applications is, however, the KS-scheme. In order to set up this scheme one first introduces auxiliary single-particle four spinors $\varphi_k(\mathbf{x})$, in terms of which the exact interacting ground state four current (2.41) is represented as[6]

$$j^\nu(x) = j^\nu_V(x) + j^\nu_D(x) \tag{3.5}$$

$$j^\nu_V(x) = \frac{1}{2}\left\{ \sum_{\varepsilon_k \leqslant -m} \bar{\varphi}_k(x)\gamma^\nu\varphi_k(x) - \sum_{-m < \varepsilon_k} \bar{\varphi}_k(x)\gamma^\nu\varphi_k(x) \right\} + \Delta j^{\nu,(0)}(x) \tag{3.6}$$

$$j^\nu_D(x) = \sum_{-m < \varepsilon_k \leqslant \varepsilon_F} \bar{\varphi}_k(x)\gamma^\nu\varphi_k(x), \tag{3.7}$$

where the counterterm[7] $\Delta j^{\nu,(0)}(x)$ is given by Eq. (C.5) (with the total RKS-potential on the right hand side) and ε_F represents the Fermi level below which all orbitals φ_k are occupied. From a field theoretical point of view the single particle RKS-approach corresponds to a problem of the type (2.28). The form (3.5–3.7) thus follows from the charge conjugation invariant current operator (2.6) via (2.39): The matrix element of $\hat{j}^\nu$ with respect to a noninteracting ground state yields in the first step

$$j^\nu(x) = \frac{1}{2}\left\{ \sum_{\varepsilon_k \leqslant \varepsilon_F} \bar{\varphi}_k(x)\gamma^\nu\varphi_k(x) - \sum_{\varepsilon_F < \varepsilon_k} \bar{\varphi}_k(x)\gamma^\nu\varphi_k(x) \right\} + \Delta j^{\nu,(0)}(x).$$

Rearrangement then gives Eqs. (3.5–3.7) in which the vacuum part $j^\nu_V(x)$ has been separated from the four current $j^\nu_D(x)$ which involves only the discrete occupied orbitals with eigenvalues between $-m$ and ε_F.

In the next step one decomposes the ground state energy functional (3.3) in the standard fashion,

$$E_{tot}[j^\nu] = T_s[j^\nu] + E_{ext}[j^\nu] + E_H[j^\nu] + E_{xc}[j^\nu], \tag{3.8}$$

where the counterterms for E_{tot} given in Eq. (2.40) are understood to be included in the individual energy components. The latter are defined as follows: The noninteracting kinetic energy functional T_s, i.e. the kinetic energy of the 'KS-particles', is

$$T_s[j^\nu] = T_{s,V}[j^\nu] + T_{s,D}[j^\nu] \tag{3.9}$$

$$T_{s,V}[j^\nu] = \frac{1}{2}\int d^3x \left\{ \sum_{\varepsilon_k \leqslant -m} \bar{\varphi}_k(x)[-i\gamma\cdot\nabla + m]\varphi_k(x) - \sum_{-m < \varepsilon_k} \bar{\varphi}_k(x)[-i\gamma\cdot\nabla + m]\varphi_k(x) \right\} - \langle 0|\hat{H}_e|0\rangle + \Delta T_s^{inhom} \tag{3.10}$$

[6] The question of noninteracting v-representability which is required for this representation has not been examined in the relativistic case. One would, however, expect analogous statements as in the nonrelativistic situation [47].

[7] This term has erroneously been omitted in Ref. [22].

$$T_{s,D}[j^\nu] = \int d^3x \sum_{-m < \varepsilon_k \leq \varepsilon_F} \bar{\varphi}_k(\mathbf{x})[-i\boldsymbol{\gamma}\cdot\nabla + m]\varphi_k(\mathbf{x}), \tag{3.11}$$

where (2.38, C.10, C.13) have to be used to obtain the counterterm ΔT_s^{inhom}. The external potential term is

$$E_{ext} = e\int d^3x j_\mu(\mathbf{x}) V^\mu(\mathbf{x}), \tag{3.12}$$

(the counterterm (C.12) has already been absorbed into the renomalised current j_μ) and the covariant Hartree energy reads

$$\begin{aligned} E_H[j^\nu] &= \frac{1}{2}\int d^3x \int d^4y j^\mu(x) D^0_{\mu\nu}(x-y) j^\nu(y) \\ &= \frac{e^2}{2}\int d^3x \int d^3y \frac{j^\nu(\mathbf{x}) j_\nu(\mathbf{y})}{|\mathbf{x}-\mathbf{y}|}, \end{aligned} \tag{3.13}$$

with $D^0_{\mu\nu}$ being the free photon propagator (A.9). The xc-energy functional is then the remainder

$$E_{xc}[j^\nu] = E_{tot}[j^\nu] - T_s[j^\nu] - E_{ext}[j^\nu] - E_H[j^\nu]. \tag{3.14}$$

Note that the ground state energy pertains to the many-electron sector. Free photons and positrons are not present in the ground state considered.

One may then use the basic variational principle, varying with respect to the orbitals (a more careful argument can be given following the lines of Ref. [47]), to obtain the complete relativistic KS-equations

$$\gamma^0\{-i\boldsymbol{\gamma}\cdot\nabla + m + e\not{V}(\mathbf{x}) + \not{v}_H(\mathbf{x}) + \not{v}_{xc}(\mathbf{x})\}\varphi_k(\mathbf{x}) = \varepsilon_k\varphi_k(\mathbf{x}), \tag{3.15}$$

where

$$v_H^\nu(\mathbf{x}) = e^2\int d^3y \frac{j^\nu(\mathbf{y})}{|\mathbf{x}-\mathbf{y}|} \tag{3.16}$$

$$v_{xc}^\nu(\mathbf{x}) = \frac{\delta E_{xc}[j^\mu]}{\delta j_\nu(\mathbf{x})}. \tag{3.17}$$

Eqs. (3.5, 3.15–3.17) have to be solved selfconsistently in order to obtain the exact $j^\nu(\mathbf{x})$ of the interacting system. The corresponding ground state energy is given by

$$\begin{aligned} E_{tot} = &\frac{1}{2}\Big\{\sum_{\varepsilon_k \leq -m} \varepsilon_k - \sum_{-m < \varepsilon_k} \varepsilon_k\Big\} - \langle 0|\hat{H}_e|0\rangle + \Delta T_s^{inhom} \\ &+ \sum_{-m < \varepsilon_k \leq \varepsilon_F} \varepsilon_k - E_H[j^\nu] + E_{xc}[j^\nu] - \int d^3x\, v_{xc}^\nu(\mathbf{x}) j_\nu(\mathbf{x}). \end{aligned}$$

The first three terms represent the Casimir energy [31], that is the energy shift induced in the vacuum by the presence of the RKS-potential. In addition, the terms of the second line contain vacuum corrections via both the form (3.5) of j^ν and the functional dependence of E_H, E_{xc} and v^μ_{xc} on j^ν.

A brief glance at the various terms involved in the RKS-scheme reveals a selfconsistency problem of considerable complexity: The evaluation of both j^ν_V and $T_{s,V}$ requires summation over all negative and positive energy solutions and appropriate renormalisation in each step of the iterative procedure. Moreover, all potentials exhibit a four component structure. As the solution of this selfconsistency problem is at best tedious (if at all possible), one is bound to consider a suitable hierarchy of (hopefully) useful approximations.

The most important, and at the same time most legitimate, simplification if one aims at electronic structure calculations in quantum chemistry and condensed matter physics is the *no-sea* (or alternatively *no-pair*) *approximation*. In this approximation all radiative contributions to the four current and T_s are neglected,

$$j^\nu_V(x) = 0; \quad T_{s,V} = 0. \tag{3.18}$$

In addition the vacuum contributions in the functional dependence of E_{xc} on j^ν are dropped[8], so that one is led to the RDFT analogue of the no-pair approximation applied in conventional relativistic many-body approaches (see e.g. [7]). An a posteriori perturbative evaluation of these corrections is possible and should be adequate, except in special circumstances as for instance the calculation of the structure of super-heavy atoms (with $Z \approx 137$ [33]). The resulting RKS-equations are then still given by Eqs. (3.15–3.17), but $j^\mu(x)$ and T_s are determined by the simpler expressions (3.7) and (3.11).

In addition to the no-sea approximation, two further simplifications may be considered. If the external potential is purely electrostatic

$$\boldsymbol{V} = \boldsymbol{0}; \quad eV^0(\boldsymbol{r}) = v_{ext}(\boldsymbol{r}), \tag{3.19}$$

which is the situation encountered for standard electronic structure calculations, the density $n(\boldsymbol{r}) = j^0(\boldsymbol{r})$ is the only quantity which is really required for the RDFT scheme (as discussed in the previous section). In this *electrostatic limit* the spatial current $\boldsymbol{j}$ and also the energy components (3.9–3.14) can be understood as functionals of n only and RDFT reduces to a pure density functional approach. The structure of the density functional representations for E_H and E_{xc} is determined by the original functionals $E_H[j^\nu]$ and $E_{xc}[j^\nu]$ by simply inserting the exact density functional for $\boldsymbol{j}$,

$$\tilde{E}_H[n] = E_H[n, \boldsymbol{j}[n]], \quad \tilde{E}_{xc}[n] = E_{xc}[n, \boldsymbol{j}[n]]. \tag{3.20}$$

The corresponding RKS-equations are then given by (3.15) with the Hartree and

[8] The definition of the no-sea approximation for E_{xc} is not completely unambiguous. As discussed in Appendix B we define it through neglect of all vacuum fermion loops in the derivation of an approximate $E_{xc}[j^\nu]$. Alternatively, one could project out all negative energy states, thus generating a direct equivalent of the standard no-pair approximation. As one would expect the differences between these two schemes to be small, we do not differentiate between these approximations here.

xc-potentials consisting only of a time-like component $\tilde{v}_{H,xc}(\boldsymbol{r})$,

$$\tilde{v}_{H,xc}(\boldsymbol{r}) = \frac{\delta \tilde{E}_{H,xc}[n]}{\delta n(\boldsymbol{r})} = \frac{\delta E_{H,xc}[j^\nu]}{\delta n(\boldsymbol{r})} + \int d^3r' \frac{\delta E_{H,xc}[j^\nu]}{\delta j^k(\boldsymbol{r}')} \frac{\delta j^k([n],\boldsymbol{r}')}{\delta n(\boldsymbol{r})},$$

so that their structure is considerably simplified.

An additional approximation can be obtained if one uses the decomposition of the electron-electron interaction mediated by the free photon propagator $D^0_{\mu\nu}$ into a longitudinal and a transverse part (according to Eq. (A.10)) to introduce the corresponding decomposition for E_H and E_{xc},

$$E_H[n] = E^L_H[n] + E^T_H[n]; \quad E_{xc}[n] = E^L_{xc}[n] + E^T_{xc}[n] \tag{3.21}$$

(in the following we shall only distinguish between the functionals in the electrostatic limit, $\tilde{E}_{H,xc}[n] \equiv E_{H,xc}[n]$, and the more general $E_{H,xc}[j^\nu]$ by their respective arguments). While this decomposition is obvious for $E_H[n]$,

$$E^L_H[n] = \frac{e^2}{2} \int d^3r \int d^3r' \frac{n(\boldsymbol{r})n(\boldsymbol{r}')}{|\boldsymbol{r}-\boldsymbol{r}'|} \tag{3.22}$$

$$E^T_H[\boldsymbol{j}[n]] = -\frac{e^2}{2} \int d^3r \int d^3r' \frac{\boldsymbol{j}([n];\boldsymbol{r}) \cdot \boldsymbol{j}([n];\boldsymbol{r}')}{|\boldsymbol{r}-\boldsymbol{r}'|}, \tag{3.23}$$

which is linear in $D^0_{\mu\nu}$, E^L_{xc} is defined by neglecting the transverse interaction to all orders in $D^0_{\mu\nu}$ (and the remainder is then called E^T_{xc}).

The *longitudinal approximation* then consists in neglecting the transverse contributions E^T_H and E^T_{xc} in the selfconsistency loop, i.e.

$$v^T_H(\boldsymbol{r}) = 0; \quad v^T_{xc}(\boldsymbol{r}) = 0. \tag{3.24}$$

The transverse contributions to the ground state energy can then be calculated a posteriori in a perturbative fashion. As most available energy functionals in RDFT include both longitudinal and transverse contributions and the selfconsistency problem is not simplified by the longitudinal approximation, this approximation is not required for an efficient application of the RKS-equations. Rather it leads to the RDFT-equivalent of the so-called Dirac-Coulomb Hamiltonian usually applied in conventional ab initio calculations, so that it is mainly useful for purposes of comparison[9]. Note that the longitudinal approximation does not automatically imply the neglect of all current contributions to the xc-potential (and vice versa) as one might expect on the basis of $E_H[j^\nu]$, for which these two approximations coincide. This point is most readily demonstrated by the current-dependence of explicit current-density functional representations of T_s (as discussed in Appendix D).

[9] A selfconsistent treatment of the transverse interaction in conventional ab initio methods is often based on the Gaunt approximation [48]. There are, however, no density functionals available within this approximation.

If one restricts oneself to the electrostatic limit of RDFT and relies on both the no-sea as well as the longitudinal approximation one arrives at the RKS-equations

$$\{-i\boldsymbol{\alpha}\cdot\nabla + \beta m + v^L_{tot}(\boldsymbol{r})\}\,\varphi_k(\boldsymbol{r}) = \varepsilon_k\,\varphi_k(\boldsymbol{r}), \tag{3.25}$$

where $\boldsymbol{\alpha}$, β are the usual Dirac matrices and

$$v^L_{tot}(\boldsymbol{r}) = v_{ext}(\boldsymbol{r}) + v^L_H([n];\boldsymbol{r}) + v^L_{xc}([n];\boldsymbol{r}) \tag{3.26}$$

$$v^L_H([n];\boldsymbol{r}) = e^2 \int d^3r' \frac{n(\boldsymbol{r}')}{|\boldsymbol{r}-\boldsymbol{r}'|} \tag{3.27}$$

$$v^L_{xc}([n];\boldsymbol{r}) = \frac{\delta E^L_{xc}[n]}{\delta n(\boldsymbol{r})} \tag{3.28}$$

$$n(\boldsymbol{r}) = \sum_{-m<\varepsilon_k\leqslant\varepsilon_F} \varphi_k^+(\boldsymbol{r})\varphi_k(\boldsymbol{r}). \tag{3.29}$$

Referring to the various approximations suggested, the corresponding ground state energy is obtained as

$$E^L_{tot} = T_{s,D} + E_{ext} + E^L_H + E^L_{xc}, \tag{3.30}$$

which may be corrected by addition of the transverse energy terms,

$$E_{tot} = E^L_{tot} + E^T_H + E^T_{xc}, \tag{3.31}$$

as well as radiative corrections. The scheme (3.25–3.30) corresponds to the no-pair Dirac-Coulomb Hamiltonian of standard relativistic many-body theory. Selfconsistent treatment of the transverse contributions, on the other hand, would lead to the RDFT-version of the no-pair Dirac-Coulomb-Breit Hamiltonian, ignoring the small difference between the full transverse and the Breit interaction.

A time-dependent generalisation of the RKS-equation (3.25) has been suggested by Parpia and Johnson [49]. While a rigorous foundation of this approach is not available to date, this method has been successfully applied to the photoionisation of Hg [50] and Xe [49] as well as the evaluation of the polarisabilities of heavy closed-shell atoms [51] (using a direct time-dependent extension of the local density approximation for $E_{xc}[n]$).

3.3 Optimised-Potential Method

The important observation leading to the transition from the original variational approach (3.4) to the KS-equations is the fact that all currently available density functional representations of the kinetic energy are not able to reproduce one of the most basic features of quantum systems, i.e. the electronic shell structure. Thus, as soon as one is interested in properties of quantum systems which are related to the shell structure (i.e. merely all), one is forced to go to the equivalent KS single-particle problem and treat its kinetic energy exactly[10]. The next important physical feature for which an accurate density functional representation is currently not at hand is the cancellation of the self-interaction effects to be provided by the exchange energy functional. A natural route to circumvent this problem is to resort to an orbital representation for the exchange part of the xc-energy and to treat the resulting extended single-particle problem exactly. In the nonrelativistic context this extension of the selfconsistency problem to include exchange on an exact level was originally introduced by Talman and Shadwick [52] (see also [53]) and has later been adopted as a DFT-method by Sahni, Gruenebaum and Perdew [54] as well as Langreth and Mehl [55] and Sham [56] under the name of the optimised-potential-method (OPM). It should be viewed as a systematic extension of the KS-scheme and thus its relativistic extension is also discussed in this section devoted to the foundations of RDFT. The OPM selfconsistency procedure is much more involved than the KS-scheme and thus applications beyond spherical systems [57, 58, 59, 60] or the atomic sphere approximation in band structure calculations [61, 62] are lacking to date. This approach has, however, attracted considerable attention (see e.g. [63, 64]) since an accurate, approximate solution of the crucial OPM equation for the exchange potential has been given by Krieger, Li and Iafrate [58, 65, 66, 67].

A relativistic extension of the OPM on the longitudinal no-pair level has been put forward by Talman and collaborators [35] (and recently been applied to atoms [36]). Further extension to a covariant exchange energy functional is straightforward on the basis of the RKS propagator G_{KS},

$$E_x = \frac{1}{2}\int d^3x \int d^4y\, D^0_{\mu\nu}(x-y)\, tr\left[G_{KS}(x,y)\gamma^\nu G_{KS}(y,x)\gamma^\mu\right], \tag{3.32}$$

(the required counterterms are not explicitly shown). The propagator G_{KS} describes the motion of the 'KS-particles' in the total RKS-potential. It can be expressed in terms of the RKS-orbitals via the standard expansion (2.29). The exchange energy functional defined in this fashion should not be confused with the relativistic HF (RHF) exchange energy (at least from a rigorous point of view), as the RKS-orbitals satisfy the local RKS-equations (3.15) rather than the nonlocal RHF-equations.

The energy (3.32) is, via the (rather involved) dependence of the RKS-orbitals on the four current density, a functional of this quantity,

$$E_x = E_x[j^\nu]. \tag{3.33}$$

The RDFT correlation energy is then defined by

$$E_c[j^\nu] = E_{xc}[j^\nu] - E_x[j^\nu]. \tag{3.34}$$

[10] Note that quite generally the rearrangement of the ground state energy functional by addition and subtraction of terms, that leads to the KS-scheme, is quite arbitrary. Instead of adding and subtracting e.g. the kinetic energy T_s any other suitable approximation of the full kinetic energy could be used, resulting in a modified selfconsistency problem.

and the individual potentials are

$$v_x^\mu([j^\nu];\boldsymbol{r}) = \frac{\delta E_x[j^\nu]}{\delta j_\mu(\boldsymbol{r})} \tag{3.35}$$

$$v_c^\mu([j^\nu];\boldsymbol{r}) = \frac{\delta E_c[j^\nu]}{\delta j_\mu(\boldsymbol{r})}. \tag{3.36}$$

If $E_c[j^\nu]$ is neglected the resulting scheme is called the exchange-only (x-only) limit of RDFT.

For the case of the longitudinal no-pair approximation and a purely electrostatic external potential $V^\mu = (V^0, \mathbf{0})$, to which we restrict further discussion of the ROPM, Eq. (3.32) reduces to (summation over the spinor indices $a, b = 1, \ldots, 4$ is implicitly understood)

$$E_x^L[n] = -\frac{e^2}{2}\int d^3r \int d^3r' \sum_{-m < \varepsilon_k, \varepsilon_l \leqslant \varepsilon_F} \times \frac{\varphi_{a,k}^+(\boldsymbol{r})\varphi_{b,l}^+(\boldsymbol{r}')\varphi_{a,l}(\boldsymbol{r})\varphi_{b,k}(\boldsymbol{r}')}{|\boldsymbol{r}-\boldsymbol{r}'|}, \tag{3.37}$$

where the φ_k now have to be interpreted as functionals of the density only. The crucial feature of $E_x^L[n]$ is its linear dependence on the electron-electron coupling constant e^2: The functional dependence of the φ_k on n is independent of e^2 as the φ_k experience a local (in the sense of multiplicative) one-body potential [54, 55, 56], where it does not matter whether the local potential is obtained selfconsistently or is just a given external potential. The correlation energy functional $E_c^L[n]$ thus contains all contributions to $E_{xc}^L[n]$ which are of higher order in e^2.

As the exact density dependence of $\varphi_k([n]; \boldsymbol{r})$ is not known, however, the corresponding x-only potential $v_x^L(\boldsymbol{r})$ cannot be directly evaluated via (3.35). Nevertheless, utilising the fact that the one-body potential which minimises $E_{tot}^L[n]$ is unique [54, 55, 56], v_x^L can be obtained by minimisation of the total ground state energy with respect to the total one-body potential (3.26) which the φ_k experience,

$$\frac{\delta E_{tot}^L[n]}{\delta v_{tot}^L(\boldsymbol{r})} = \int d^3r' \sum_{-m < \varepsilon_k \leqslant \varepsilon_F} \frac{\delta E_{tot}^L[n]}{\delta\varphi_k(\boldsymbol{r}')}\frac{\delta\varphi_k(\boldsymbol{r}')}{\delta v_{tot}^L(\boldsymbol{r})} + \text{c.c.} = 0. \tag{3.38}$$

Decomposing v_{tot}^L according to (3.26) and

$$v_{xc}^L(\boldsymbol{r}) = v_x^L(\boldsymbol{r}) + v_c^L(\boldsymbol{r}), \tag{3.39}$$

one obtains an integral equation for $v_x^L(\boldsymbol{r})$,

$$\int d^3r' \, K(\boldsymbol{r}, \boldsymbol{r}')v_x^L(\boldsymbol{r}') = Q(\boldsymbol{r}). \tag{3.40}$$

The kernel of the integral equation is given by

$$K(\boldsymbol{r},\boldsymbol{r}') = \sum_{-m<\varepsilon_k\leqslant\varepsilon_F} \varphi_k^+(\boldsymbol{r})G_k(\boldsymbol{r},\boldsymbol{r}')\varphi_k(\boldsymbol{r}'), \tag{3.41}$$

which the inhomogeneous term reads

$$Q(\boldsymbol{r}) = -e^2 \sum_{-m<\varepsilon_k,\varepsilon_l\leqslant\varepsilon_F} \int d^3r' \int d^3r'' \times \varphi_k^+(\boldsymbol{r})\, G_k(\boldsymbol{r},\boldsymbol{r}') \frac{\varphi_l(\boldsymbol{r}')\varphi_l^+(\boldsymbol{r}'')}{|\boldsymbol{r}'-\boldsymbol{r}''|} \varphi_k(\boldsymbol{r}''). \tag{3.42}$$

The quantity $G_k(\boldsymbol{r},\boldsymbol{r}')$ represents the Greens function

$$G_k(\boldsymbol{r},\boldsymbol{r}') = \sum_{-m<\varepsilon_l<m, l\neq k} \frac{\varphi_l(\boldsymbol{r})\varphi_l^+(\boldsymbol{r}')}{\varepsilon_l-\varepsilon_k}, \tag{3.43}$$

which satisfies

$$[-i\boldsymbol{\alpha}\cdot\boldsymbol{\nabla} + \beta m + v_{tot}^L(\boldsymbol{r}) - \varepsilon_k]\,G_k(\boldsymbol{r},\boldsymbol{r}') = \delta^{(3)}(\boldsymbol{r}-\boldsymbol{r}') - \varphi_k(\boldsymbol{r})\varphi_k^+(\boldsymbol{r}'). \tag{3.44}$$

The longitudinal no-pair ROPM thus requires the simultaneous selfconsistent solution of (3.40)[11] and the RKS-equation (3.25), either including some correlation potential or in the x-only limit with $v_c^L = 0$.

The advantage of the ROPM lies in the fact that due to the Fock form of $E_x^L[n]$ the self-interaction of the 'KS-particles' is cancelled exactly. This also manifests itself in the asymptotic form of v_x^L for finite systems,

$$v_x^L(r\to\infty)\to -\frac{1}{r}, \tag{3.45}$$

so that the total potential is free of self-interaction effects, e.g. for neutral systems one obtains

$$v_{tot}^L(r\to\infty)\to -\frac{1}{r}. \tag{3.46}$$

In order to provide a quantitative comparison of this DF-concept for exchange and the standard HF approach we list in Table 3.1 the corresponding x-only ground state energies and eigenvalues of the highest occupied orbitals for spherical (closed subshell) atoms [36]. As is obvious from Table 3.1 the differences between ROPM and RHF ground state energies are rather small from a numerical point of view: They are below 60 mhartree even for the largest atoms, with the ROPM-energies always being somewhat less attractive (consistent with the reduced variational freedom of the ROPM-orbitals) – apart from

[11] The integral equation (3.40) determines v_x^L up to a trivial constant which is determined by the boundary condition $v_x^L(r\to\infty) = 0$.

Table 3.1. Longitudinal ground state energies ($-E^L_{tot}$) and highest occupied eigenvalues ($-\varepsilon^L_{mk}$) for closed subshell atoms from nonrelativistic OPM (NROPM [59]), relativistic OPM (ROPM [36]) and relativistic HF (RHF [68]) calculations [69] (all energies are in hartree).

Atom	$-E^L_{tot}$			$-\varepsilon^L_{mk}$		
	NROPM	ROPM	RHF	NROPM	ROPM	RHF
He (1S1/2)	2.862	2.862	2.862	0.918	0.918	0.918
Be (2S1/2)	14.572	14.575	14.576	0.309	0.309	0.309
Ne (2P3/2)	128.545	128.690	128.692	0.851	0.848	0.848
Mg (3S1/2)	199.611	199.932	199.935	0.253	0.253	0.253
Ar (3P3/2)	526.812	528.678	528.684	0.591	0.587	0.588
Ca (4S1/2)	676.751	679.704	679.710	0.196	0.196	0.196
Zn (4S1/2)	1777.828	1794.598	1794.613	0.293	0.299	0.299
Kr (4P3/2)	2752.028	2788.848	2788.861	0.523	0.515	0.514
Sr (5S1/2)	3131.514	3178.067	3178.080	0.179	0.181	0.181
Pd (4D5/2)	4937.858	5044.384	5044.400	0.335	0.319	0.320
Cd (5S1/2)	5465.056	5593.299	5593.319	0.266	0.282	0.281
Xe (5P3/2)	7232.018	7446.876	7446.895	0.456	0.439	0.440
Ba (6S1/2)	7883.404	8135.625	8135.644	0.158	0.163	0.163
Yb (6S1/2)	13391.070	14067.621	14067.669	0.182	0.196	0.197
Hg (6S1/2)	18408.313	19648.826	19648.865	0.262	0.329	0.328
Rn (6P3/2)	21865.826	23601.969	23602.005	0.427	0.382	0.384
Ra (7S1/2)	23093.258	25028.027	25028.061	0.149	0.167	0.166
No (7S1/2)	32787.471	36740.625	36740.682	0.171	0.209	0.209

He where both approaches coincide. Also the corresponding highest occupied eigenvalues are essentially identical. The same holds for the longitudinal x-only energies (see Table 5.1). Here the maximum difference of 106 mhartree is found for No. The x-only ROPM thus demonstrates explicitly that one can obtain RHF-level results for all interesting atomic properties with a local exchange potential.

3.4 Nonrelativistic Limit

The weakly relativistic limit of the Hamiltonian (2.20) for fermions in external electric and magnetic fields can be derived with standard techniques, either by direct expansion or by a low order Foldy-Wouthuysen transformation. One obtains

$$\hat{H} = \int d^3x \hat{\varphi}^+(\boldsymbol{x}) \left\{ \frac{1}{2m} \left[(-i\hbar\nabla)^2 + 2i\hbar\frac{e}{c} \boldsymbol{V}(\boldsymbol{x}) \cdot \nabla + \frac{e^2}{c^2} \boldsymbol{V}(\boldsymbol{x})^2 \right] \right. \tag{3.47}$$

$$\left. - \frac{e\hbar}{2mc} \boldsymbol{\sigma} \cdot (\nabla \times \boldsymbol{V}(\boldsymbol{x})) + eV_0(\boldsymbol{x}) \right\} \hat{\varphi}(\boldsymbol{x}) + \hat{H}_{ee}.$$

In Eq. (3.47) $\hat{\varphi}(\boldsymbol{x})$ is a nonrelativistic field operator of two-component structure, $\boldsymbol{\sigma}$ are the Pauli matrices and the electron-electron interaction reduces to the Coulomb interaction, denoted by $\hat{H}_{ee}$. As usual, the gauge term proportional to

$(e^2/c^2)\,V(x)^2$ has been kept, although it is of order $1/c^2$ and terms of this order are not included consistently. The Hamiltonian (3.47) is invariant under the gauge transformation

$$\hat{\varphi}'(x) = e^{-ie\lambda(x)/\hbar}\,\hat{\varphi}(x), \quad V'(x) = V(x) - c\nabla\lambda(x), \tag{3.48}$$

i.e.

$$H(\hat{\varphi}', V') = H(\hat{\varphi}, V).$$

The density operator is defined in terms of the field operators as

$$\hat{n}(x) = \hat{\varphi}^+(x)\,\hat{\varphi}(x), \tag{3.49}$$

and by a weakly relativistic expansion of the current operator (2.6) (together with the appropriate redefinition of the vacuum) the following expression for the nonrelativistic current operator can be extracted,

$$\hat{j}(x) = \hat{j}_p(x) - \frac{c}{e}\nabla\times\hat{m}(x) - \frac{e}{mc}V(x)\,\hat{n}(x). \tag{3.50}$$

Here the paramagnetic current,

$$\hat{j}_p(x) = -\frac{i\hbar}{2m}\left[\hat{\varphi}^+(x)(\nabla\hat{\varphi}(x)) - (\nabla\hat{\varphi}^+(x))\,\hat{\varphi}(x)\right], \tag{3.51}$$

and the magnetisation-density,

$$\hat{m}(x) = -\frac{e\hbar}{2mc}\,\hat{\varphi}^+(x)\,\sigma\,\hat{\varphi}(x), \tag{3.52}$$

are defined as usual. It is important to note that $j_p(x)$ is not invariant under the gauge transformation (3.48), only the combination

$$\hat{j}_p(x) - \frac{e}{mc}V(x)\,\hat{n}(x)$$

has this property.

If one re-expresses the Hamiltonian (3.47) in terms of the density and current operators in order to exhibit the coupling to the external fields more explicitly, one finds

$$\hat{H} = \int d^3x \left\{\hat{\varphi}^+(x)\frac{(-i\hbar\nabla)^2}{2m}\hat{\varphi}(x) + eV_0(x)\hat{n}(x)\right\} + \hat{H}_{ee} \tag{3.53}$$
$$- \int d^3x \frac{e}{c}V(x)\cdot\left\{\hat{j}_p(x) - \frac{c}{e}\nabla\times\hat{m}(x) - \frac{e}{2mc}V(x)\hat{n}(x)\right\}.$$

This expression indicates that it is the sum of the paramagnetic current operator and the curl of the magnetisation density,

$$\hat{j}_p(x) - \frac{c}{e}\nabla\times\hat{m}(x),$$

which couples to the external vector potential, but the situation is somewhat ambiguous concerning the gauge term, as one may rearrange the Hamiltonian (3.53) either in the form

$$\hat{H} = \hat{T} + \hat{H}_{ee} + \int d^3x \left\{ -\frac{e}{c} V(x) \cdot \hat{j}(x) + \left[eV_0(x) - \frac{e^2}{2mc^2} V(x)^2 \right] \hat{n}(x) \right\}, \tag{3.54}$$

or

$$\hat{H} = \hat{T} + \hat{H}_{ee} - \int d^3x \frac{e}{e} V(x) \cdot \left[\hat{j}_p(x) - \frac{c}{e} \nabla \times \hat{m}(x) \right] \tag{3.55}$$
$$+ \int d^3x \left[eV_0(x) + \frac{e^2}{2mc^2} V(x)^2 \right] \hat{n}(x) .$$

It is directly possible to prove a HK-theorem for the form (3.55) using the density n and the gauge-dependent current $j_p - (c/e)\nabla \times m$ as basic DFT variables, but not for the form (3.54) which would suggest to use n and the full current j. One is thus led to the statement that the first set of variables can legitimately be used to set up nonrelativistic current density functional theory, indicating at first glance a conflict with the fully relativistic DFT approach.

It is important to notice, however, that consistent neglect of all terms of the order $1/c^2$ (which has not been treated consistently in the weakly relativistic expansion) in the Hamiltonian allows a proof of a HK-theorem on the basis of the variables n and j. In other words: Only a fully relativistic approach combines consistency in $1/c$ with gauge invariance. It remains to be investigated explicitly, whether inclusion of all relevant terms to order $1/c^2$ allows to reinstate the physical current $j(x)$ as basic variable also in this order as one would expect from the fully relativistic theory.

4 Explicit Relativistic Exchange-Correlation Functionals

The derivation of explicit energy functionals in *nonrelativistic* DFT follows a variety of avenues. The present day "standard" is the LDA,

$$E_{xc}^{NRLDA}[n] = \int d^3r \, e_{xc}^{NRHEG}(n(r)), \tag{4.1}$$

in which the density dependence of the xc-energy density e_{xc}^{NRHEG} of the nonrelativistic HEG is used with the density $n(\mathbf{r})$ of the actual inhomogeneous system. In particular, accurate representations of the correlation part e_c^{NRHEG} [70, 71, 72, 73] are obtained by a parametrisation of Monte Carlo results for the

NRHEG [74, 73]. Gradient corrections in low order,

$$E_{xc}^{NRGE}[n] = \int d^3r\, e_{xc}^{NRGE}(n(\mathbf{r}), (\nabla n(\mathbf{r}))^2, \nabla^2 n(\mathbf{r})), \tag{4.2}$$

were first obtained by painstaking analysis of the relevant diagrammatic contributions (see e.g. [75, 76, 77, 78, 79, 80, 81, 60]), but turned out not to be adequate [75, 76, 82]. The situation was improved by the introduction of generalised gradient approximations (GGAs) [10, 11, 83, 84, 85, 86], in which $(\nabla n)^2$-terms beyond the lowest order contribution to the gradient expansion (GE) are included. These types of functionals yield the most accurate values for atomic ground state and ionisation energies, molecular geometries and dissociation energies as well as cohesive properties of solids that have been obtained to date with nonrelativistic density functionals (see e.g. [2, 3, 82, 87, 88, 89, 90, 91]).

An alternative approach under the heading of weighted density approximation (WDA) attempts to model the density dependence of the pair correlation function of inhomogeneous systems,

$$E_{xc}^{NRWDA}[n] = \frac{e^2}{2}\int d^3r \int d^3r' \frac{n(\mathbf{r})n(\mathbf{r}')}{|\mathbf{r}-\mathbf{r}'|}[\tilde{g}([n];\mathbf{r},\mathbf{r}')-1], \tag{4.3}$$

where $\tilde{g}$ indicates that a coupling constant integration over the pair correlation function is involved. The WDA is characterised by the feature that self-interaction effects are corrected to a large degree, though not as completely as in the OPM.

Compared with the nonrelativistic case, the derivation of explicit relativistic functionals is not as fully developed. Concerning the RLDA both the x-only limit and the correlation contribution in the so-called random phase approximation (RPA) are available. We discuss the RLDA in Section 4.1. Relativistic gradient corrections for E_{xc}, on the other hand, have not been evaluated at all, although the basic technique for their derivation can be extended to the relativistic regime. In view of the absence of explicit results we only illustrate this method for the case of T_s in Appendix D. An extension of the WDA scheme to relativistic systems (RWDA) [92, 36] is summarised in Section 4.2. However, no information on the RWDA beyond the longitudinal x-only limit is available. Moreover, it should be emphasised at the very outset that on the present level of sophistication neither the RLDA nor the RWDA contain radiative corrections. The issue of vacuum corrections in $E_{xc}[n]$ is discussed in detail in Appendix B and will not be addressed in this section.

4.1 Relativistic Local Density Approximation

In complete analogy to (4.1) the RLDA for $E_{xc}[j^\nu]$ is based on the xc-energy density $e_{xc}^{RHEG}(n)$ of the relativistic homogeneous electron gas (RHEG), which automatically reduces the full j^ν-dependence of the exact xc-energy functional to a pure density dependence,

$$E_{xc}^{RLDA}[n] = \int d^3r\, e_{xc}^{RHEG}(n_0 = n(\boldsymbol{r})). \tag{4.4}$$

The derivation of its lowest order contribution, i.e. the exchange energy, is discussed in some detail in Appendix B, illustrating in particular the UV-renormalisation required[12]. The final result, that has been obtained by a number of authors [93, 94, 95, 96, 18, 19], can be expressed as the nonrelativistic exchange energy density e_x^{NRHEG} multiplied by a relativistic correction factor. Separation into its longitudinal and transverse part according to (3.21) yields,

$$e_x^{RHEG,L/T}(n) = e_x^{NRHEG}(n)\,\Phi_x^{L/T}(\beta), \tag{4.5}$$

where β represents the basic relativistic density variable,

$$\beta = \frac{(3\pi^2 n)^{\frac{1}{3}}}{mc}, \tag{4.6}$$

and $e_x^{NRHEG}(n)$ and the $\Phi_x^{L/T}$ are explicitly given in (B.51, B.54, B.55). The variation of the $\Phi_x^{L/T}$ with β is shown in Fig. 4.1. One notices that the longitudinal contribution dominates in the low density limit and that it depends only weakly on β. The transverse part shows a stronger dependence on β and dominates in the high density regime, in which e_x^{RHEG} even changes its sign.

The transverse part can be analysed further if one separates it into a magnetic (or current-current) and a retardation component or if one restricts oneself to its weakly relativistic limit, the so-called Breit contribution [48, 97],

$$e_x^{RHEG,T} = e_x^{mag} + e_x^{ret} = e_x^{T,Breit} + \ldots,$$

where the individual terms follow from the structure of the transverse photon propagator $D_{\mu\nu}^{0,T}$,

$$D_{00}^{0,T}(k) = \frac{-4\pi e^2}{k^2 + i\varepsilon} - \frac{4\pi e^2}{\boldsymbol{k}^2} = \frac{4\pi e^2 (k^0)^2}{\boldsymbol{k}^4} + \ldots$$

$\Rightarrow$ retardation of charge-charge interaction

$$D_{ij}^{0,T}(k) = \delta_{ij}\frac{4\pi e^2}{k^2 + i\varepsilon} = \delta_{ij}\frac{-4\pi e^2}{\boldsymbol{k}^2} + \ldots$$

$\Rightarrow$ magnetic interaction (= Gaunt interaction + ...),

inserted into (3.32). The density dependence of these terms is illustrated in Fig. 4.2, which shows that the behaviour of e_x^T is largely dominated by the magnetic contribution. The relative minor role of higher order retardation effects is emphasised by the similarity between e_x^T and $e_x^{T,Breit}$.

[12] Note that in the case of e_x^{RHEG} vacuum corrections do not contribute after renormalisation, so that the complete result is identical with its no-sea/pair approximation.

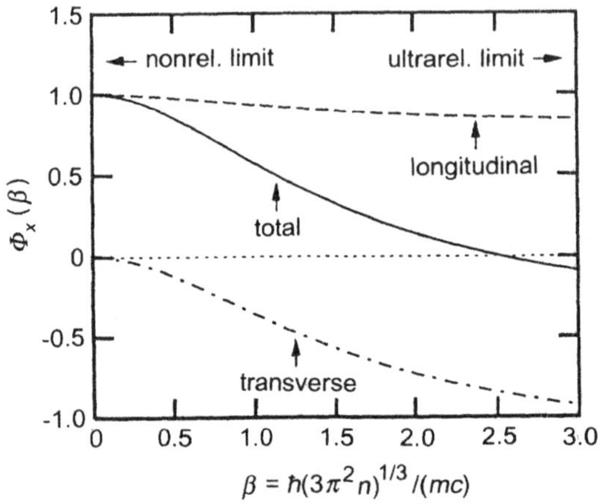

Fig. 4.1. Relativistic correction factor for the LDA exchange energy density: longitudinal contribution (B.54), transverse contribution (B.55) and total correction $\Phi_x^L + \Phi_x^T$

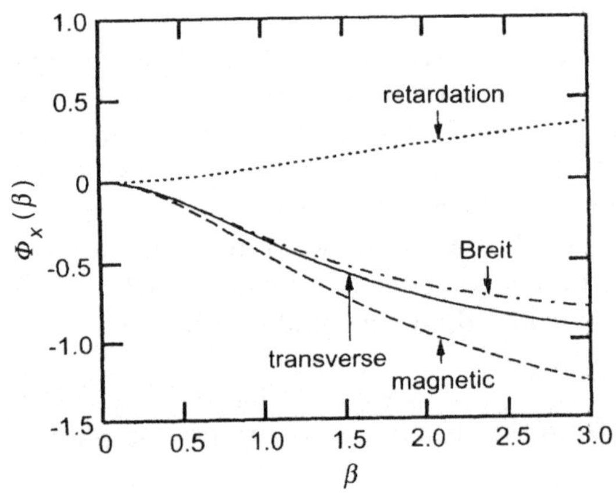

Fig. 4.2. Relativistic correction factor for the LDA exchange energy density: magnetic, retardation and Breit contribution ($e_x^{T,Breit}$ from Ref. [97] has been corrected for a typographical error)

In Fig. 4.3 we plot the density dependence of the resulting exchange potentials. The relevant range of density values for electronic structure calculations is indicated by the β-values at the origin and the expectation value of the radial coordinate of the $1S1/2$-orbital for the Kr and Hg atoms. One finds that relativistic effects are somewhat more pronounced for v_x than for e_x and are definitely relevant for inner shell features of high Z-atoms.

Relativistic correlation contributions in the LDA have so far only been considered on the basis of a partial resummation of those terms in the perturbation expansion in e^2 which are the most relevant in the high density limit. This contribution is either called the ring approximation (in accordance with its diagrammatic form) or, most often, the random phase approximation (RPA – which we shall use here) to e_c^{RHEG}. The detailed discussion of $e_c^{RHEG,RPA}$ is again relegated to Appendix B. In contrast to the case of exchange no closed analytical expression can be given for $e_c^{RHEG,RPA}$. Numerical results (within the

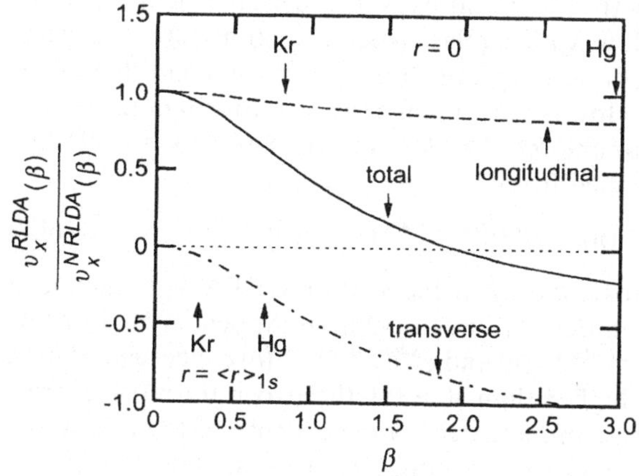

Fig. 4.3. Relativistic correction factor for the LDA exchange potential. The values of the densities of Kr and Hg at the origin ($r = 0$) and the r-expectation values of the $1s$-orbitals ($r = <r>_{1s}$) from RLDA-calculations using finite nuclei are also indicated

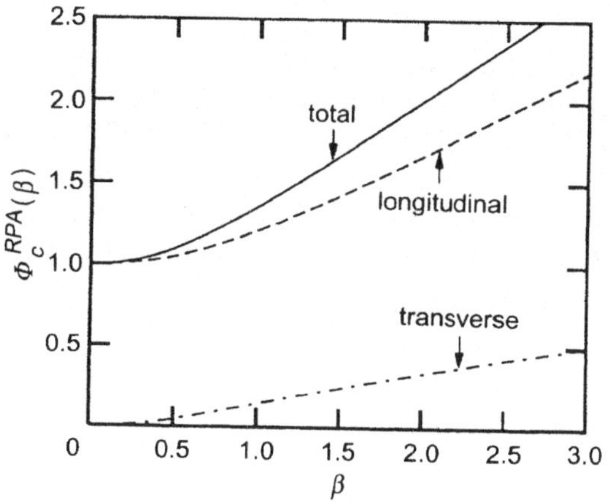

Fig. 4.4. Relativistic corrections to the LDA correlation energy density within the RPA [99]

no-sea approximation) have been obtained by Ramana and Rajagopal [98] and by Müller and Serot [99]. The resulting relativistic correction factor,

$$e_c^{RHEG,RPA}(n) = e_c^{NRHEG,RPA}(n)\,\Phi_c^{RPA}(\beta), \tag{4.7}$$

(again decomposed into its longitudinal and transverse contributions) is shown in Fig. 4.4. As for the exchange energy the relativistic correction for the RPA to e_c^{RHEG} is substantial (a similar correction factor is found for the corresponding correlation potential [22,98]).

One may ask whether the correction factor Φ_c^{RPA}, Eq. (4.7), can be of any use in actual applications. In this respect it is important to note that for the nonrelativistic HEG the RPA, which only contains one of the diagrammatic contributions to e_c^{NRHEG} of order e^4, is not an accurate approximation to the full e_c^{NRHEG} even for the highest relevant densities inside the nuclei of high-Z atoms ($\beta \approx 10 \Leftrightarrow r_s \approx 0.001$). On the other hand, the combination of $e_c^{RHEG,RPA}$ with the remaining contributions of order e^4 (here abbreviated by RPA +) agrees within an accuracy of 3% with e_c^{NRHEG} already for densities with $r_s \approx 0.1$ ($\beta \approx 0.1$), i.e. for densities below which relativistic corrections are irrelevant (note that all relativistic corrections are proportional to β^2 for small β). Consequently in this intermediate and low density regime the well known nonrelativistic e_c^{NRHEG} can also be used for the RLDA. An accurate form for the RLDA correlation energy is thus obtained from

$$e_e^{RHEG}(n) = e_c^{RHEG,RPA+}(n) - e_c^{NRHEG,RPA+}(n) + e_c^{NRHEG}(n), \tag{4.8}$$

where the relativistic corrections are only included via the RPA + (which also dominates in the relativistic high-density limit – compare Appendix B). For low and intermediate densities $e_c^{RHEG,RPA+}(n)$ and $e_c^{NRHEG,RPA+}(n)$ cancel each other, while for relativistic densities higher than $\beta \approx 0.1$ the two right-most expressions cancel. Unfortunately, the complete density dependence of the relativistic RPA + is not known. In the high density limit, however, one finds [96]

$$e_c^{RHEG,RPA+}(n) \underset{\beta \gg 1}{\approx} 1.4\, e_c^{RHEG,RPA}(n), \tag{4.9}$$

so that $\Phi_c^{RPA}(n)$ represents a lower bound for the total $\Phi_c(n)$. The error for the relevant density range $0.1 < \beta < 1$ is difficult to estimate, but should be less than the 30% indicated by (4.9) as on the nonrelativistic level $e_c^{NRHEG,RPA+}$ and $e_c^{NRHEG,RPA}$ differ by roughly 20% for $\beta = 0.1$. In view of the fact that the RPA + has not been fully evaluated, however, only the relativistic correction due to the RPA can presently be used instead of the more accurate form (4.8),

$$e_c^{RHEG}(n) \simeq e_c^{RHEG,RPA}(n) - e_c^{NRHEG,RPA}(n) + e_c^{NRHEG}(n). \tag{4.10}$$

4.2 Relativistic Weighted Density Approximation

In the nonrelativistic LDA one finds partial, but by no means satisfactory cancellation of self-interaction effects between E_{H} and E_x^{LDA}. The WDA [100, 101] constitutes a relatively direct approach, in which one attempts to improve on this situation by a density functional representation of the pair correlation function.

In order to apply the WDA in the relativistic regime, a fully covariant extension of the concept of the pair correlation function would be desirable. To our knowledge this is, however, not available. Nevertheless, if one restricts the discussion to the (instantaneous) longitudinal limit, one can express $E_{xc}^L[n]$, Eq. (3.21), via a relativistic pair correlation function defined in analogy to the nonrelativistic case as

$$g(\boldsymbol{r}, \boldsymbol{r}') = \frac{\langle \Phi_0 | \hat{n}(\boldsymbol{r}) \hat{n}(\boldsymbol{r}') | \Phi_0 \rangle}{n(\boldsymbol{r}) n(\boldsymbol{r}')} - \frac{\delta^{(3)}(\boldsymbol{r} - \boldsymbol{r}')}{n(\boldsymbol{r})}. \tag{4.11}$$

By definition, g is symmetric,

$$g(\boldsymbol{r}, \boldsymbol{r}') = g(\boldsymbol{r}', \boldsymbol{r}), \tag{4.12}$$

and satisfies the sum rule

$$\int d^3r' \, n(\mathbf{r}') [g(\mathbf{r}, \mathbf{r}') - 1] = -1 . \tag{4.13}$$

In fact, (4.13) is also satisfied by the x-only limit g_x of g, i.e. its lowest order contribution in e^2. In the relativistic case only this limit of the pair correlation function of the RHEG, $g_x^{RHEG}(k_F|\mathbf{r}-\mathbf{r}'|, k_F)$, specified in Eq. (B.68), is known (within the no-pair approximation [19, 102]), so that we restrict the subsequent discussion to the x-only limit.

For the transition from the x-only RLDA to the x-only RWDA one replaces the constant k_F inside $g_x^{RHEG}(k_F|\mathbf{r}-\mathbf{r}'|, k_F)$ by a local screening momentum $\tilde{k}_F(\mathbf{r})$, which is determined by the requirement, that the basic sum rule (4.13) be satisfied in the form

$$\int d^3r' \, n(\mathbf{r}') [g_x^{RHEG}(\tilde{k}_F(\mathbf{r})|\mathbf{r}-\mathbf{r}'|, \tilde{k}_F(\mathbf{r})) - 1] = -1 \tag{4.14}$$

for any given $\mathbf{r}$. One should note that this prescription leads to a truly non-local functional,

$$E_x^{L,RWDA}[n] = \frac{e^2}{2} \int d^3r \int d^3r' \, \frac{n(\mathbf{r}) n(\mathbf{r}')}{|\mathbf{r}-\mathbf{r}'|} \times [g_x^{RHEG}(\tilde{k}_F(\mathbf{r})|\mathbf{r}-\mathbf{r}'|, \tilde{k}_F(\mathbf{r})) - 1], \tag{4.15}$$

but violates the general symmetry (4.12) so that one obtains different x-only potentials depending on whether the WDA scheme is applied before or after variation with respect to the density. Moreover, using g_x^{RHEG} as a kernel the x-only RWDA reduces to a pure density functional, without any $\mathbf{j}$-dependence.

The main advantage of this approximation is that it is exact for two-electron systems (if the correct $\tilde{k}_F(\mathbf{r}) = 0$ is utilised in (4.15) before performing the functional differentiation (3.17) required for its application) and also correctly accounts for the self-interaction energies of individual closed shells if a shell-partitioning scheme is used [71]. Furthermore, the RWDA reproduces the asymptotic r^{-1} proportionality of the exact x-only potential (although with the incorrect prefactor of 1/2 [103]).

5 Relativistic DFT Results for Atoms

So far only few applications of the RKS-equations (3.25–3.29) utilising a relativistic form for $E_{xc}[n]$ have been reported (and none for the 'field theoretical' KS-equations (3.5, 3.15–3.17)). MacDonald and Vosko [19] as well as Das et al. [104] analysed the x-only RLDA for high-Z atoms and ions, emphasising the importance of relativistic corrections to $E_x[n]$. This work has been extended by inclusion of the RPA limit of $E_c[n]$ in the RLDA by Ramana et al. [97]. These authors also point out the need for nonlocal corrections, in particular in the case of the x-only energy. The influence of relativistic corrections to $E_{xc}[n]$ on the band structures of Pd and Pt has been examined by MacDonald et al. [105], who find a significant effect on the Fermi surface of Pt. In particular, they conclude that relativistic corrections to $v_x[n]$ can be as important as the nonrelativistic $v_c[n]$ even for the valence levels of high-Z systems.

On the other hand, a large number of relativistic Slater calculations [23] (Dirac-Fock-Slater – DFS), in which the RKS-equations are used with the nonrelativistic x-only LDA, can be found in the literature (see e.g. [8,9]). However, no attempt is made to review this extensive body of literature here.

In this section we summarise the properties of the approximations to $E_{xc}[n]$ discussed in Section 4 in applications to atoms. All results presented in the following [36] are based on the direct numerical solution of Eqs. (3.25–3.29) using a nuclear potential which corresponds to a homogeneously charged sphere [69][13]. Only spherical, i.e. closed subshell, atoms and ions are considered. Whenever suitable we use Hg as a prototype of all high-Z atoms.

In the following we investigate both the importance of relativistic corrections to $E_{xc}[n]$ and the adequacy of the RLDA and RWDA to reproduce them, using ROPM and conventional relativistic ab initio results as reference standards. Consequently it is not so much the accuracy of the total relativistic xc-energies which is of interest, but rather their intrinisically relativistic ingredients, i.e. the differences between the selfconsistent relativistic and the self consistent non-relativistic longitudinal xc-energies,

$$\Delta E^L_{x,c} = E^L_{x,c}[n^R] - E^{NR}_{x,c}[n^{NR}] \tag{5.1}$$

and the transverse xc-energies $E^T_{x,c}$. Here n^R represents the selfconsistent density obtained by solving the RKS-equations with a given (approximate) form for $E^L_{x,c}[n]$. Its nonrelativistic counterpart n^{NR} is obtained by solution of the nonrelativistic KS-equations with the nonrelativistic limit $E^{NR}_{x,c}[n]$ of $E^L_{x,c}[n]$. In the x-only limit one can examine relativistic corrections on a local level via

$$\Delta v^L_x(r) = \frac{v^L_x([n^R];r) - v^{NR}_x([n^{NR}];r)}{v^{NROPM}_x([n^{NROPM}];r)} . \tag{5.2}$$

Δv^L_x represents the percentage deviation of the selfconsistent relativistic potential $v^L_x([n^R];\ r)$ from the corresponding selfconsistent nonrelativistic potential $v^{NR}_x([n^{NR}];\ r)$. The selfconsistent $v^L_x([n^R];\ r)$ is calculated by insertion of the self-consistent $n^R(r)$ into the functional derivative (3.28) for that $E^L_x[n]$ which has been used to determine $n^R(r)$. In particular, the ROPM x-only potential can, in principle, be obtained by insertion of the exact x-only density $n^{ROPM}(r)$ into the exact $v^L_x([n];r) = v^{ROPM}_x([n];r)$ and thus can be used as a comparative standard:

[13] All our calculations have been performed without a Latter-type correction for $v_x(r)$ in the large-r regime (compare [104]).

$\Delta v_x^{L,ROPM}(r)$ constitutes a direct local measure of relativistic corrections in the x-only limit with which $\Delta v_x^L(r)$ from various approximations can be compared (for the corresponding deviations of approximate $v_x^{NR}(r)$ from $v_x^{NROPM}(r)$ see [59, 60, 106]).

5.1 Longitudinal x-only Functionals

Before considering approximations to the exact $E_x^L[n]$, Eq. (3.37), some general remarks on the importance of relativistic corrections on the ROPM-level seem appropriate. First of all one gleans from Table 5.1 that the relativistic contributions in E_x^L are quite substantial: For instance for Hg one obtains a 20 hartree shift when going from the nonrelativistic OPM- to the ROPM-value which represents 5.8% of E_x^L, very similar to the 6.8% relativistic correction which is found for E_{tot}^L. Clearly, relativistic effects are most important for the innermost orbitals. This is obvious from Table 5.2 showing the single-particle spectrum of Hg. Note that the $2P1/2$-eigenvalue is modified by 17% and that even the outermost $6S1/2$-eigenvalue experiences a 26% shift (reflecting the 'gold maximum'). Table 5.2 also demonstrates that apart from the physically relevant highest occupied eigenvalue and in spite of the very similar ground state energies ROPM and RHF single particle energies differ substantially (compare [59]), in consistency with their auxiliary character.

Table 5.1. Longitudinal (Coulomb) x-only energies ($-E_x^L$) for closed subshell atoms from NROPM-, ROPM-, RHF-, DFS-, RLDA-, and RWDA-calculations [36, 69] (all energies are in hartree).

Atom	NROPM	ROPM	RHF	DFS	RLDA	RWDA
He	1.026	1.026	1.026	0.853	0.853	1.026
Be	2.666	2.667	2.668	2.278	2.278	2.706
Ne	12.105	12.120	12.123	10.952	10.944	12.843
Mg	15.988	16.017	16.023	14.564	14.550	17.093
Ar	30.175	30.293	30.303	27.897	27.844	32.419
Ca	35.199	35.371	35.383	32.702	32.627	37.967
Zn	69.619	70.245	70.269	66.107	65.834	37.967
Kr	93.833	95.048	95.072	89.784	89.293	75.604
Sr	101.926	103.404	103.429	97.836	97.251	102.095
Pd	139.113	141.898	141.930	134.971	133.887	111.133
Cd	148.879	152.143	152.181	144.931	143.687	152.275
Xe	179.062	184.083	184.120	175.926	174.102	163.321
Ba	189.065	194.804	194.841	186.417	184.363	197.564
Yb	276.143	288.186	288.265	278.642	274.386	209.171
Hg	345.240	365.203	365.277	354.299	347.612	310.268
Rn	387.445	414.082	414.151	402.713	394.102	392.339
Ra	401.356	430.597	430.664	419.218	409.871	444.584
No	511.906	564.309	564.415	554.242	538.040	462.365
						606.216

Table 5.2. Single particle energies ($-\varepsilon_{nlj}$) for Hg from NROPM-, ROPM- and RHF-calculations in comparison with DFS-, RLDA- and RWDA-results (all energies are in hartree).

Level	NROPM	ROPM	RHF	DFS	RLDA	RWDA
1S1/2	2756.925	3047.430	3074.228	3047.517	3044.410	3051.995
2S1/2	461.647	540.056	550.251	539.713	539.250	540.530
2P1/2	444.015	518.061	526.855	518.164	517.746	519.244
2P3/2	444.015	446.682	455.157	446.671	446.399	447.469
3S1/2	108.762	128.272	133.113	128.001	127.905	128.292
3P1/2	100.430	118.350	122.639	118.228	118.148	118.592
3P3/2	100.430	102.537	106.545	102.397	102.346	102.691
3D3/2	84.914	86.201	89.437	86.085	86.060	86.364
3D5/2	84.914	82.807	86.020	82.690	82.668	82.959
4S1/2	23.522	28.427	30.648	28.067	28.046	28.200
4P1/2	19.895	24.161	26.124	23.871	23.854	24.023
4P3/2	19.895	20.363	22.189	20.039	20.030	20.167
4D3/2	13.222	13.411	14.797	13.148	13.146	13.271
4D5/2	13.222	12.700	14.053	12.434	12.432	12.553
4F5/2	4.250	3.756	4.473	3.556	3.559	3.665
4F7/2	4.250	3.602	4.312	3.402	3.404	3.509
5S1/2	3.501	4.403	5.103	4.290	4.286	4.349
5P1/2	2.344	3.012	3.538	2.898	2.896	2.955
5P3/2	2.344	2.363	2.842	2.219	2.218	2.265
5D3/2	0.538	0.505	0.650	0.363	0.363	0.397
5D5/2	0.538	0.439	0.575	0.296	0.296	0.328
6S1/2	0.262	0.329	0.328	0.222	0.222	0.254

The percentage relativistic contribution in the exact v_x^L is plotted in Fig. 5.1. One clearly observes an 'oscillatory' behavior between 0.01a.u. and 3a.u. reflecting the shell structure of the atom: Most relativistic single particle orbitals are shifted towards the nucleus with respect to the nonrelativistic ones[14], resulting in an oscillating structure of $n^R(r) - n^{NR}(r)$ and thus also of Δv_x^L. This leads to an additional attraction of about 20% in the region close to the nucleus. Even in the valence regime the relativistic corrections amounts to roughly 10%.

In view of these results there is little doubt that the density dependence of explicit density functionals for the x-only energy has to be modified in the relativistic regime.

On the basis of exact OPM results one can now examine approximate forms for $E_x^L[n]$. Here we consider the two relativistic functionals available, i.e. the RLDA and the RWDA[15] discussed in Section 4, as well as the nonrelativistic LDA (NRLDA), i.e. the DFS-approach with $\alpha = 2/3$. For comparison with the ROPM-results in Table 3.1 the corresponding longitudinal x-only ground state energies and highest occupied eigenvalues are listed in Table 5.3. Two features are apparent from this comparison: On the one hand, the total ground state

[14] All $S1/2$ and $P1/2$ orbitals are considerably contracted, the $P3/2$, $3D3/2$ and $4D3/2$ orbitals only slightly. While the $5D3/2$ as well as the $3D5/2$ and $4D5/2$ r-expectation values remain essentially unchanged, the $5D5/2$, $4F5/2$ and $4F7/2$ orbitals are expanded.

[15] All numerical results from the RWDA have been obtained with the fit (B.73) to the exact pair correlation function.

Table 5.3. Longitudinal x-only ground state energies ($-E^L_{tot}$) and highest occupied eigenvalues ($-\varepsilon^L_{mk}$) for closed subshell atoms from DFS, RLDA and RWDA calculations [36] (all energies are in hartree). Note that for Yb the DFS-, RLDA- and RWDA-calculations incorrectly predict the 4F7/2-orbital to be most weakly bound. The values given (*), however, correspond to the 6S1/2-orbital.

Atom	$-E^L_{tot}$			$-\varepsilon^L_{mk}$		
	DFS	RLDA	RWDA	DFS	RLDA	RWDA
He (1S1/2)	2.724	2.724	2.862	0.517	0.517	0.599
Be (2S1/2)	14.226	14.226	14.609	0.170	0.170	0.204
Ne (2P3/2)	127.635	127.628	129.417	0.441	0.441	0.511
Mg (3S1/2)	198.569	198.556	200.963	0.142	0.142	0.181
Ar (3P3/2)	526.387	526.337	530.747	0.331	0.331	0.369
Ca (4S1/2)	677.118	677.047	682.204	0.112	0.112	0.140
Zn (4S1/2)	1790.721	1790.458	1799.949	0.191	0.191	0.232
Kr (4P3/2)	2783.758	2783.282	2795.778	0.291	0.291	0.321
Sr (5S1/2)	3172.638	3172.071	3185.631	0.104	0.104	0.129
Pd (4D5/2)	5037.733	5036.677	5054.707	0.111	0.111	0.151
Cd (5S1/2)	5586.299	5585.086	5604.337	0.181	0.181	0.218
Xe (5P3/2)	7438.858	7437.076	7460.124	0.250	0.250	0.275
Ba (6S1/2)	8127.344	8125.336	8149.714	0.095	0.095	0.116
Yb (6S1/2)	14058.528	14054.349	14089.603	0.119*	0.119*	0.146*
Hg (6S1/2)	19638.195	19631.622	19675.706	0.222	0.222	0.254
Rn (6P3/2)	23590.763	23582.293	23632.105	0.214	0.214	0.237
Ra (7S1/2)	25016.763	25007.568	25059.377	0.097	0.097	0.117
No (7S1/2)	36730.804	36714.839	36782.219	0.128	0.128	0.156

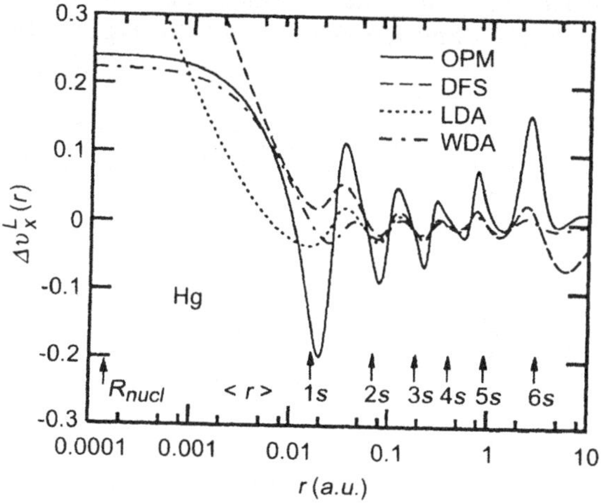

Fig. 5.1. Percentage relativistic correction to the longitudinal x-only potential $\Delta v^L_x(r)$, Eq. (5.2), for neutral Hg from OPM-, DFS-, LDA- and WDA-calculations [69]. Also shown are the nuclear radius R_{nucl} and the r-expectation values of the individual s-orbitals

energies of all three approximations differ considerably from the corresponding ROPM values (in Table 3.1). While the RWDA overestimates the binding energy substantially, both the RLDA and the DFS energies are too small. This deviation can be directly traced to the corresponding E^L_x (given in Table 5.1): For instance for Hg the error of 17.204 hartree for E^L_{tot} matches rather well with the corresponding error of 17.591 hartree for E^L_x. The same holds for the DFS- and RWDA-results (and throughout the periodic table).

On the other hand, the highest occupied eigenvalues which reflect the asymptotic form of the density, and thus even in the x-only limit should be close to the ionisation potential, are too small by roughly 50% for the DFS and RLDA and by 30–40% for the RWDA[16]. Thus while the latter improves on the asymptotic form of v_x^L in principle, the effect of this formal improvement on the physically relevant part of the asymptotic regime is rather limited. Note further, that for Yb all three schemes incorrectly predict the 4F7/2-orbital to be most weakly bound instead of the 6S1/2-orbital. The same deficiency has been observed for Cr and Cu in the nonrelativistic case [59]. These difficulties to reproduce the size and the ordering of the outermost eigenvalues are well known from the nonrelativistic case and are not related to the relativistic corrections in $E_x^L[n]$.

The percentage deviation of the relativistic correction ΔE_x^L, Eq. (5.1), obtained for the three approximations with respect to the OPM standard is plotted in Fig. 5.2. While both the DFS and the WDA overestimate the exact ΔE_x^L by about 10%, the LDA underestimates them by roughly 20% in the relevant Z-regime. One notes, however, that the deviation of the LDA seems to be decreasing with increasing Z in contrast to the errors of the DFS and the WDA. In any case, all three approximations are not satisfactory from a quantitative point of view.

The fact that the longitudinal x-only energies obtained from some approximate functional are rather insensitive to the density inserted into the functional [19] is demonstrated explicitly in Table 5.4. Here the E_x^L obtained by insertion of the selfconsistent ROPM-, DFS-, RLDA-, and RWDA-densities for Hg into the

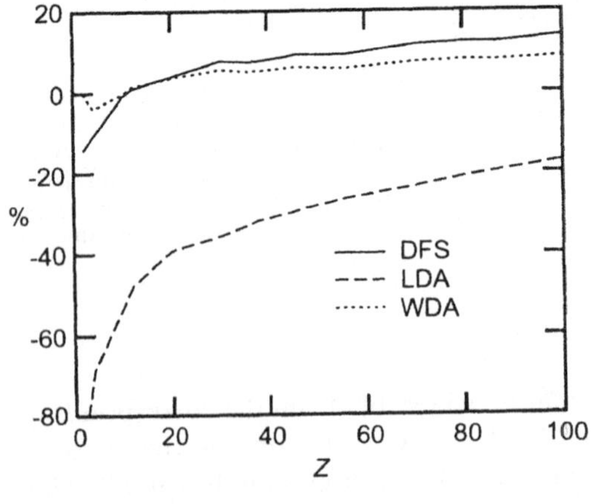

Fig. 5.2. Percentage deviation of ΔE_x^L for neutral atoms with respect to the exact OPM results obtained from DFS, LDA and WDA

[16] For He the selfconsistent RWDA-potential has also been evaluated by insertion of $\tilde{k}_F$ into the complete functional derivative of (4.15).

Table 5.4. Longitudinal (Coulomb) exchange energies ($-E_x^L$) for Hg obtained by insertion of selfconsistent ROPM-, DFS-, RLDA- and RWDA-densities into the corresponding $E_x^L[n]$ (all energies are in hartree).

Hg $E_x^L[n]$	Densities from ROPM	 DFS	 RLDA	 RWDA
ROPM	365.203	364.865	364.745	365.366
DFS	354.508	354.299	354.181	354.805
RLDA	347.928	347.722	347.612	348.216
RWDA	391.988	391.785	391.656	392.339

corresponding functionals are listed[17]: While the energy differences between the four functionals (for any given density) are substantial, the differences generated by the four densities for a given functional are of the order of a few 100 mhartree. These latter differences originate from the selfconsistency procedure, i.e. from the corresponding v_x^L. In other words: The selfconsistent n^R is dominated by relativistic kinematics and the nuclear attraction, the x-potential only plays a minor role. Nevertheless the impact of the relativistic corrections in v_x^L on atomic exchange energies is larger than the differences between more refined nonlocal functionals on the nonrelativistic level. Consequently relativistic corrections to the longitudinal x-only energy should not just be taken into account perturbatively, but rather be included in the selfconsistency loop.

The relativistic corrections in v_x^L are also apparent for the innermost RKS-eigenvalues. Notwithstanding their auxiliary nature, these single particle levels are a direct measure of the quality of any approximation to the relativistic $v_x^L[n]$. Focusing on the 1S1/2-eigenvalue of Hg Table 5.2 shows that the values for both the RLDA and the RWDA deviate by several hartree from the ROPM eigenvalue, indicating that $v_x^{L,RLDA}$ is not sufficiently attractive, while $v_x^{L,RWDA}$ overestimates the actual ROPM-potential. The innermost DFS-eigenvalues, on the other hand, seem to agree rather well with the ROPM-results, which however, must be regarded as fortuitious.

This last point is emphasised if one looks at the relativistic correction to the x-only potential, Δv_x^L, shown for Hg in Fig. 5.1. While the shell oscillations of the exact v_x^L are at least partially reproduced by all three approximations (the OPM-amplitudes being larger by more than a factor of 2), both the DFS- and the RLDA-potentials are far from the exact $\Delta v_x^L(r)$ in the region close to the nucleus. For these approximations v_x^L is proportional to $n^{1/3}$ for high densities [19], with the prefactor being larger for DFS. In spite of the fact that we use extended nuclei this proportionality starts to show up at about the r-expectation value of the 1S1/2-orbital. On the other hand, the RWDA follows the exact

[17] Note that insertion of a given density into the exact x-only functional corresponds to insertion of the RKS-orbitals which yield this density into (3.37).

$\Delta v_x^L(r)$ rather closely in the small-r regime. This can be understood from the fact that this part of $v_x^L(r)$ is dominated by the self-interaction correction of the 1S1/2-orbital, for which the RWDA-concept is particularly suitable. This result suggests to search for a relativistic extension of the model pair correlation functions which are more accurate than $g_x^{(R)HEG}$ in the nonrelativistic regime [36, 100].

In order to anlayse the origin of the failure of the RLDA it is useful to separate the two effects leading to the relativistic correction in the x-only energies [36],

$$\Delta E_x^L = \Delta E_x(dens) + \Delta E_x(\text{fctl}) \tag{5.3}$$

$$\Delta E_x(dens) = E_x^{NR}[n^R] - E_x^{NR}[n^{NR}] \tag{5.4}$$

$$\Delta E_x(fctl) = E_x^L[n^R] - E_x^{NR}[n^R], \tag{5.5}$$

i.e. the correction $E_x^{NR}[n^R] - E_x^{NR}[n^{NR}]$ resulting from the difference between the selfconsistent relativistic density n^R and the selfconsistent non-relativistic density n^{NR} and the contribution $E_x^L[n^R] - E_x^{NR}[n^R]$ arising from the relativistic modification of the functional form of the x-only functional. Using the LDA one finds for Hg $\Delta E_x(dens) = -22.303$ hartree and $\Delta E_x(fctl) = 6.569$ hartree. Thus the dominating density contribution increases the x-only energy, while the functional correction leads to a reduction (as is immediately obvious from Φ_x^L, Fig. 4.1) and both are of the same order of magnitude. Moreover, $\Delta E_x(dens)$ is rather insensitive to the precise form of $E_x^L[n]$ due to the insensitivity of n^R to v_x^L noted earlier. For instance the values of $\Delta E_x(dens)$ from the LDA are almost identical with the complete ΔE_x^L from the DFS approach (-22.421 hartree for Hg). $\Delta E_x(fctl)$, on the other hand, is exclusively determined by the form of $E_x^L[n]$. The need to balance the contributions $\Delta E_x(dens)$ and $\Delta E_x(fctl)$ becomes even more obvious if a more accurate nonrelativistic $E_x^{NR}[n]$ like a GGA is used: For Hg Becke's GGA [10] (applied without any functional correction) yields $\Delta E_x^L = \Delta E_x(dens) = 24.978$ hartree which overshoots the exact value by 5.015 hartree to be compensated by $\Delta E_x(fctl)$.

For an evaluation of the exact $\Delta E_x(dens)$ and $\Delta E_x(fctl)$ one needs the set of nonrelativistic KS-orbitals which generates the ROPM density (required for $E_x^{NR}[n^R]$). While the calculation of these orbitals is a rather complicated task and thus will not be addressed here, there is one type of system for which the exact $\Delta E_x(dens)$ is identical with the complete ΔE_x^L as the functional contribution $\Delta E_x(fctl)$ vanishes: For all two-electron systems the exact $E_x^L[n]$,

$$E_x^L[n] = -\frac{e^2}{4}\int d^3r \int d^3r' \frac{n(\boldsymbol{r})n(\boldsymbol{r}')}{|\boldsymbol{r}-\boldsymbol{r}'|}, \tag{5.6}$$

is identical with the exact $E_x^{NR}[n]$. The OPM values for Hg^{78+} can thus be compared with the corresponding LDA results,

$$\Delta E_x^{OPM}(dens) = -7.963\,\text{hartree}, \quad \Delta E_x^{OPM}(fctl) = 0$$

$$\Delta E_x^{LDA}(dens) = -7.198\,\text{hartree}, \quad \Delta E_x^{LDA}(fctl) = 3.157\,\text{hartree}.$$

The error of 765 mhartree for $\Delta E_x(dens)$ reflects the fact that the LDA is missing important nonlocal contributions responsible for the cancellation of the self interaction energy already on the nonrelativistic level. The much larger LDA error for $\Delta E_x(fctl)$, on the other hand, directly shows that nonlocal corrections are even more relevant for the relativistic correction $E_x^L[n^R] - E_x^{NR}[n^R]$. Furthermore, the total deviation of 3.922 hartree in the LDA found for Hg^{78+} agrees rather well with the deviation of 4.229 hartree for neutral Hg, suggesting that the major contribution to ΔE_x^L comes from the $1S1/2$-electrons, while contributions of the order of 300 mhartree are due to the relativistic rearrangement of other orbitals.

The last point becomes more transparent if one decomposes $\Delta v_x^L(r)$ into its density and functional components (in analogy to Eqs. (5.3–5.5)),

$$\Delta v_x^L(r) = \Delta v_x(r, dens) + \Delta v_x(r, fctl) \tag{5.7}$$

$$\Delta v_x(r, dens) = \frac{v_x^{NR}([n^R];r) - v_x^{NR}([n^{NR}];r)}{v_x^{NROPM}([n^{NROPM}];r)} \tag{5.8}$$

$$\Delta v_x(r, fctl) = \frac{v_x^L([n^R];r) - v_x^{NR}([n^R];r)}{v_x^{NROPM}([n^{NROPM}];r)}\,. \tag{5.9}$$

One finds that the oscillatory structure of $\Delta v_x^L(r)$ observed in Fig. 5.1 is completely due to the density correction $\Delta v_x(r, dens)$, as can be gleaned from Fig. 5.3[18]. As expected, the functional correction $\Delta v_x(r, fctl)$ is only relevant for the high density small-r regime, i.e. the innermost orbitals. In this regime, however, the $\Delta v_x^L(r)$ from the LDA is rather different from the exact OPM result. In order to demonstrate that both components of $\Delta v_x^L(r)$ contribute to the error of the LDA it is again advantageous to consider Hg^{78+}: Similar to the situation for ΔE_x^L the exact $\Delta v_x(r, fctl)$ vanishes. As is clear from Fig. 5.4 the LDA does not reproduce the exact $\Delta v_x^L(r)$. In particular, $|\Delta v_x^{LDA}(r, fctl)|$ does not vanish but is of the same order of magnitude as the exact $\Delta v_x(r, dens)$.

Consequently the origin of the failure of the RLDA for neutral atoms is twofold: The error is dominated by the self interaction of the $1S1/2$-electrons which manifests itself mainly in the functional correction $E_x^L[n] - E_x^{NR}[n]$. The smaller (but chemically equally relevant) error in the density components $\Delta E_x(dens)$ and $\Delta v_x(r, dens)$, on the other hand, is a result of the difficulties of the LDA to reproduce the shell structure of the exact density already in the nonrelativistic regime [59, 106]: The amplitude (not the location) of the relativistic shift of the individual shells becomes larger if the shells are more pronounced. While the first source of errors calls for improved, i.e., nonlocal, relativistic corrections to the x-only energy functional, the second problem can only be resolved by using a more accurate nonrelativistic starting point for $E_x^L[n]$ than the NRLDA.

[18] The LDA-result for $\Delta v_x(r, dens)$ is almost identical with the complete selfconsistent $\Delta v_x^L(r)$ from the DFS-scheme.

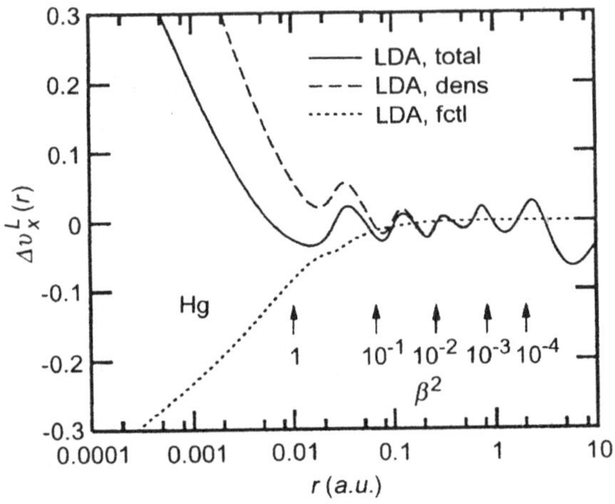

Fig. 5.3. Decomposition of Δv_x^L (total) into its density component (5.8) (dens) and the functional contribution (5.9) (fctl) : LDA-results for neutral Hg. Also shown are some characteristic β^2-values

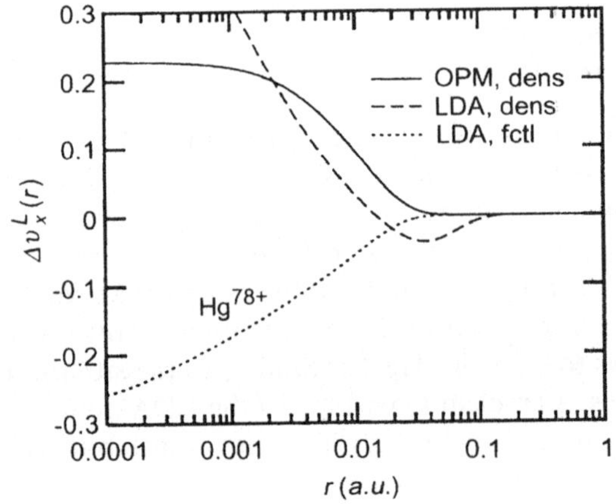

Fig. 5.4. Decomposition of $\Delta v_x^L(r)$ for Hg^{78+}: Exact $\Delta v_x(r, dens)$ (OPM, dens) in comparison with $\Delta v_x^{LDA}(r, dens)$ (LDA, dens) and $\Delta v_x^{LDA}(r, fctl)$ (LDA, fctl)

5.2 Relativistic LDA for $E_x^T[n]$

We now turn to a brief discussion of the RLDA for $E_x^T[n]$, which is the only density functional approximation for this quantity available to date. The transverse x-only energies calculated from the selfconsistent RLDA densities are compared with RHF results [68, 48] in Table 5.5. The RHF orbitals used for these energies [68] could be replaced by the ROPM or RLDA orbitals without significantly changing the resulting E_x^T's: For Hg, for example, one obtains 22.145 hartree by inserting the selfconsistent RLDA-orbitals into the transverse Fock term. As pointed out by MacDonald and Vosko [19] the RLDA overestimates the exact E_x^T by about a factor of 1.5. Moreover, in contrast to the RLDA's error for ΔE_x^L the error for E_x^T does not decrease with increasing Z. Thus the RLDA can only serve a qualitative measure of transverse x-only energies.

Table 5.5. Transverse exchange energies (E_x^T) and its Breit approximation ($E_x^{T,Breit}$) from RHF- [68] and RLDA-calculations [36]. Also the total correction $\Delta E_x^L + E_x^T$ is given (all energies are in hartree).

Atom	RHF			RLDA		
	E_x^T	$E_x^{T,Breit}$	$\Delta E_x^L + E_x^T$	E_x^T	$E_x^{T,Breit}$	$\Delta E_x^L + E_x^T$
He	0.000	0.000	0.000	0.000	0.000	0.000
Be	0.001	0.001	0.000	0.002	0.002	0.001
Ne	0.017	0.017	0.002	0.035	0.035	0.028
Mg	0.032	0.032	0.003	0.065	0.064	0.050
Ar	0.132	0.132	0.014	0.249	0.248	0.180
Ca	0.191	0.191	0.019	0.353	0.352	0.249
Zn	0.759	0.761	0.131	1.322	1.312	0.920
Kr	1.419	1.427	0.203	2.401	2.373	1.586
Sr	1.710	1.720	0.231	2.867	2.831	1.862
Pd	3.290	3.318	0.503	5.358	5.263	3.399
Cd	3.808	3.842	0.541	6.162	6.045	3.840
Xe	5.711	5.775	0.687	9.089	8.877	5.431
Ba	6.473	6.552	0.733	10.255	10.001	6.038
Yb	13.897	14.148	1.846	21.557	20.778	12.355
Hg	22.166	22.665	2.192	34.201	32.654	18.444
Rn	28.676	29.397	2.028	44.313	42.046	22.964
Ra	31.148	31.957	1.899	48.202	45.636	24.638
No	53.576	55.248	1.154	84.987	79.083	41.085

Furthermore, the rather subtle cancellation observed for the exact ΔE_x^L and E_x^T is not found in the RLDA. The RLDA thus completely misrepresents the total relativistic correction to atomic exchange energies [36] (see Table 5.5).

As is well known the Breit approximation to E_x^T is rather close to the full E_x^T throughout the periodic table for both RHF [48] and the RLDA [97]. Taking into account the subtle cancellation between E_x^T and ΔE_x^L, however, the difference between E_x^T and $E_x^{T,Breit}$ amounts to an appreciable percentage of the total $\Delta E_x^L + E_x^T$, at least for high-Z atoms. Furthermore, the RLDA for the Breit approximation leads to energies smaller than the $E_x^{T,RLDA}$ in contrast to the correct relationship. Thus while the error of the RLDA for ΔE_x^L, i.e. in the nonrelativisitic (Coulomb) limit of the full photon exchange, is roughly 20%, it increases to about 50% for $E_x^{T,Breit}$, i.e. the first order weakly relativistic correction to the Coulomb interaction, and finally the RLDA even misses the correct sign for $E_x^T - E_x^{T,Breit}$, i.e. all higher order relativistic corrections to the Coulomb interaction.

The failure of the RLDA for $E_x^T[n]$ is, however, not surprising in view of its RHEG origin. The finite speed of light plays a much more important role for an infinite system like the RHEG than for atoms: In the RHEG extremely distant points in space are interacting with each other so that the travelling time of photons between these points modifies the form of the long range electromagnetic forces (similar to the difference between Casimir-Polder and van der Waals forces [26]). The electronic density of atoms, on the other hand, is rather localised so that the actual propagation of photons as compared with the instantaneous Coulomb interaction can not have the same impact.

5.3 Relativistic LDA for $E_c[n]$

Our review of the properties of explicit relativistic functionals is concluded by an examination of the RLDA for $E_c[n]$ (using the RPA for the relativistic corrections, Eq. (4.10), and the parametrisation of Vosko et al. [70] for $e_c^{NRHEG}(n)$). The RLDA correlation energies obtained by solution of (3.25) (using the RLDA for the complete v_{xc}^L) for a number of atoms are listed in Table 5.6 and compared with the usual quantum chemical correlation energy E_c^{RCI},

$$E_c^{RCI} = E_{tot} - E_{tot}^{RHF} - \Delta E^{QED} . \qquad (5.10)$$

Here the RHF ground state energy E_{tot}^{RHF} is understood to include the transverse exchange energy contribution as well as the finite nuclear size correction and ΔE^{QED} represents all quantum electrodynamical corrections (when extracting E_{tot} from experimental data also the finite nuclear mass has to be taken into account [107]). Consequently E_c^{RCI} contains all relativistic correlation contributions. Unfortunately, except for two-electron systems, E_{tot}^{RHF} is not identical with the corresponding ROPM ground state energy so that the exact density functional E_c does not exactly agree with E_c^{RCI} but is somewhat larger in magnitude. This small difference is, however, irrelevant for the present purpose, so that one may use E_c^{RCI} as a reference standard[19]. The same holds for the nonrelativistic quantum chemical correlation energy,

$$E_c^{NRCI} = E_{tot}^{NR} - E_{tot}^{NRHF} , \qquad (5.11)$$

with respect to the nonrelativistic E_c. Moreover, full CI results are only available for rather small atoms so that we resort to data of second order many-body perturbation theory (MBPT2) for E_c^{NRCI} and for E_c^{RCI} [108] in the case of the more interesting large atoms. On the level of accuracy required for the present purpose the differences between MBPT2 and full CI energies are not relevant.

Table 5.6 demonstrates once more the well known fact that the nonrelativistic LDA overestimates the exact atomic correlation energies by about a factor of 2. Here, however, not the accuracy of the complete functional (4.10) is of interest, but rather the relativistic corrections ΔE_c^L and E_c^T, as the correction scheme (4.10) could be combined with more accurate nonrelativistic $E_c[n]$ like GGAs. Table 5.6 shows that both ΔE_c^L and E_c^T are much smaller than their x-only counterparts. On the other hand, ΔE_c^L and E_c^T add up constructively so that the total correction $\Delta E_c^L + E_c^T$ is somewhat closer to $\Delta E_x^L + E_x^T$ than the individual components: For Hg one obtains $\Delta E_c^L + E_c^T = -0.49$ hartree within MBPT2 compared with the exact $\Delta E_x^L + E_x^T$ of about 2.19 hartree. Nevertheless, in absolute values the relativistic corrections to $E_c[n]$ are clearly less important than those to $E_x[n]$.

[19] Note that the presently remaining numerical uncertainty in the available CI-correlation energies is much smaller than the error of the RLDA discussed here.

Table 5.6. Comparison of LDA [36], CI (estimated from nonrelativistic CI-calculations for the three innermost electrons and the experimental ionisation potentials of all other electrons [109]) and MBPT2 [108] correlation energies for neutral atoms: E_c^{NR} – nonrelativistic correlation energy, ΔE_c^L – relativistic contribution in the longitudinal correlation energy, E_c^T – transverse correlation energy (in the case of the MBPT2 only the dominating Breit contribution to E_c^T is given – all energies are in mhartrees).

Atom	$-E_c^{NR}$			$-\Delta E_c^L$		$-E_c^T$	
	MBPT2	CI	LDA	MBPT2	LDA	MBPT2	LDA
He	37.14	42.04	111.47	0.00	0.00	0.04	0.00
Be		94.34	224.44		0.02		0.02
Ne	383.19	390.47	743.38	0.20	0.38	1.87	0.32
Mg		438.38	891.42		0.75		0.57
Ar	697.28	722.16	1429.64	0.84	2.60	7.92	1.89
Zn	1650.61		2665.20	10.51	10.97	26.43	7.92
Kr	1835.43		3282.95	11.39	19.61	41.07	13.10
Cd	2618.11		4570.56	35.86	44.79	82.32	28.58
Xe	2921.13		5200.19	37.57	64.73	108.75	39.27
Hg	5086.24		8355.68	203.23	200.87	282.74	113.08
Rn	5392.07		9026.90	195.36	257.00	352.60	138.43

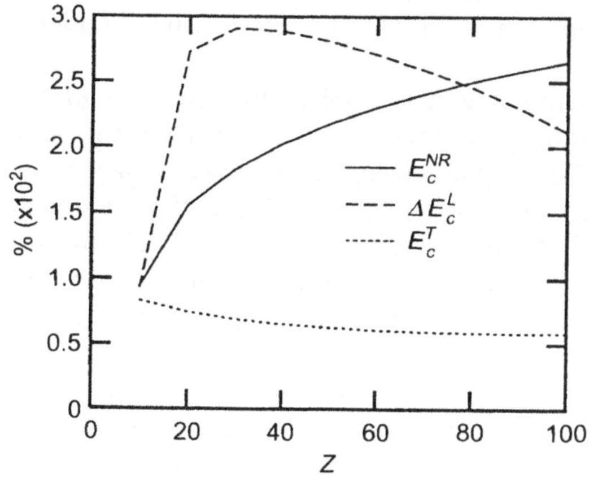

Fig. 5.5. Percentage deviation of the LDA for $E_c[n]$ from MBPT2 results [108] for the Ne isoelectronic series. For E_c^T the absolute value of the error has been plotted (note the take out scale)

Concerning the accuracy of $\Delta E_c^{L,LDA}$ and $E_c^{T,RLDA}$ two features are immediately clear from Table 5.6: On the one hand, the MBPT2 energies do not increase with Z as smoothly as the corresponding RLDA results. On the other hand, while the LDA results for ΔE_c^L agree with the MBPT2 values at least for some atoms, the magnitude of E_c^T from the MBPT2 is not reproduced by the RLDA. It seems unlikely that the drastic deviation of the RLDA for E_c^T can be explained by the fact that in contrast to the RLDA in the case of the MBPT2 only the Breit limit of the full transverse interaction has been used, in particular in view of the similarity of $E_x^{T,Breit}$ and E_x^T shown in Table 5.5. Thus the overall relative error for ΔE_c^L and E_c^T in the RLDA is even larger than that for ΔE_x^L and E_x^T. Note, however, that the underestimation of E_c^T in the RLDA might be reduced by inclusion of the relativistic corrections of order e^4 which are neglected in the RPA.

The inappropriate scaling of the RLDA with Z, and thus also with β, becomes particularly obvious for fixed electron number. In Fig. 5.5 the percentage deviations of the RLDA for ΔE_c^L and E_c^T are shown for the Ne isoelectronic series. The error for the correlation energy in the RLDA shows little tendency to approach zero with increasing Z, indicating that the relativistic correction factor Φ_c^{RPA} plotted in Fig. 4.4 is inadequate for electronic structure calculations.

In summary, the RLDA addresses relativistic corrections to $E_c[n]$ on the same limited level of sophistication as the NRLDA does for the nonrelativistic correlation energy functional. Even more than in the case of exchange, nonlocal corrections seem to be required for a really satisfactory description of (relativistic) correlation effects in atoms.

6 Concluding Remarks

On the basis of the rather limited applications of RDFT the following comments can be offered.

The ROPM can be readily formulated and applied for the case of the longitudinal exchange in the no-pair approximation. It yields results for ground state and exchange energies as well as highest occupied eigenvalues which are very close to the corresponding RHF results. This demonstrates the possibility of approaching the exchange contributions (including the necessary self-interaction corrections) in terms of a local (i.e. multiplicative) rather than a nonlocal potential. The next goal is the inclusion of transverse exchange contributions in the ROPM scheme.

Concerning the RLDA it is found that the errors already present in the nonrelativistic case are enhanced in the relativistic regime: The error of the total longitudinal exchange energy, which is dominated by its nonrelativistic ingredients, is of the order of 10% in the small-Z regime and reduces to about 5% for heavier atoms. The relativistic corrections in the longitudinal exchange energy are, however, underestimated in magnitude by about 20% and the transverse contribution to the exchange energy is overestimated by more than 50%. As both errors add up rather than cancel, the total relativistic correction to atomic exchange energies is rather poorly reproduced by the RLDA. This picture is confirmed by the local errors found for the longitudinal exchange potential. In order to improve the situation, it seems necessary to work towards a relativistic extension of the generalised gradient approximation, which proved to be quite successful in the nonrelativistic regime.

The relative success of the nonrelativistic LDA is to some extent due to a fortunate partial cancellation of errors between the exchange and the correlation contributions. As the importance of correlation effects decreases with increasing nuclear charge, this situation does not occur in the relativistic regime. For the present, the discussion of correlation effects was based on the RLDA form suggested by Eq. (4.10) which essentially contains a nonrelativistic functional in the LDA augmented by relativistic corrections within the RPA. The results obtained with this functional cannot be considered to be satisfactory. It remains to be investigated whether the two corrective options offered, (i) use of a more accurate nonrelativistic correlation functional, and (ii) inclusion of all relativistic correlation contributions of the order e^4 besides the RPA, will improve this situation. The next possible step, the evaluation of relativistic gradient corrections for the case of the correlation energy, would certainly constitute an extensive task, even on the lowest possible level.

The investigation of the RWDA, be it in the simplest form on the basis of the x-only pair correlation function of the RHEG, demonstrated that the problems of the RLDA with the cancellation of the self-interaction and with the (related) asymptotic form of the x-only potential are only corrected in part in this approximation. On the other hand, the performance of this RWDA is definitely superior with respect to the relativistic corrections (near the nucleus). Thus further improvement might be possible if a refined relativistic pair correlation function is used.

Quite generally, it must be stated that some additional effort is required to develop the RDFT towards the same level of sophistication that has been achieved in the nonrelativistic regime. In particular, all exchange-correlation functionals, which are available so far, are functionals of the density alone. An appropriate extension of the nonrelativistic spin density functional formalism on the basis of either the time reversal invariance or the assembly of current density contributions (which are e.g. accessible within the gradient expansion) is one of the tasks still to be undertaken.

In addition, there is interest in further extending the discussion to a variety of situations, that have recently gained much attention in the nonrelativistic case, as time-dependent systems [49], excited states [45] or finite temperature ensembles [110]. As an example of work along these lines we mention the gradient expansion of the noninteracting, relativistic free energy [110], leading to a temperature-dependent relativistic extended Thomas-Fermi model.

Acknowledgements. We wish to thank A. Facco Bonetti, R.N. Schmid and Dr. H. Müller for very valuable discussions. Financial support by the Deutsche Forschungsgemeinschaft (projects En 265/1–1 and Dr 113/20–1) and the BMFT (project 06 OF 705) is gratefully acknowledged.

Note added: After this manuscript was completed, a contribution presenting an orbital-dependent self-interaction correction for the relativistic LDA xc-energy functional was published [111].

7 Appendices

A Relevant Aspects of Vacuum QED

In the first appendix we outline some basic features of vacuum QED (without external potential) in order to establish our notation (adjusted to allow a direct comparison with nonrelativistic DF results) and to provide details of the renormalisation procedure for Greens functions and ground state energies utilised in Sections 2, 3 and the Appendices B, C.

We consider the Lagrangian specified by Eqs. (2.1–2.6) without external sources,

$$V_\mu(x) = 0. \tag{A.1}$$

Due to the fact that the Lagrangian incorporates the creation and destruction of field quanta, not even the time-development of a single particle is a simple matter. The time development can be expressed in terms of the electron (fermion) and photon propagators, which are defined as the vacuum expectation values of the time-ordered product of field operators. For the fermions one has

$$G_V(x, y) = -i\langle 0|\mathbf{T}\hat{\psi}(x)\hat{\bar{\psi}}(y)|0\rangle, \tag{A.2}$$

and for the photons

$$D_V^{\mu\nu}(x, y) = -ie^2\langle 0|\mathbf{T}\hat{A}^\mu(x)\hat{A}^\nu(y)|0\rangle. \tag{A.3}$$

As a consequence of translational invariance both quantities are functions of the difference of the Minkowski coordinates only, so that their four-dimensional Fourier transform can be written as

$$G_V(x-y) = \int\frac{d^4p}{(2\pi)^4}e^{-ip\cdot(x-y)}G_V(p) \tag{A.4}$$

$$D_V^{\mu\nu}(x-y) = \int\frac{d^4q}{(2\pi)^4}e^{-iq\cdot(x-y)}D_V^{\mu\nu}(q). \tag{A.5}$$

The standard approach to the calculation of the propagators (A.4, A.5) is perturbation theory with respect to the electron-electron coupling constant $\alpha = e^2/(\hbar c)$ on the basis of the interaction picture. Technically this results in an expansion of expectation values of interacting field operators in powers of expectation values of the free (or noninteracting) field operators $\hat{\psi}_0$ and $\hat{A}_0^\mu$. The structure of this expansion can be summarised in a set of formal rules, the Feynman rules. For instance for the electron propagator one obtains[20]

---

[20] The occurence of the denominator expression leads, speaking in diagrammar, to the elimination of all unconnected diagrams, i.e. it reflects the proper normalisation of $G_V(x-y)$. $|0_0\rangle$ represents the free vacuum, to be distinguished from the vacuum $|0\rangle$ of the interacting theory.

$$G_V(x-y) = -i\frac{\langle 0_0|\mathrm{T}\hat{\psi}_0(x)\hat{\bar{\psi}}_0(y)\exp[-ie\int d^4z\hat{\bar{\psi}}_0(z)\hat{A}\!\!\!/_0(z)\hat{\psi}_0(z))]|0_0\rangle}{\langle 0_0|\mathrm{T}\exp[-ie\int d^4z\hat{\bar{\psi}}_0(z)\hat{A}\!\!\!/_0(z)\hat{\psi}_0(z))]|0_0\rangle},$$

which allows an immediate expansion in powers of e. Each power introduces an additional integration $\int d^4z$ (called loop integration), while (using Wick's theorem) the corresponding integrands consist of three basic elements, i.e. the expectation values of the free field operators,

$$G_V^0(x-y) = -i\langle 0_0|\mathrm{T}\hat{\psi}_0(x)\hat{\bar{\psi}}_0(y)|0_0\rangle \tag{A.6}$$

$$D^{0,\mu\nu}(x-y) = -ie^2\langle 0_0|\mathrm{T}\hat{A}_0^{\mu}(x)\hat{A}_0^{\nu}(y)|0_0\rangle \tag{A.7}$$

and their link, the vertex,

$$\gamma^{\mu}\delta^{(4)}(z_1-z_3)\delta^{(4)}(z_2-z_3),$$

where the z_i represent the coordinates of the three field operators linked at the vertex. For homogeneous systems this expansion is most easily performed in momentum space, leading to loop integrations over four momenta rather than space-time coordinates. In momentum space the free (or noninteracting) electron propagator G_V^0, which represents the $e^2 \to 0$ limit of G_V, reads

$$G_V^0(p) = \frac{p\!\!\!/ + m}{p^2 - m^2 + i\varepsilon}. \tag{A.8}$$

The form of the free photon propagator $D_{\mu\nu}^0$ depends on the choice of gauge, that is the parameter λ in the Lagrangian (2.1). In Landau gauge, corresponding to the choice $\lambda = \infty$, one has

$$D_{\mu\nu}^0(q) = \left(g_{\mu\nu} - \frac{q_\mu q_\nu}{q^2}\right)D^0(q); \quad D^0(q) = \frac{-4\pi e^2}{q^2 + i\varepsilon}. \tag{A.9}$$

The standard decomposition of $D_{\mu\nu}^0(x-y)$ into the nonretarded Coulomb (longitudinal) and the transverse contribution [29],

$$D_{\mu\nu}^0(x-y) = D_{\mu\nu}^{0,L}(x-y) + D_{\mu\nu}^{0,T}(x-y) \tag{A.10}$$

$$D_{\mu\nu}^{0,L}(x-y) = e^2 g_{0\mu}g_{0\nu}\frac{\delta(x^0-y^0)}{|\boldsymbol{x}-\boldsymbol{y}|}, \tag{A.11}$$

implies that the longitudinal part of $D_{\mu\nu}^0(q)$ is identical with the Coulomb interaction,

$$D_{\mu\nu}^{0,L}(q) = g_{0\mu}g_{0\nu}D_L^0(\boldsymbol{q}); \quad D^{0L}(\boldsymbol{q}) = \frac{4\pi e^2}{\boldsymbol{q}^2}. \tag{A.12}$$

The third basic element of the perturbation expansion, the vertex, which describes the emission or absorption of a photon by a fermion (in lowest order), is given by γ_μ in our notation (as usual, four momentum conservation is automatically implied at the vertices). In diagrammar the free propagators and the simple vertex will be represented by

$G^0_{V,\alpha\beta}(p) \;=$ 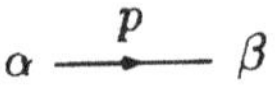 (A.13)

$iD^0_{\mu\nu}(q) \;=$ 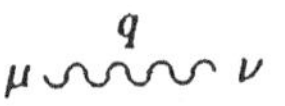 (A.14)

$\gamma^{\mu}_{\alpha\beta} \;=$ 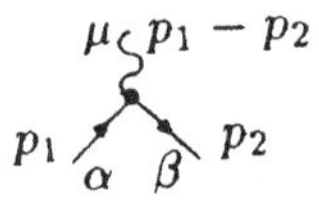 (A.15)

where α, β are the spinor indices.

In the subsequent discussion we shall need the irreducible 2- and 3-point functions[21], i.e. the electron self energy $\Sigma_V(p)$, the vacuum polarisation $\Pi_{V,\mu\nu}(q)$ and the full vertex function $\Gamma_{V,\mu}(p_1, p_2)$, as knowledge of $\Sigma_V(p)$ and $\Pi_{V,\mu\nu}(q)$ determines the corresponding propagators $G_V(p)$ and $D_{V,\mu\nu}(q)$ completely and $\Gamma_{V,\mu}(p_1, p_2)$ represents the perturbative corrections to the free vertex γ_μ. The connection between these quantities is established by the Dyson equations (see e.g. [26])

$$G_V(p) = G_V^0(p) + G_V^0(p)\Sigma_V(p)G_V(p) \tag{A.16}$$

$$D_{V,\mu\nu}(q) = D^0_{\mu\nu} q) + D^0_{\mu\rho}(q)\Pi_V^{\rho\lambda}(q)D_{V,\lambda\nu}(q). \tag{A.17}$$

The relations (A.16–A.17) separate the nontrivial higher order contributions in the perturbation expansions for $G_V(p)$ and $D_{V,\mu\nu}(q)$ from trivial multiples of lower order terms, thus isolating the essential information contained in higher orders. These relations become particularly simple if (A.16) is rewritten in terms of inverse propagators,

$$G_V(p)^{-1} = G_V^0(p)^{-1} - \Sigma_V(p) = \not{p} - m - \Sigma_V(p), \tag{A.18}$$

and if the tensor structure of $\Pi_V^{\mu\nu}(q)$,

$$\Pi_V^{\mu\nu}(q) = (q^2 g^{\mu\nu} - q^\mu q^\nu)\omega_V(q^2), \tag{A.19}$$

which results from gauge invariance, is used in (A.17),

$$D_{V,\mu\nu}(q) = \left(g_{\mu\nu} - \frac{q_\mu q_\nu}{q^2}\right)\frac{-4\pi e^2}{q^2[1 + 4\pi e^2\omega_V(q^2)]}. \tag{A.20}$$

Moreover, $\Sigma_V(p)$ and $\Gamma_{V,\mu}(p_1, p_2)$ satisfy the Ward–Takahashi identity (see e.g. [26])

$$(p_\mu - p'_\mu)\Gamma_V^\mu(p, p') = \Sigma_V(p') - \Sigma_V(p), \tag{A.21}$$

which also is a manifestation of gauge invariance.

[21] n-point functions are called irreducible if their diagrammatic expansions only consist of graphs which do not split into two pieces if one internal electron or photon line is cut.

One encounters ultraviolet (UV) divergent contributions to the electron and photon propagators as well as the vertex function already in the first order of the perturbation expansion. These UV-divergent first order contributions (1-loop contributions) correspond to the following diagrams for the irreducible 2- and 3-point functions:
The electron self-energy,

$$\Sigma_V^{(1)}(p) \quad = \qquad\qquad\qquad\qquad \text{(A.22)}$$

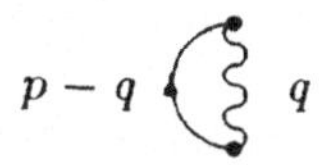

the vacuum polarisation,

$$-i\Pi^{(0)}_{V,\mu\nu}(q) \quad = \qquad\qquad , \qquad\qquad \text{(A.23)}$$

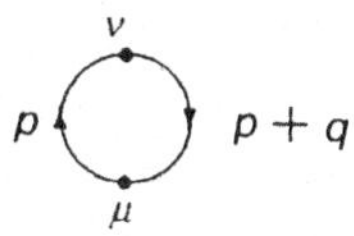

and the vertex function,

$$\Gamma^{(1)}_{V,\mu}(p_1 . p_2) \quad = \qquad\qquad . \qquad\qquad \text{(A.24)}$$

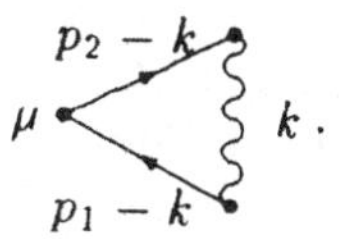

Transcription of the diagrammar into explicit equations using standard Feynman rules (within the notation chosen here) yields

$$\Sigma_V^{(1)}(p) = i\int\frac{d^4q}{(2\pi)^4}D^0_{\mu\nu}(q)\gamma^\mu G^0_V(p-q)\gamma^\nu \qquad \text{(A.25)}$$

$$\Pi^{(0)}_{V,\mu\nu}(q) = -i\int\frac{d^4p}{(2\pi)^4}\, tr\left[\gamma_\mu G^0_V(p)\gamma_\nu G^0_V(p-q)\right], \qquad \text{(A.26)}$$

where the extra minus sign in (A.26) arises from the Feynman rule concerning closed fermion loops and the trace is to be taken in spinor space, as well as

$$\Gamma^{(1)}_{V,\mu}(p_1, p_2) = i\int\frac{d^4k}{(2\pi)^4}D^0_{\rho\nu}(k)\gamma^\rho G^0_V(p_1-k)\gamma_\mu G^0_V(p_2-k)\gamma^\nu . \qquad \text{(A.27)}$$

If one evaluates the integrals over the loop momenta one finds that neither of the quantities is well-defined. One method to endow these quantities with a meaning is the counterterm technique, based e.g. on dimensional regularisation (see e.g. [112, 26]). In this method one first evaluates the integrals (A.25–A.27) not in (1,3)-dimensional Minkowski space but in a (1, d-1)-dimensional space. The results evaluated for integer dimension can then be analytically continued to noninteger d, allowing to consider the physically interesting limit $d \to 4$. Using the abbreviation $\Delta = (4 - d)/2$ one finds for the integrals[22] (A.25–A.27)

[22] For brevity we suppress all problems related to infrared divergencies in the following.

$$\Sigma_V^{(1)}(p) = \frac{e^2}{16\pi^2}\Gamma(\Delta)(-\not{p} + 4m) + \Sigma_{V,finite}^{(1)}(p) \tag{A.28}$$

$$\Pi_{V,\mu\nu}^{(0)}(q) = (q^2 g_{\mu\nu} - q_\mu q_\nu)\left(\frac{1}{12\pi^2}\Gamma(\Delta) + \omega_{V,finite}^{(0)}(q^2)\right) \tag{A.29}$$

$$\Gamma_{V,\mu}^{(1)}(p_1, p_2) = \frac{e^2}{16\pi^2}\Gamma(\Delta)\gamma_\mu + \Gamma_{V,\mu,finite}^{(1)}(p_1, p_2), \tag{A.30}$$

where $\Gamma(\Delta)$ is Euler's Γ-function in which the UV-divergences of the loop-integrals are isolated,

$$\Gamma(\Delta) \underset{\Delta \to 0}{\to} \frac{1}{\Delta} + \ldots,$$

while inside the finite remainders the limit $d \to 4$ can be taken directly. These finite parts, which are not of immediate interest in the present context, correspond to rather lengthy expressions (for the static limit of $\omega_{V,finite}^{(0)}(q^2)$ see Eq. (B.38)). The general feature that emerges from dimensional regularisation is the isolation of the divergent contributions to the integrals in question: In all cases one finds a simple pole structure in the deviation of the space-time dimensionality from $d = 4$.[23] For the discussion of the next step, the actual renormalisation, one starts with the unrenormalised Lagrangian density, that is the Lagrangian of the form (2.1–2.4) with the (finite) physical mass m and charge e and physical (interacting) field operators

$$\mathscr{L}_{unren} = \mathscr{L}(\hat{\psi}, \hat{A}_\mu, m, e). \tag{A.31}$$

As the 2- and 3-point functions $\Sigma_V^{(1)}$, $\Pi_{V,\mu\nu}^{(0)}$ and $\Gamma_{V,\mu}^{(1)}$ calculated on the basis of this Lagrangian are divergent, it is necessary to add a so-called counterterm Lagrangian

$$\mathscr{L}_{CT} = \mathscr{L}(\hat{\psi}, \hat{A}_\mu, A, B, \ldots) \tag{A.32}$$

in order to eliminate the divergencies. The crucial observation for both the physical interpretation as well as the technical success of this *at first glance* somewhat artificial step is the fact that the divergent contributions to the three relevant functions, Eqs. (A.28–A.30), essentially repeat the forms of the corresponding free propagators and the free vertex: the divergent part of $\Sigma_V^{(1)}$ is just proportional to $\not{p}$ and m, but not e.g. to p^2, the divergent part of $\Pi_{V,\mu\nu}^{(0)}$ repeats the tensor structure of $D_{\mu\nu}^0$, Eq. (A.9), and the divergent part of $\Gamma_{V,\mu}^{(1)}$ is proportional to the free vertex γ_μ (but not dependent on p_μ). Thus the contributions (counterterms) to be generated by (A.32) have the form of free propagators and the free vertex with modified physical constants m and e as well as modified normalisation. As a consequence the operator structure of the counterterm

[23] It is perhaps interesting to note that the divergent contributions to $\Sigma_V^{(1)}$ and $\Gamma_{V,\mu}^{(1)}$ satisfy the identity (A.21).

Lagrangian is identical to that of the initial Lagrangian (A.31), so that the sum $\mathscr{L}_{unren} + \mathscr{L}_{CT}$ can be combined in the form

$$\mathscr{L}_R = \mathscr{L}_{unren} + \mathscr{L}_{CT} = \mathscr{L}_{unren}\left(\sqrt{Z_2}\hat{\psi}, \sqrt{Z_3}\hat{A}_\mu, m - \delta m, \frac{Z_1 e}{Z_2\sqrt{Z_3}}\right). \tag{A.33}$$

Thus the renormalisation procedure simply amounts to a redefinition of the physical constants and operators in the initial Lagrangian, leaving the physical implications of this Lagrangian unmodified.

This procedure is readily demonstrated (on the 1-loop level) for the case of the "mass renormalisation". Here one has with (A.18) and (A.28)

$$G_V^{(1)}(p)^{-1} = \not{p}\left(1 + \frac{e^2}{16\pi^2}\Gamma(\Delta)\right) - m\left(1 + \frac{e^2}{4\pi^2}\Gamma(\Delta)\right) - \Sigma^{(1)}_{V,finite}(p). \tag{A.34}$$

In order to compensate the divergent terms (in the limit $d \to 4$), one adds to the electron part of the unrenormalised Lagrangian the counterterm Lagrangian

$$\mathscr{L}_{e,CT} = \bar{\hat{\psi}}(x)(iA\not{\partial} - B)\hat{\psi}(x), \tag{A.35}$$

so that the renormalised electron Lagrangian reads

$$\mathscr{L}_{e,R} = \bar{\hat{\psi}}(x)(i(1 + A)\not{\partial} - (m + B))\hat{\psi}(x). \tag{A.36}$$

Evaluating the electron propagator on the one loop level as before (using the same $\mathscr{L}_{int}$) one finds to first order

$$G_{V,R}^{(1)}(p)^{-1} = \not{p}\left(1 + A + \frac{e^2}{16\pi^2}\Gamma(\Delta)\right) - m\left(1 + \frac{B}{m} + \frac{e^2}{4\pi^2}\Gamma(\Delta)\right) - \Sigma^{(1)}_{V,finite}(p), \tag{A.37}$$

so that the choice

$$A = -\frac{e^2}{16\pi^2}\Gamma(\Delta) \qquad B = -\frac{e^2 m}{4\pi^2}\Gamma(\Delta) \tag{A.38}$$

leads to the finite result

$$G_{V,R}^{(1)}(p)^{-1} = \not{p} - m - \Sigma^{(1)}_{V,finite}(p). \tag{A.39}$$

In diagrammar the corresponding statement derived from the Dyson equation is

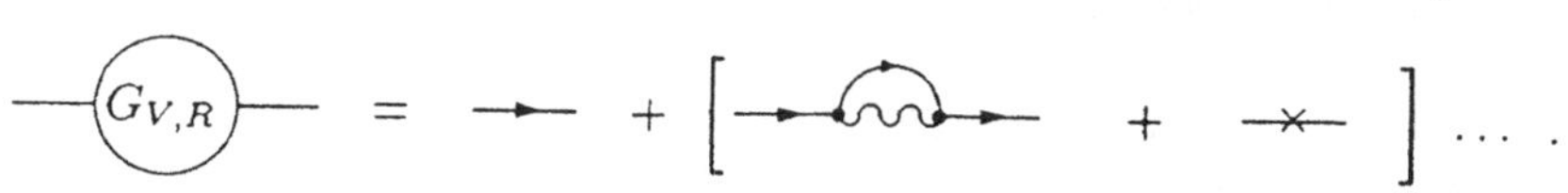

The additional diagram represents the counterterm contribution

$$-\!\!\times\!\!- = \frac{(1+A)\not{p} + m + B}{(1+A)^2 p^2 - (m+B)^2 + i\varepsilon} - \frac{\not{p} + m}{p^2 - m^2 + i\varepsilon}$$

$$= \frac{\not{p} + m}{p^2 - m^2 + i\varepsilon}[-A\not{p} + B]\frac{\not{p} + m}{p^2 - m^2 + i\varepsilon} + \dots .$$

In fact, the choice (A.38) is not only motivated by technical necessity but rather results from an underlying physical requirement on the electron propagator: In order to describe real fermions, which satisfy the standard dispersion relation $p^2 = (p^0)^2 - \boldsymbol{p}^2 = m^2$ (with the finite experimental mass m) in the presence of the virtual photon cloud, the full electron propagator $G_{V,R}(p)$ obtained from some Lagrangian of the form (A.33) with a priori arbitrary coefficients should reduce to the form of the free propagator $G_V^0(p)$ with physical mass m for on-shell momentum $p^2 = m^2$. Consequently the physical, i.e. renormalised, self-energy $\Sigma_{V,R}(p)$ has to satisfy the normalisation conditions

$$\Sigma_{V,R}(p)|_{\not{p}=m} = 0; \quad \frac{d}{d\not{p}}\Sigma_{V,R}(p)|_{\not{p}=m} = 0, \tag{A.40}$$

which (to first order) is achieved by the counterterm Lagrangian (A.32) with the coefficients (A.38).

The form invariance of the Lagrangian under this renormalisation procedure, indicated in Eq. (A.33), can now be implemented by defining the wavefunction (more correctly the field operator) renormalisation constant

$$Z_2 = 1 + A, \tag{A.41}$$

and the corresponding bare field operator

$$\hat{\psi}_b(x) = \sqrt{Z_2}\,\hat{\psi}(x), \tag{A.42}$$

as well as the bare mass

$$m_b = \frac{1}{Z_2}(m + B) = m - \delta m. \tag{A.43}$$

The renormalised electron Lagrangian (A.33) can then be expressed as

$$\mathscr{L}_{e,R} = \hat{\bar{\psi}}_b(x)(i\not{\partial} - m_b)\hat{\psi}_b(x), \tag{A.44}$$

i.e. the counterterms required for keeping $\Sigma_{V,R}^{(1)}$ finite have been completely absorbed in a redefinition (or renormalisation) of the ingredients of the free electron part of $\mathscr{L}$. In terms of Z_2 and δm the resulting renormalised self-energy then reads quite generally

$$\Sigma_{V,R}(p) = Z_2[\Sigma_V(p) - \delta m] + (1 - Z_2)(\not{p} - m), \tag{A.45}$$

so that the first order counterterm is given by

$$\begin{aligned}\Delta\Sigma_V^{(1)}(p) &= -\delta m^{(1)} + (1 - Z_2^{(1)})(\not p - m) \\ &= \frac{e^2}{16\pi^2}\Gamma\left(\frac{4-d}{2}\right)\not p - \frac{e^2}{4\pi^2}\Gamma\left(\frac{4-d}{2}\right)m. \end{aligned} \tag{A.46}$$

An equivalent procedure can be carried through for $\Pi^{(0)}_{V,\mu\nu}$ addressing $\mathscr{L}_\gamma$ as well as $\Gamma^{(1)}_{V,\mu}$ addressing $\mathscr{L}_{int}$. The argument can be extended to any order, after discussion of overlapping divergencies, which occur for instance in the diagram.

For the subsequent discussion we also note the normalisation condition for the vacuum polarisation kernel $\omega_{V,R}(q^2)$: In order to obtain the physical charge e for real electrons, i.e. to reproduce the Coulomb interaction for well separated electrons, $\omega_{V,R}(q^2)$ must vanish for vanishing momentum,

$$\omega_{V,R}(q^2 = 0) = 0, \tag{A.47}$$

which is immediately obvious from the Dyson equation (A.20) for the full photon propagator. Consistent with (A.29) the counterterm for $\Pi^{(0)}_{V,\mu\nu}(q)$ reads

$$\Delta\Pi^{(0)}_{V,\mu\nu}(q) = -(q^2 g_{\mu\nu} - q_\mu q_\nu)\frac{1}{12\pi^2}\Gamma\left(\frac{4-d}{2}\right). \tag{A.48}$$

The final form of the fully renormalised Lagrangian of QED is

$$\begin{aligned}\mathscr{L}_R = &\frac{1}{4}\{[\hat{\bar\psi}_b(x), (i\overrightarrow{\not\partial} - m_b - e_b \not V_b(\boldsymbol{x}) - e_b \hat{\not A}_b(x))\hat\psi_b(x)] \\ &+ [\hat{\bar\psi}_b(x)(-i\overleftarrow{\not\partial} - m_b - e_b \not V_b(\boldsymbol{x}) - e_b \hat{\not A}_b(x)), \hat\psi_b(x)]\} \\ &- \frac{1}{16\pi}\hat F_{b,\mu\nu}(x)\hat F_b^{\mu\nu}(x) - \frac{\lambda}{8\pi}(\partial_\nu \hat A_b^\nu(x))^2 \\ = &\frac{Z_2}{4}\{[\hat{\bar\psi}(x), (i\overrightarrow{\not\partial} - m + \delta m - e\not V(\boldsymbol{x}) - e\hat{\not A}(x))\hat\psi(x)] \\ &+ [\hat{\bar\psi}(x)(-i\overleftarrow{\not\partial} - m + \delta m - e\not V(\boldsymbol{x}) - e\hat{\not A}(x)), \hat\psi(x)]\} \\ &- \frac{Z_3}{16\pi}\hat F_{\mu\nu}(x)\hat F^{\mu\nu}(x) - \frac{Z_3\lambda}{8\pi}(\partial_\nu \hat A^\nu(x))^2, \end{aligned} \tag{A.49}$$

where we have included the static external potential in order to indicate that it has to be renormalised in the same spirit as the quantised photon field. The physical fields and constants are related to the bare quantities by

$$\hat{\psi}_b(x) = \sqrt{Z_2}\hat{\psi}(x) \tag{A.50}$$

$$\hat{A}_b^\nu(x) = \sqrt{Z_3}\,\hat{A}^\nu(x) \tag{A.51}$$

$$V_b^\nu(\boldsymbol{x}) = \sqrt{Z_3}\,V^\nu(\boldsymbol{x}) \tag{A.52}$$

$$e_b = \frac{Z_1}{Z_2\sqrt{Z_3}}e \tag{A.53}$$

$$m_b = m - \delta m \quad . \tag{A.54}$$

The renormalisation constants Z_1, Z_2, Z_3 and δm have to be understood as functions of the finite physical charge e and mass m of the electrons which can be constructed order by order in the perturbation series. It is important to notice that these constants are uniquely determined by vacuum QED without any external potential. They do not depend on the specific external potential present. If one bases the perturbation expansion on the Lagrangian (A.49) all Greens and n-point functions of the theory (defined in terms of the physical fields $\hat{\psi}$ *and* $\hat{A}^\mu$) are finite.

B Relativistic Homogeneous Electron Gas

In this appendix we summarise some properties of the relativistic homogeneous electron gas (RHEG) in order to illustrate the renormalisation of ground state energies (indicated in Section 2) and to provide the details for the RLDA and the relativistic gradient expansion (RGE), which are discussed in Section 4 and Appendix D, respectively. For simplicity we restrict the discussion to the unpolarised RHEG.

We start by noting the basic difference between the perturbative approaches to the RHEG and the QED vacuum (discussed in Appendix A). As a consequence of the difference between the ground state $|\Phi_0>$ of the RHEG and the homogeneous vacuum $|0>$ the fermion propagator

$$G(x,y) = -i\langle\Phi_0|\mathrm{T}\hat{\psi}(x)\hat{\bar{\psi}}(y)|\Phi_0\rangle, \tag{B.1}$$

already differs from G_V on the noninteracting level. In our notation the noninteracting fermion propagator of the RHEG is given by

$$G^0(p) = G_V^0(p) + G_D^0(p) = G_-(p) + G_+(p) \tag{B.2}$$

$$G_D^0(p) = 2\pi i\delta(p^0 - E_p)\frac{\not{p}+m}{2E_p}\Theta(k_F - |\boldsymbol{p}|) \tag{B.3}$$

$$G_-(p) = \frac{\not{p}_- + m}{2E_p}\frac{-1}{p^0 + E_p - i\varepsilon} \tag{B.4}$$

$$G_+(p) = \frac{\not{p}_+ + m}{2E_p}\left[\frac{\Theta(|\boldsymbol{p}| - k_F)}{p^0 - E_p + i\varepsilon} + \frac{\Theta(k_F - |\boldsymbol{p}|)}{p^0 - E_p - i\varepsilon}\right], \tag{B.5}$$

where $E_p = \sqrt{p^2 + m^2}$, $p^\mu_\pm = (\pm E_p, p^i)$ and the Fermi momentum k_F is related to the electron density n_0 of the RHEG as usual, $n_0 = k_F^3/(3\pi^2)$. Here two alternative forms have been listed, the first one emphasising the relation between $G^0(p)$ and the vacuum propagator $G_V^0(p)$, Eq. (A.8), the second one visualising its decomposition into the electron (G_+) and positron (G_-) contributions. Note that due to charge conservation the density of the RHEG is not changed by switching on the electron-electron interaction, so that n_0 also represents the density of the interacting RHEG and $n_0 = k_F^3/(3\pi^2)$ then relates the interacting density to the noninteracting k_F. Diagrammatically the full $G^0(p)$, Eq. (B.2), will be represented by

$G^0(p) \quad =$  (B.6)

in the following.

While the other two basic elements of perturbation theory, the free photon propagator (A.9) and the simple vertex, remain unchanged, it seems worth pointing out that the full photon propagator

$$D_{\mu\nu}(x, y) = -ie^2 \langle \Phi_0 | \mathbf{T} \hat{A}_\mu(x) \hat{A}_\nu(y) | \Phi_0 \rangle, \tag{B.7}$$

and the full vertex function do not: In the case of the RHEG not only virtual electron-positron pairs screen the bare interaction but also virtual electron-hole pairs.

B.1 Response Functions

Most information concerning the RHEG required in the present context is contained in the response functions of the RHEG. In our notation the time-ordered current response functions (n-point functions) are defined as

$$\chi^{(n)}_{\mu_1 \ldots \mu_n}(x_1, \ldots x_n) = (-i)^{n-1} \langle \Phi_0 | \mathbf{T} \hat{\tilde{j}}_{\mu_1}(x_1) \ldots \hat{\tilde{j}}_{\mu_n}(x_n) | \Phi_0 \rangle, \tag{B.8}$$

where the current deviation operator $\hat{\tilde{j}}_\mu$ is given by

$$\hat{\tilde{j}}_\mu(t, \boldsymbol{r}) = \hat{j}_\mu(t, \boldsymbol{r}) - \langle \Phi_0 | \hat{j}_\mu(t, \boldsymbol{r}) | \Phi_0 \rangle = \hat{j}_\mu(t, \boldsymbol{r}) - j_\mu(\boldsymbol{r}). \tag{B.9}$$

For the time-independent systems of interest here a partial Fourier transformation of $\chi^{(n)}_{\mu_1 \ldots \mu_n}$ is advantageous,

$$\chi^{(n)}_{\mu_1 \ldots \mu_n}(t_1, \boldsymbol{r}_1; \ldots t_n, \boldsymbol{r}_n) = \int \frac{d\omega_1}{2\pi} \ldots \int \frac{d\omega_n}{2\pi} e^{-i\omega_1 t_1 - \ldots - i\omega_n t_n} \times 2\pi\delta(\omega_1 + \ldots + \omega_n) \chi^{(n)}_{\mu_1 \ldots \mu_n}(\omega_1, \boldsymbol{r}_1; \ldots \omega_n, \boldsymbol{r}_n). \tag{B.10}$$

The static response functions utilised in Appendix C are then obtained by taking the zero-frequency limit,

$$\chi^{(n)}_{\mu_1 \ldots \mu_n}(\boldsymbol{r}_1, \ldots \boldsymbol{r}_n) \equiv \chi^{(n)}_{\mu_1 \ldots \mu_n}(\omega_1 = 0, \boldsymbol{r}_1; \ldots \omega_n = 0, \boldsymbol{r}_n). \tag{B.11}$$

For the case of the RHEG further Fourier transformations is useful,

$$\chi^{(n)}_{\mu_1 \ldots \mu_n}(q_1^0, \boldsymbol{r}_1; \ldots q_n^0, \boldsymbol{r}_n) = \int \frac{d^3 q_1}{(2\pi)^3} \cdots \int \frac{d^3 q_n}{(2\pi)^3} e^{i\boldsymbol{r}_1 \cdot \boldsymbol{q}_1 + \ldots + i\boldsymbol{r}_n \cdot \boldsymbol{q}_n} \times (2\pi)^3 \delta^{(3)}(\boldsymbol{q}_1 + \ldots + \boldsymbol{q}_n) \chi^{(n)}_{\mu_1 \ldots \mu_n}(q_1, \ldots q_n). \tag{B.12}$$

Gauge invariance then implies the transversality of the $\chi^{(n)}_{\mu_1 \ldots \mu_n}$ with respect to all arguments [113],

$$q_i^{\mu_i} \chi^{(n)}_{\mu_1 \ldots \mu_i \ldots \mu_n}(q_1, \ldots, q_i, \ldots q_n) = 0. \tag{B.13}$$

In the following the connected contributions of the $\chi^{(n)}$, for which all external vertices are in some way linked to each other, will be denoted by $\chi_c^{(n)}$, while the linear response function of the RHEG will be abbreviated by $\chi_{\mu\nu}(q)$ for brevity.

As a consequence of (B.13) and the Dyson equation for $\chi^{\mu\nu}(q)$,

$$\chi^{\mu\nu}(q) = \Pi^{\mu\nu}(q) + \Pi^{\mu\rho}(q) D^0_{\rho\lambda}(q) \chi^{\lambda\nu}(q). \tag{B.14}$$

the irreducible 2-point function $\Pi^{\mu\nu}(q)$ also satisfies the transversality relation,

$$q_\mu \Pi^{\mu\nu}(q) = 0, \tag{B.15}$$

which essentially determines its tensor structure. Introducing the (4×4) polarisation tensors

$$P_L^{\mu\nu}(q) = \frac{-1}{\boldsymbol{q}^2 q^2} \begin{pmatrix} (\boldsymbol{q}^2)^2 & \boldsymbol{q}^2 q^0 q^j \\ \boldsymbol{q}^2 q^0 q^i & (q^0)^2 q^i q^j \end{pmatrix} \tag{B.16}$$

$$P_T^{\mu\nu}(q) = \frac{1}{\boldsymbol{q}^2} \begin{pmatrix} 0 & 0 \\ 0 & \boldsymbol{q}^2 g^{ij} + q^i q^j \end{pmatrix}, \tag{B.17}$$

$(g^{ij} = -\delta_{ij})$ $\Pi^{\mu\nu}(q)$ can be written as

$$\Pi^{\mu\nu}(q) = P_L^{\mu\nu}(q) \Pi_L(q) - P_T^{\mu\nu}(q) \Pi_T(q). \tag{B.18}$$

For convenience, we note some useful properties of $P_{L/T}^{\mu\nu}(q)$,

$$P^\lambda_{L,\nu}(q) P^\nu_{L,\mu}(q) = P^\lambda_{L,\mu}(q) \tag{B.19}$$

$$P^\lambda_{T,\nu}(q) P^\nu_{T,\mu}(q) = P^\lambda_{T,\mu}(q) \tag{B.20}$$

$$P^\lambda_{T,\nu}(q) P^\nu_{L,\mu}(q) = 0 \tag{B.21}$$

$$P^\lambda_{L,0}(q) P^0_{L,\mu}(q) = -\frac{\boldsymbol{q}^2}{q^2} P^\lambda_{L,\mu}(q) \tag{B.22}$$

$$P_L^{\mu\nu}(q) + P_T^{\mu\nu}(q) = g^{\mu\nu} - \frac{q^\mu q^\nu}{q^2} \tag{B.23}$$

$$P^\mu_{L,\mu}(q) = 1 \tag{B.24}$$

$$P^\mu_{T,\mu}(q) = 2. \tag{B.25}$$

Decomposing $\Pi^{\mu\nu}(q)$ into an electron gas (D) and the vacuum (V) component,

$$\Pi^{\mu\nu}(q) = \Pi_D^{\mu\nu}(q) + \Pi_V^{\mu\nu}(q) \tag{B.26}$$

$$\Pi_L(q) = \Pi_{L,D}(q) + \Pi_V(q) \tag{B.27}$$

$$\Pi_T(q) = \Pi_{T,D}(q) - \Pi_V(q), \tag{B.28}$$

the vacuum contribution can be recast in the form (A.19), with the polarisation function $\Pi_V(q)$ given by

$$\Pi_V(q) = q^2 \omega_V(q). \tag{B.29}$$

Using the polarisation tensors (B.16, B.17), the free photon propagator (A.9) and the longitudinal and transverse polarisation functions $\Pi_{L/T}(q)$ the Dyson equation for $\chi^{\mu\nu}(q)$ reads

$$\chi^{\mu\nu}(q) = \frac{\Pi_L(q)}{1 - D^0(q)\Pi_L(q)} P_L^{\mu\nu}(q) - \frac{\Pi_T(q)}{1 + D^0(q)\Pi_T(q)} P_T^{\mu\nu}(q). \tag{B.30}$$

The full photon propagator $D^{\mu\nu}(q)$ can be directly obtained from the Dyson equation (A.17) (which also holds for the RHEG) by insertion of (B.18). For a discussion of the renormalisation of $D^{\mu\nu}(q)$ it is instructive to rewrite the form corresponding to (B.30) by introducing the full vacuum photon propagator,

$$D_V(q) = \frac{D^0(q)}{1 - D^0(q)\Pi_V(q)}, \tag{B.31}$$

which allows to 'decouple' the screening effects entirely due to the vacuum from those originating from the actual electron gas,

$$D^{\mu\nu}(q) = \frac{D_V(q)}{1 - D_V(q)\Pi_{L,D}(q)} P_L^{\mu\nu}(q) + \frac{D_V(q)}{1 + D_V(q)\Pi_{T,D}(q)} P_T^{\mu\nu}(q). \tag{B.32}$$

At first glance this form seems to suggest that $D^{\mu\nu}(q)$ is UV-finite as soon as $D_V(q)$ has been renormalised, which amounts to replacing $D_V(q)$ by $D_{V,R}(q)$. However, $\Pi_{L/T,D}(q)$ also contains UV-divergent subgraphs. The following 2-loop contribution may illustrate this point,

Replacing G^0 by $G_V^0 + G_D^0$ one recognises that besides the obvious pure vacuum loop absorbed in D_V also mixtures between the first order vacuum vertex correction loop (A.24) and G_D^0-type propagators occur. The counterterms required to keep such subgraphs finite are, however, completely determined by vacuum QED: Similar to the renormalisation of overlapping divergencies each vacuum subgraph in a (larger) non-vacuum diagram has to be supplemented individually by the corresponding counterterm (this also holds for multi-loop vacuum subgraphs). The same procedure applies to $\chi^{\mu\nu}(q)$.

We also indicate the purely longitudinal limit of $\chi^{\mu\nu}(q)$,

$$\chi_L^{\mu\nu}(q) = \frac{\Pi_{LL}(q)}{1 - D^0(q)\Pi_{LL}(q)} P_L^{\mu\nu}(q) - \Pi_{TL}(q) P_T^{\mu\nu}(q), \tag{B.33}$$

where $\Pi_{LL/TL}(q)$ is obtained from $\Pi_{L/T}(q)$ by replacing all internal photon propagators by $D_{\mu\nu}^{0,L}(q)$ (note that in the denominator $D^0(q)$ appears rather than $D_L^0(q)$ due to the properties of $P_L^{\mu\nu}(q)$, i.e. Eq. (B.22)).

For the discussion of inhomogeneity corrections to the RLDA one also needs the inverse response function $\chi_{\mu\nu}^{-1}(q)$, at least in the static limit,

$$\chi^{-1,\mu\nu}(q^0 = 0, \boldsymbol{q}) = -g^{\mu\nu} D_L^0(\boldsymbol{q}) + \Pi^{-1,\mu\nu}(q^0 = 0, \boldsymbol{q}) \tag{B.34}$$

$$\Pi^{-1,\mu\nu}(q^0 = 0, \boldsymbol{q}) = \begin{pmatrix} \dfrac{1}{\Pi_L(0,\boldsymbol{q})} & 0 & 0 & 0 \\ 0 & & & \\ 0 & & \left(g^{ij} + \dfrac{q^i q^j}{\boldsymbol{q}^2}\right) \dfrac{-1}{\Pi_T(0,\boldsymbol{q})} & \\ 0 & & & \end{pmatrix}. \tag{B.35}$$

As far as explicit approximations for the polarisation functions $\Pi_{L/T}(q)$ are concerned only very little is known, even in the static limit. The complete frequency dependence is available for the noninteracting limit $\Pi_{L/T}^{(0)}(q)$, i.e. the relativistic generalisation of the Lindhard function [95, 114]. In addition to its vacuum part (A.26) one has

$$\begin{aligned} \Pi_{D,\mu\nu}^{(0)}(q) = & -i \int \frac{d^4p}{(2\pi)^4} tr [\gamma_\mu G_D^0(p) \gamma_\nu G_D^0(p-q)] \\ & -i \int \frac{d^4p}{(2\pi)^4} tr [\gamma_\mu G_V^0(p) \gamma_\nu G_D^0(p-q)] \\ & -i \int \frac{d^4p}{(2\pi)^4} tr [\gamma_\mu G_D^0(p) \gamma_\nu G_V^0(p-q)] \\ \neq & -i \int \frac{d^4p}{(2\pi)^4} tr [\gamma_\mu G_+^0(p) \gamma_\nu G_+^0(p-q)], \end{aligned}$$

where it has been emphasised that the no-sea approximation is not just obtained by completely neglecting the 'positron propagator' $G_-(p)$ in the perturbation expansion (which would constitute the no-pair approximation for $\Pi_{L/T}^{(0)}$). Rather the no-sea approximation for any given diagram contributing to $\Pi_{\mu\nu}$ is defined by neglecting all those contributions to each fermion loop inside the diagram which do not vanish in the limit $k_F \to 0$. The static limit of $\Pi_{D,\mu\nu}^{(0)}$ and its small-$\boldsymbol{q}^2$

expansion are given by ($Q = |\boldsymbol{q}|/(2k_F)$),

$$\Pi^{(0)}_{L,D}(0,\boldsymbol{q}) \tag{B.36}$$

$$= -\frac{mk_F}{2\pi^2}\left\{\frac{4}{3}\eta - \frac{4}{3}\beta Q^2 \operatorname{arsinh}(\beta) + \frac{1}{3}\frac{\eta}{\beta^2 Q}(\eta^2 - 3\beta^2 Q^2)\ln\left|\frac{1+Q}{1-Q}\right|\right.$$

$$\left. - \frac{1}{3}\frac{(1-2\beta^2Q^2)(1+\beta^2Q^2)^{1/2}}{\beta^2 Q}\ln\left|\frac{(1+\beta^2Q^2)^{1/2}+\eta Q}{(1+\beta^2Q^2)^{1/2}-\eta Q}\right|\right\}$$

$$= -\frac{mk_F\eta}{\pi^2}\left\{1 - \frac{1}{3}\left[1 + 2\frac{\beta}{\eta}\operatorname{arsinh}(\beta)\right]Q^2 - \frac{1}{15}[1-4\beta^2]Q^4 + \ldots\right\}$$

$$\Pi^{(0)}_{T,D}(0,\boldsymbol{q}) = -\Pi^{(0)}_{L,D}(0,\boldsymbol{q}) - \frac{mk_F\eta}{2\pi^2}\left\{1 + \frac{1-Q^2}{2Q}\ln\left|\frac{1+Q}{1-Q}\right|\right\} \tag{B.37}$$

$$= \frac{mk_F\eta}{\pi^2}\left\{-\frac{2}{3}\frac{\beta}{\eta}\operatorname{arsinh}(\beta)Q^2 + \frac{4}{15}\beta^2 Q^4 + \ldots\right\}$$

$$\Pi^{(0)}_{V,R}(0,\boldsymbol{q}) = \frac{\boldsymbol{q}^2}{12\pi^2}\left\{\frac{1}{3} + \left(\frac{1-2\beta^2Q^2}{\beta^2Q^2}\right)\left[1 - \frac{(1+\beta^2Q^2)^{1/2}}{2\beta Q}\right.\right. \tag{B.38}$$

$$\left.\left.\times \ln\left|\frac{(1+\beta^2Q^2)^{1/2}+\beta Q}{(1+\beta^2Q^2)^{1/2}-\beta Q}\right|\right]\right\}$$

$$= \frac{1}{60\pi^2}\frac{\boldsymbol{q}^4}{m^2} + \ldots,$$

where

$$\beta = \frac{(3\pi^2 n_0)^{\frac{1}{3}}}{m}; \quad \eta = (1+\beta^2)^{\frac{1}{2}}. \tag{B.39}$$

Note that quite generally one has

$$\lim_{q^2\to 0}\frac{\Pi_{V,R}(0,\boldsymbol{q})}{\boldsymbol{q}^2} = 0 \tag{B.40}$$

due to the on-shell normalisation condition for $\omega_{V,R}$, Eq.(A.47). Beyond the noninteracting limit only the vacuum part of the first order (2-loop) contribution to the polarisation function [115, 116] has been evaluated. Moreover, the screening length $\Pi_{L,D}(\mathbf{0},0)$ is related to the energy density via the compressibility sum rule [117],

$$\frac{d^2}{dn_0^2}[t_s(n_0) + e_{xc}(n_0)] = -\frac{1}{\Pi_{L,D}(0,\mathbf{0})},$$

so that the long wavelength limit of higher orders of $\Pi_{L,D}$ can be obtained from the associated contributions to $e_{xc}(n_0)$. Finally, in the context of the quark-gluon gas the high temperature limits of certain classes of higher order diagrams have also been examined (see e.g. [118]). These results are, however, only of limited value in the present context aiming at $T = 0$ and $m \neq 0$.

B.2 Ground State Energy

The ground state energy of the RHEG constitutes the basis for the RLDA and also provides an instructive example for the renormalisation procedure described in Section 2 and Appendix A. We start by emphasising that the ground state energy is defined with respect to the vacuum energy as [96, 113]

$$E_{tot}^{RHEG} = \langle \Phi_0 | \hat{H}^{RHEG} | \Phi_0 \rangle - \langle 0 | \hat{H}^{RHEG} | 0 \rangle + \Delta E_{tot}^{RHEG}, \tag{B.41}$$

where ΔE_{tot}^{RHEG} represents the additional counterterms beyond the energy of the homogeneous vacuum, $\langle 0 | \hat{H}^{RHEG} | 0 \rangle$, which are required to keep E_{tot}^{RHEG} UV-finite. Moreover, in the case of the RHEG one deals with the energy density corresponding to (B.41), rather than the energy itself.

The kinetic energy density of the noninteracting RHEG can be evaluated without addressing the UV-renormalisation procedure [12],

$$\begin{aligned} t_s(n_0) &= \langle \Phi_0 | [\hat{\bar{\psi}}(x), (-i\gamma\cdot\nabla + (1-\gamma^0)m)\hat{\psi}(x)] | \Phi_0 \rangle \\ &\quad - \langle 0 | [\hat{\bar{\psi}}(x), (-i\gamma\cdot\nabla + (1-\gamma^0)m)\hat{\psi}(x)] | 0 \rangle \\ &= i \lim_{y\to x} tr[(-i\gamma\cdot\nabla + (1-\gamma^0)m) G_D^0(x-y)] \\ &= t_s^{NRHEG}(n_0) \Phi_{s,0}(\beta) \end{aligned} \tag{B.42}$$

$$t_s^{NRHEG}(n_0) = \frac{1}{10\pi^2 m} k_F^5 \tag{B.43}$$

$$\Phi_{s,0}(\beta) = \frac{10}{\beta^5}\left[\frac{1}{8}(\beta\eta^3 + \beta^3\eta - \operatorname{arsinh}(\beta)) - \frac{1}{3}\beta^3\right], \tag{B.44}$$

where the electron rest mass has been subtracted. Taking into account the homogeneous positive charge background the (direct) Hartree energy of the RHEG vanishes. The exchange-correlation energy can be written in terms of a coupling constant integral over the current-current response function [93, 98],

$$e_{xc}(n_0) = \frac{i}{2}\int_0^1 d\lambda \int \frac{d^4q}{(2\pi)^4} D_{\mu\nu}^0(q)[\chi_\lambda^{\mu\nu}(q) - \chi_{V,\lambda}^{\mu\nu}(q)] + \Delta E_{xc}^{RHEG}, \tag{B.45}$$

where $\chi_\lambda^{\mu\nu}(q)$ is given by (B.30) with the electron-electron coupling strength e^2 replaced by λe^2, $\chi_{V,\lambda}^{\mu\nu}(q)$ represents its vacuum limit and ΔE_{xc}^{RHEG} is identical with the total counterterm ΔE_{tot}^{RHEG}.

The first order term (in e^2) in (B.45), i.e. the exchange energy of the RHEG (according to Eq.(3.32)), is the most simple energy contribution for which the UV-renormalisation is nontrivial. The basic problem associated with the renormalisation of energies (rather than Greens functions) is that energy contributions can not be rewritten entirely in terms of renormalised n-point functions: At least one overall loop integration remains to be treated separately (as e.g. the q-integration in (B.45)), even if one takes into account the subtraction of the pure vacuum energy. As an additional complication, this outermost loop integration often leads to overlapping divergencies. In fact, visualising the exchange energy graphically

$-2ie_x(n_0) =$ 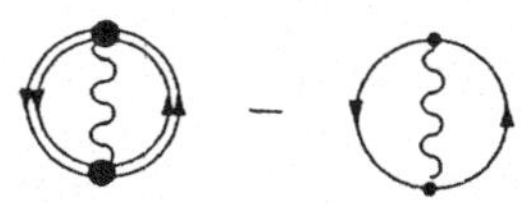 (B.46)

one realises that three divergent 1-loop subgraphs contribute to the (left) electron gas loop,

$-i\Pi^{(0)}_{\mu\nu} =$  $\qquad \Sigma^{(1)} =$  . (B.47)

While the UV-divergence of the $\Pi^{(0)}_{\mu\nu}$-subgraph is eliminated by the subtraction of the vacuum exchange energy, the two (identical) self energy subgraphs require additional counterterms. As one is facing overlapping divergencies in (B.46) each divergent subgraph has to be renormalised individually. Of course, only the vacuum contribution to $\Sigma^{(1)}$ requires renormalisation, the corresponding counterterms being well known from vacuum QED,

$$e_x(n_0) = \frac{i}{2}\int \frac{d^4q}{(2\pi)^4} D^0_{\mu\nu}(q)[\Pi^{(0),\mu\nu}(q) - \Pi^{(0),\mu\nu}_V(q)] - i\int \frac{d^4p}{(2\pi)^4} tr[G^0_D(p)\Delta\Sigma^{(1)}_V(p)]. \tag{B.48}$$

As discussed in detail in Appendix A the self-energy counterterm $\Delta\Sigma^{(1)}_V(p)$ is defined so that the renormalised vacuum self-energy $\Sigma_{V,R}(p)$ (A.45) satisfies the standard on-shell normalisation condition (A.40), i.e. on the 1-loop level one obtains (A.46). Using (B.2) $e_x(n_0)$ can thus be rewritten as

$$e_x(n_0) = \frac{1}{2}\int \frac{d^4q}{(2\pi)^4}\int \frac{d^4p}{(2\pi)^4} D^0_{\mu\nu}(q) tr[G^0_D(p+q)\gamma^\mu G^0_D(p)\gamma^\nu] - \int \frac{d^4p}{(2\pi)^4} tr[G^0_D(p)\Sigma^{(1)}_{V,R}(p)].$$

The second term vanishes according to Eqs. (B.3, A.40),

$$[(\not p + m)\Sigma_{V,R}(p)]_{p^2=\not p^2=m^2} = 0, \tag{B.49}$$

so that the standard renormalisation scheme eliminates all vacuum corrections to $e_x(n_0)$ without the need to apply either the no-sea or the no-pair approximation. The first term, which is also obtained within both the no-sea and the no-pair approximation for $e_x(n_0)$, can be evaluated straight-forwardly [93, 94, 95],

$$e_x(n_0) = e_x^{NRHEG}(n_0)\Phi_x(\beta) \tag{B.50}$$

$$e_x^{NRHEG}(n_0) = -\frac{e^2}{4\pi^3}k_F^4 \tag{B.51}$$

$$\Phi_x(\beta) = 1 - \frac{3}{2}\left[\frac{\eta}{\beta} - \frac{1}{\beta^2}\text{arsinh}(\beta)\right]^2 . \tag{B.52}$$

Using the decomposition (A.10) $e_x(n_0)$ can be split into a longitudinal and a transverse contribution [19, 95],

$$e_x^{L/T}(n_0) = e_x^{NRHEG}(n_0)\Phi_x^{L/T}(\beta) \tag{B.53}$$

$$\Phi_x^L(\beta) = \frac{5}{6} + \frac{1}{3\beta^2} + \frac{2\eta}{3\beta}\text{arsinh}(\beta) \tag{B.54}$$

$$- \frac{2\eta^4}{3\beta^4}\ln(\eta) - \frac{1}{2}\left(\frac{\eta}{\beta} - \frac{\text{arsinh}(\beta)}{\beta^2}\right)^2$$

$$\Phi_x^T(\beta) = \frac{1}{6} - \frac{1}{3\beta^2} - \frac{2\eta}{3\beta}\text{arsinh}(\beta) \tag{B.55}$$

$$+ \frac{2\eta^4}{3\beta^4}\ln(\eta) - \left(\frac{\eta}{\beta} - \frac{\text{arsinh}(\beta)}{\beta^2}\right)^2 .$$

The UV-renormalisation procedure is particularly involved for the correlation energy e_c. Most of the relevant counterterms provided by ΔE_{xc}^{RHEG}, however, are already included if the basic expression (B.45) is understood as being written in terms of the renormalised response function $\chi_{R,\lambda}^{\mu\nu}(q)$,

$$e_{xc}(n_0) = \frac{i}{2}\int_0^1 d\lambda \int \frac{d^4q}{(2\pi)^4} D_{\mu\nu}^0(q)[\chi_{R,\lambda}^{\mu\nu}(q) - \chi_{V,R,\lambda}^{\mu\nu}(q)] + \Delta\tilde{E}_{xc}^{RHEG}, \tag{B.56}$$

where the exchange energy has not been subtracted for brevity. The only remaining divergence (to be eliminated by $\Delta\tilde{E}_{xc}^{RHEG}$) originates from the outermost loop integration in (B.56). It is most easily discussed within the so-called random phase (or ring) approximation (RPA) in which $\Pi_{L/T}$ is approximated by its 1-loop contribution [93, 119, 96],

$$e_{xc}^{RPA}(n_0) = \frac{i}{2}\int_0^1 d\lambda \int \frac{d^4q}{(2\pi)^4}\left[\frac{D^0(q)\Pi_L^{(0)}(q)}{1 - \lambda D^0(q)\Pi_L^{(0)}(q)} - 2\frac{D^0(q)\Pi_T^{(0)}(q)}{1 + \lambda D^0(q)\Pi_T^{(0)}(q)}\right.$$

$$\left. - 3\frac{D^0(q)\Pi_{V,R}^{(0)}(q)}{1 - \lambda D^0(q)\Pi_{V,R}^{(0)}(q)}\right] + \Delta\tilde{E}_{xc}^{RHEG,RPA},$$

i.e. in addition to the exchange graph (B.46) included for brevity, $e_{xc}^{RPA}(n_0)$ consists of the following diagrams,

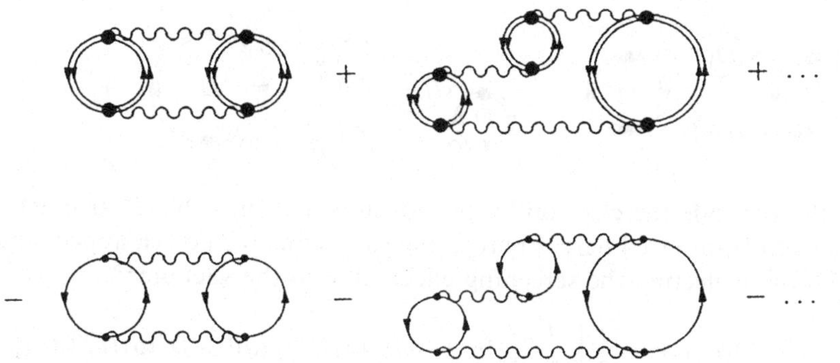

where the coupling constant integration is understood implicitly. The latter, however, can be easily performed to yield

$$e_{xc}^{RPA}(n_0) = -\frac{i}{2}\int \frac{d^4q}{(2\pi)^4}\big[\ln[1 - D_{V,R}(q)\Pi_{L,D}^{(0)}(q)] + 2\ln[1 + D_{V,R}(q)\Pi_{T,D}^{(0)}(q)]\big] + \Delta\tilde{E}_{xc}^{RHEG,RPA}. \tag{B.57}$$

At this point one usually defines the vacuum-screened exchange energy,

$$e_{x,s}(n_0) = \frac{i}{2}\int \frac{d^4q}{(2\pi)^4}\big[D_{V,R}(q)\Pi_{L,D}^{(0)}(q) - 2D_{V,R}(q)\Pi_{T,D}^{(0)}(q)\big] + \Delta\tilde{E}_{xc}^{RHEG,RPA}, \tag{B.58}$$

graphically given by

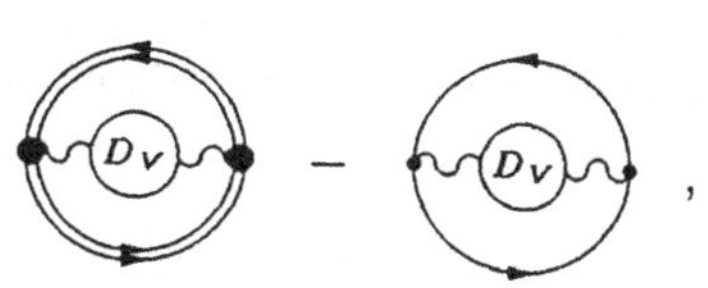

which requires renormalisation similar to its unscreened counterpart and, from a rigorous DFT point of view, is no longer a pure exchange contribution. Subtracting $e_{x,s}(n_0)$ from $e_{xc}^{RPA}(n_0)$ one ends up with a UV-convergent correlation energy [119],

$$e_{c,s}^{RPA}(n_0) = -\frac{i}{2}\int \frac{d^4q}{(2\pi)^4}\big[\ln[1 - D_{V,R}(q)\Pi_{L,D}^{(0)}(q)] + 2\ln[1 + D_{V,R}(q)\Pi_{T,D}^{(0)}(q)] + D_{V,R}(q)\Pi_{L,D}^{(0)}(q) - 2D_{V,R}(q)\Pi_{T,D}^{(0)}(q)\big], \tag{B.59}$$

as the lowest order diagram in (B.59) already contains $\Pi^{(0)}_{L/T,D}$ two times.[24] Diagrammatically (B.59) is given by

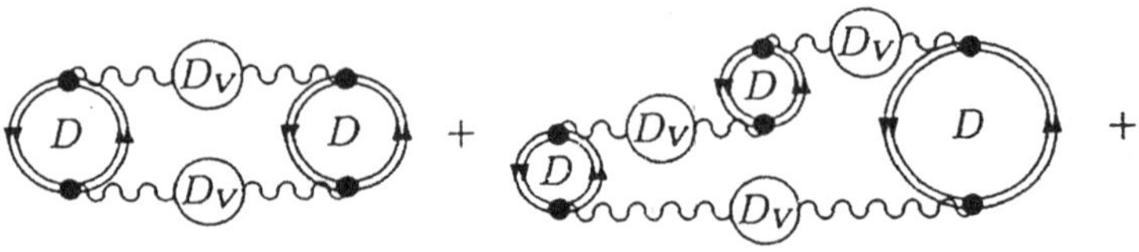

where the D inside the electron-loops indicates that only the electron gas part $\Pi^{(0)}_{D,\mu\nu}$ has to be inserted here. Finally, one can define the no-sea approximation of (B.59) by neglecting the screening effects due to the vacuum[25],

$$e_c^{L,RPA}(n_0) = -\frac{i}{2}\int \frac{d^4q}{(2\pi)^4}\{\ln|1 - D^0(q)\Pi^{(0)}_{L,D}(q)| + D^0(q)\Pi^{(0)}_{L,D}(q)\} \tag{B.60}$$

$$e_c^{T,RPA}(n_0) = -i\int \frac{d^4q}{(2\pi)^4}\{\ln|1 + D^0(q)\Pi^{(0)}_{T,D}(q)| - D^0(q)\Pi^{(0)}_{T,D}(q)\}, \tag{B.61}$$

where $e_c^{RPA}(n_0)$ has also been decomposed into the longitudinal and the transverse parts using (B.33). $e_c^{L/T,RPA}(n_0)$ has been evaluated by Ramana and Rajagopal [98] as well as Müller and Serot [99]. The high-density (ultra-relativistic) limit e_c^{RPA} is [93, 96, 99]

$$e_c^{RPA}(n_0) \underset{\beta \gg 1}{\rightarrow} \frac{e^4}{12\pi^4} k_F^4(-7.796). \tag{B.62}$$

In order to arrive at the complete the RPA + , which we understand as the combination of the RPA with the remaining second order (e^4) correlation contributions, the two second order exchange diagrams,

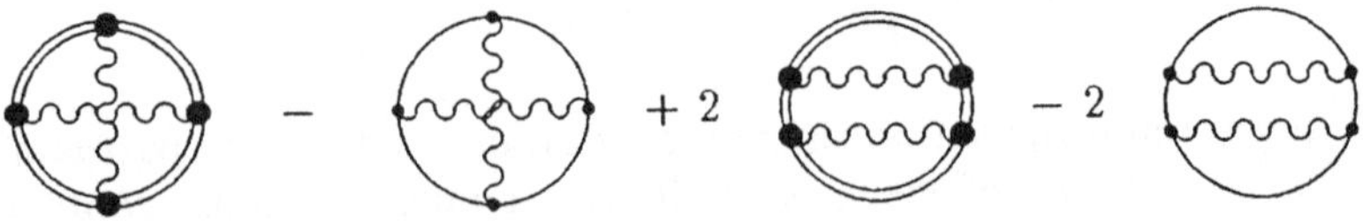

(B.63)

[24]After Wick-rotation of q_0 in (B.59) one e.g. finds $\Pi^{(0)}_{L,D}(\boldsymbol{q}, iq_0) \sim (\boldsymbol{q}^2 + q_0^2)^{-1}$ and $\Pi^{(0)}_{V,R}(\boldsymbol{q}, iq^0) \sim (\boldsymbol{q}^2 + q_0^2)\ln|\boldsymbol{q}^2 + q_0^2|$ for large q^0 and $|\boldsymbol{q}|$ so that two factors of $\Pi^{(0)}_{L,D}(\boldsymbol{q}, iq^0)$ together with the required photon propagators are sufficient to insure UV-convergence of the outermost loop integral.

[25]Note that the definition of the no-sea approximation for (B.45) is somewhat arbitrary. As the guiding principle we have chosen to neglect all closed pure vacuum electron loops wherever they occur inside a diagram.

have to be added to e_c^{RPA}. In contrast to the nonrelativistic situation the two right hand diagrams do not vanish. Both types of diagrams require renormalisation beyond the subtraction of their vacuum limit shown in Eq.(B.63). The complete density dependence of these diagrams is not known. In the ultrarelativistic limit one finds for the sum of both graphs (the individual contributions are not gauge invariant) [96]

$$e_c^{(x2)}(n_0) \underset{\beta \gg 1}{\rightarrow} \frac{e^4}{12\pi^4} k_F^4(-3.18 \pm 0.12), \tag{B.64}$$

so that their contribution constitutes about 40% of the RPA in this limit.

One can also analyse the 2-loop contribution to the screened exchange energy (B.58),

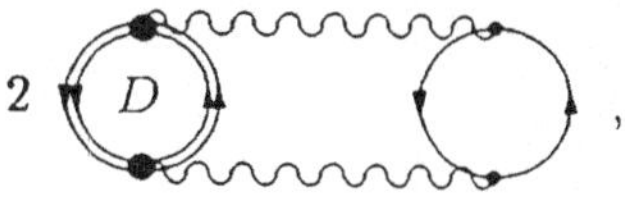

which in our definition is beyond the no-sea approximation. Its ultrarelativistic limit has been calculated by Freedman and McLerran [96],

$$e_{x,s}^{(2)}(n_0) \underset{\beta \gg 1}{\rightarrow} \frac{e^4}{12\pi^4} k_F^4\left(\ln(2\beta) - \frac{11}{6}\right). \tag{B.65}$$

In the extreme high density limit $e_{x,s}^{(2)}$ thus dominates over all other known xc-energy contributions. Note, however, that the densities required for $e_{x,s}^{(2)}$ to be of the same order of magnitude as e_c^{RPA+}, i.e. $\beta \approx 10^3$, are not relevant for electronic structure calculations.

No calculations of e_c beyond the RPA + are found in the literature. There is, however, a rather direct argument [99], which shows that the diagrammatic contributions discussed here are at least dominant in the high-density limit: As the relevant parameter for this limit is β, Eq. (B.39), it is completely equivalent to the zero mass limit. Taking the zero mass limit already from the very outset, i.e. basing the perturbation expansion for E_{xc} on an electron propagator G^0, Eq. (B.2), with $m = 0$, one can scale all loop-momenta in any given diagrammatic contribution to (B.45) by k_F and thereby extract an overall factor of k_F^4. The remaining Feynman amplitudes are dimensionless and consequently *all* diagrams give contributions which are proportional to k_F^4. The relative importance of the individual diagrams is thus essentially determined by their proportionality to α, apart from factors of $\ln|k_F/m|$ introduced by the renormalisation procedure (using m as natural renormalisation point) for vacuum parts of the diagrams. Therefore one would expect the RPA + contributions together with $e_{x,s}^{(2)}$, Eq. (B.65), to be a rather accurate representation of E_{xc} for high densities $1 \ll \beta \ll 10^3$.

B.3 Pair Correlation Function

The relativistic pair correlation function g is defined in complete analogy to its nonrelativistic counterpart,

$$\lim_{y^0 \to x^0} i\chi^{00}(x, y) - n_0 \delta^{(3)}(\boldsymbol{x} - \boldsymbol{y}) = n_0 [g(\boldsymbol{x}, \boldsymbol{y}) - 1] n_0, \tag{B.66}$$

which is an alternative form of Eq. (4.11) for the case of the RHEG. For relativistic homogeneous systems g in general depends on two variables, i.e. k_F and $|\boldsymbol{x} - \boldsymbol{y}|$, and two parameters, i.e. the coupling constant e^2 and the energy scale m. In the case of the RHEG, however, only its x-only limit (zeroth order in e^2) has been evaluated so far (compare the recent Monte Carlo study of g for the NRHEG [73]). Thus while the nonrelativistic $g_x^{NRHEG}(\boldsymbol{x}, \boldsymbol{y})$ is a function of $z = k_F|\boldsymbol{x} - \boldsymbol{y}|$ only,

$$g_x^{NRHEG}(z) = 1 - \frac{9}{2}\left[\frac{j_1(z)}{z}\right]^2, \tag{B.67}$$

the corresponding x-only pair correlation function of the RHEG [102, 19],

$$g_x^{RHEG}(z, k_F) = 1 - \frac{9}{4}\left[\frac{1}{\eta^2}\left(\sum_{i=0}^{\infty} \frac{j_{i+1}(z)}{z^{i+1}} \left(\frac{\beta}{\eta}\right)^{2i} \frac{(2i+1)!!}{2i+1}\right)^2 + \frac{\beta^2}{\eta^2}\left(\sum_{i=0}^{\infty} \frac{j_{i+2}(z)}{z^{i+1}} \left(\frac{\beta}{\eta}\right)^{2i} \frac{(2i+1)!!}{2i+1}\right)^2 + \left(\frac{j_1(z)}{z}\right)^2\right] \tag{B.68}$$

in addition depends on β. The result (B.68) has been obtained within the no-pair approximation, i.e.

$$g_x^{RHEG} = 1 - \frac{i^2}{n_0^2} \int \frac{d^4 q}{(2\pi)^4} e^{iq \cdot (x-y)} \int \frac{d^4 p}{(2\pi)^4} tr[\gamma_0 G_D^0(p) \gamma_0 G_D^0(p-q)].$$

The series representation (B.68) of g_x^{RHEG} is not very suitable for actual applications of the RWDA. A simple and reasonably accurate fit to g_x^{RHEG} [92] is based on its form for $z = 0$,

$$g_x^{RHEG}(0, k_F) = \frac{3}{4} - \frac{9}{16}\left[\frac{\eta}{\beta^2} - \frac{\operatorname{arsinh}(\beta)}{\beta^3}\right]^2, \tag{B.69}$$

with the limits

$$\lim_{\beta \to 0} g_x^{RHEG}(0, k_F) = \frac{1}{2} \tag{B.70}$$

$$\lim_{\beta \to \infty} g_x^{RHEG}(0, k_F) = \frac{3}{4}. \tag{B.71}$$

(B.71) may be interpreted as a consequence of the relativistic mixing of spin states [102]. The decreasing depth of the exchange hole $g_x^{RHEG} - 1$ with increasing β apparent from (B.70, B.71) is compensated by a broadening of the 'width' of the exchange hole (compare [102, 19]) in order to satisfy (4.13) in the homogeneous limit,

$$\frac{4}{3\pi}\int_0^\infty z^2 dz\,[g_x^{RHEG}(z, k_F) - 1] = -1. \tag{B.72}$$

The scaling behavior required for (B.72) suggests the ansatz

$$g_x^{fit}(z, k_F) = 1 - \frac{9}{2}C_1(\beta)\left[\frac{j_1(C_2(\beta)z)}{C_2(\beta)z}\right]^2 \tag{B.73}$$

$$C_1(\beta) = \frac{1}{2} + \frac{9}{8}\left[\frac{\eta}{\beta^2} - \frac{\operatorname{arsinh}(\beta)}{\beta^3}\right]^2 \tag{B.74}$$

$$C_2(\beta) = C_1(\beta)^{\frac{1}{3}}, \tag{B.75}$$

which not only approaches $g_x^{NRHEG}(z)$, Eq. (B.67), for vanishing β, but also satisfies Eqs. (B.69, B.72). The percentage deviation of $g_x^{fit}(z, k_F)$ from $g_x^{RHEG}(z, k_F)$ is plotted in Fig. B.1. In fact, even in the limit $\beta \to \infty$ in which

$$g_x^{RHEG}(z, \infty) = 1 - \frac{9}{4}\left[\left(2\frac{j_1(z)}{z^2} - \frac{j_0(z)}{z^3}(2 + z^2) + \frac{2}{z^3}\right)^2 + \left(\frac{j_1(z)}{z^2}\right)^2\right] \tag{B.76}$$

the maximal error does not exceed 1%, while for the more relevant moderately high densities ($\beta < 0.5$ – at the r-expectation value of Hg^{78+} one finds $\beta = 0.6$) the error reduces to less than 0.2%.

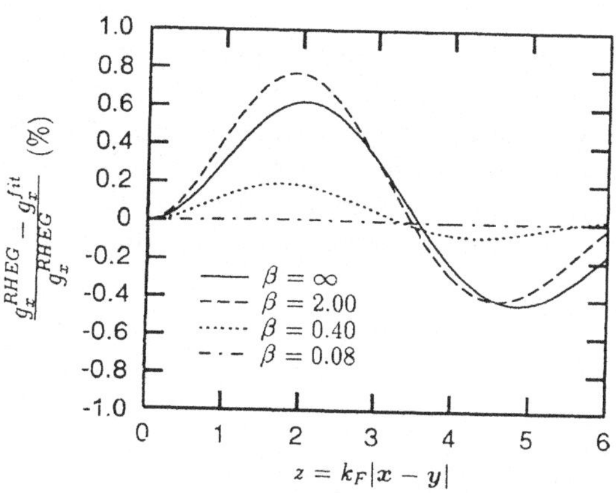

Fig. B.1. Percentage deviation of $g_x^{fit}(z, \beta)$, Eq. (B.73), from the exact $g_x^{RHEG}(z,\beta)$, Eq. (B.68), as a function of z for various values of β

C. Weakly Inhomogeneous Electron Gas

Here we provide some details on inhomogeneous systems from the viewpoint of a weakly perturbed electron gas, i.e. we consider the external potential in the Hamiltonian (2.20) as a small perturbation. In this case one can expand the four current $\delta j^\mu(\boldsymbol{r})$ induced by a static perturbation in a power series with respect to $V^\mu(\boldsymbol{r})$,

$$\delta j_\mu(\boldsymbol{r}) = \sum_{n=1}^{\infty} \frac{e^n}{n!} \int d^3r_1 \ldots \int d^3r_n \chi^{(n+1)}_{c,\mu\mu_1\ldots\mu_n}(\boldsymbol{r}, \boldsymbol{r}_1, \ldots \boldsymbol{r}_n) \times V^{\mu_1}(\boldsymbol{r}_1) \ldots V^{\mu_n}(\boldsymbol{r}_n) + \Delta j_\mu(\boldsymbol{r}), \tag{C.1}$$

where $\chi^{(n)}_{c,\mu_1\ldots\mu_n}$ represents the connected static response functions[26] defined by (B.11) [120]. Of course, δj^μ satisfies current conservation,

$$\partial_\mu \delta j^\mu(\boldsymbol{r}) = \boldsymbol{\nabla} \cdot \delta \boldsymbol{j}(\boldsymbol{r}) = 0 \Rightarrow \int d^3r\, \delta j^0(\boldsymbol{r}) = 0, \tag{C.2}$$

which is directly related to the transversality of $\chi^{(n)}_{c,\mu_1\ldots\mu_n}$ displayed in Eq. (B.13).

The induced current obtained from (C.1) is automatically UV-finite if the expansion is based on renormalised response functions, i.e. $\Delta j_\mu(\boldsymbol{r})$ just sums up the terms required for the transition from the $\chi^{(n)}_{c,\mu_1\ldots\mu_n}$ to their renormalised counterparts. It is instructive to analyse the corresponding counterterms for the noninteracting limit of (C.1) given graphically by

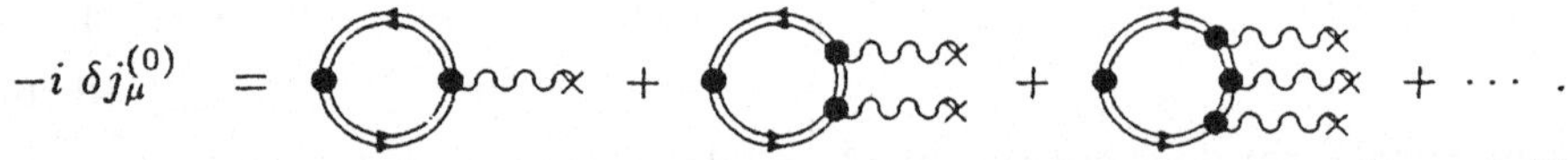

(C.3)

(note the multiplicities contained in $\chi^{(n)}_{c,\mu_1\ldots\mu_n}$) where the external potential has been represented by

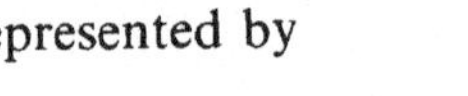

(C.4)

While the noninteracting 3-point function, i.e. the second graph, is UV-finite due to Furry's theorem, the noninteracting 4-point function (third diagram) is UV-finite due to gauge invariance and all higher order response functions are overall convergent. The only divergent term to be examined results from the relativistic Lindhard function. The corresponding counterterm has been discussed in Appendix A. Using again dimensional regularisation one finds

$$\Delta j^{(0)}_\mu(\boldsymbol{r}) = -\frac{e}{12\pi^2} \Gamma\left(\frac{4-d}{2}\right) \boldsymbol{\nabla}^2 V_\mu(\boldsymbol{r}), \tag{C.5}$$

if the gauge $\boldsymbol{\nabla} \cdot \boldsymbol{V}(\boldsymbol{r}) = 0$ is used.

---

[26] Note, that (C.1) could be used as alternative definition for the static response functions.

The corresponding energy shift can be evaluated by a coupling constant integration with respect to the external potential. Scaling the external potential Hamiltonian by λ,

$$\hat{H}_{ext}(\lambda) = \lambda e \int d^3r \hat{j}^{\mu}(\boldsymbol{r}) V_{\mu}(\boldsymbol{r}), \tag{C.6}$$

one obtains for the corresponding ground state energy

$$\begin{aligned} E_{tot}(\lambda) = {} & \langle \Phi_0(\lambda) | \hat{H}^{RHEG} + \hat{H}_{ext}(\lambda) | \Phi_0(\lambda) \rangle \\ & - \langle 0 | \hat{H}^{RHEG} | 0 \rangle + \Delta E_{tot}^{RHEG} + \Delta E_{tot}^{inhom}(\lambda), \end{aligned} \tag{C.7}$$

by a coupling constant integration approach (using a normalised ground state $|\Phi_0(\lambda)>$ for given λ),

$$\begin{aligned} E_{tot}(\lambda = 1) &= E_{tot}^{RHEG} + e \int_0^1 d\lambda \int d^3r j^{\mu}(\lambda, r) V_{\mu}(r) + \Delta E_{tot}^{inhom}(\lambda = 1) \\ &= E_{tot}^{RHEG} + e \int d^3r j^{\mu}(\lambda = 0, \boldsymbol{r}) V_{\mu}(\boldsymbol{r}) + \Delta E_{tot}^{inhom} \\ &\quad + \sum_{n=2}^{\infty} \frac{e^n}{n!} \int d^3r_1 \ldots \int d^3r_n \chi_{c,\mu_1 \ldots \mu_n}^{(n)}(\boldsymbol{r}_1, \ldots \boldsymbol{r}_n) \\ &\quad \times V^{\mu_1}(\boldsymbol{r}_1) \ldots V^{\mu_n}(\boldsymbol{r}_n), \end{aligned} \tag{C.8}$$

where (C.1) has been utilised for any given λ and $j^{\mu}(\lambda = 0, \boldsymbol{r}) = g^{\mu 0} n_0$ represents the current of the unperturbed system, i.e. the RHEG.

As for the induced four current the renormalisation of the inhomogeneity corrections to E_{tot}^{RHEG} reduces to the renormalisation of the $\chi_{c,\mu_1 \ldots \mu_n}^{(n)}$. The counterterms are thus closely related to those for δj^{μ},

$$\Delta E_{tot}^{inhom} = e \int_0^1 d\lambda \int d^3r \, \Delta j^{\mu}(\lambda, \boldsymbol{r}) V_{\mu}(\boldsymbol{r}). \tag{C.9}$$

The only counterterm on the noninteracting level, corresponding to (C.5), is e.g. given by

$$\Delta E_{tot}^{(0),inhom} = -\frac{e^2}{24\pi^2} \Gamma\left(\frac{4-d}{2}\right) \int d^3r \, V_{\mu}(\boldsymbol{r}) \nabla^2 V^{\mu}(\boldsymbol{r}). \tag{C.10}$$

The counterterm (C.9) can be decomposed into contributions to the individual energy components. Again this is most easily demonstrated for the noninteracting case. Here the total energy is just a sum of the external potential energy,

$$E_{ext} = \int d^3r \, V^{\mu}(\boldsymbol{r}) [g^{\mu 0} n_0 + \delta j^{\mu}(\boldsymbol{r})] + \Delta E_{ext}^{(0),inhom}, \tag{C.11}$$

which requires the counterterm

$$\Delta E_{ext}^{(0),inhom} = -\frac{e^2}{12\pi^2} \Gamma\left(\frac{4-d}{2}\right) \int d^3r \, V_{\mu}(\boldsymbol{r}) \nabla^2 V^{\mu}(\boldsymbol{r}), \tag{C.12}$$

and the noninteracting kinetic contribution T_s which absorbs the remainder of (C.10),

$$\Delta T_s^{inhom} = \frac{e^2}{24\pi^2}\,\Gamma\left(\frac{4-d}{2}\right)\int d^3r\, V_\mu(\boldsymbol{r})\nabla^2 V^\mu(\boldsymbol{r}). \tag{C.13}$$

D Linear Response Corrections to the Relativistic LDA

While the RLDA for $E_{xc}[j^\nu]$ is based on the xc-energy density of the RHEG, Eq. (B.45), the response expansions (C.1, C.8) allow the derivation of systematic corrections to the RLDA. We restrict ourselves here to the case of linear response for which the starting point is given by (C.1) to lowest order,

$$\delta j^\mu(\boldsymbol{q}) = e\chi^{\mu\nu}(q^0 = 0, \boldsymbol{q})\, V_\nu(\boldsymbol{q}), \tag{D.1}$$

($\chi^{\mu\nu}$ is always understood to be renormalised in this appendix, so that counter-terms are not displayed explicitly). Insertion of (D.1) into the lowest order contribution to (C.8) then leads to

$$\delta E_{tot}^{LR} = e\int\frac{d^3q}{(2\pi)^3}\,\delta j^\mu(\boldsymbol{q})\, V_\mu(\boldsymbol{q}) - \frac{1}{2}\int\frac{d^3q}{(2\pi)^3}\,\delta j^\mu(\boldsymbol{q})\,\chi_{\mu\nu}^{-1}(0,\boldsymbol{q})\,\delta j^\nu(-\boldsymbol{q}).$$

Inserting the result (B.34) for the inverse response function,

$$\begin{aligned}\delta E_{tot}^{LR} = &\int\frac{d^3q}{(2\pi)^3}\left\{e\,\delta j^\mu(\boldsymbol{q})\, V_\mu(\boldsymbol{q}) + \frac{1}{2}\delta j^\mu(\boldsymbol{q})\, D_L^0(\boldsymbol{q})\,\delta j_\mu(-\boldsymbol{q})\right\}\\ &- \frac{1}{2}\int\frac{d^3q}{(2\pi)^3}\,\delta j^\mu(\boldsymbol{q})\,\Pi_{\mu\nu}^{-1}(0,\boldsymbol{q})\,\delta j^\nu(-\boldsymbol{q}),\end{aligned} \tag{D.2}$$

one can identify the first term as the coupling to the external potential, the second as the induced Hartree energy. The third term involves the change in the kinetic energy δT_s^{LR} and the xc-energy δE_{xc}^{LR}, where the former is given by the noninteracting limit of $\Pi_{\mu\nu}^{-1}(\boldsymbol{q}, 0)$ so that the two contributions can be separated. Utilising the tensor structure of $\Pi_{\mu\nu}^{-1}(\boldsymbol{q}, 0)$, Eq. (B.35), one finally arrives at

$$\delta T_s^{LR} = -\frac{1}{2}\int\frac{d^3q}{(2\pi)^3}\left\{\frac{\delta j^0(\boldsymbol{q})\,\delta j^0(-\boldsymbol{q})}{\Pi_L^{(0)}(0,\boldsymbol{q})} + \frac{[\hat{\boldsymbol{q}}\times\delta\boldsymbol{j}(\boldsymbol{q})]\cdot[\hat{\boldsymbol{q}}\times\delta\boldsymbol{j}(-\boldsymbol{q})]}{\Pi_T^{(0)}(0,\boldsymbol{q})}\right\} \tag{D.3}$$

$$\begin{aligned}\delta E_{xc}^{LR} = \frac{1}{2}\int\frac{d^3q}{(2\pi)^3}&\left\{\delta j^0(\boldsymbol{q})\left[\frac{1}{\Pi_L^{(0)}(0,\boldsymbol{q})} - \frac{1}{\Pi_L(0,\boldsymbol{q})}\right]\delta j^0(-\boldsymbol{q})\right.\\ &\left.+\left[\frac{1}{\Pi_T^{(0)}(0,\boldsymbol{q})} - \frac{1}{\Pi_T(0,\boldsymbol{q})}\right][\hat{\boldsymbol{q}}\times\delta\boldsymbol{j}(\boldsymbol{q})]\cdot[\hat{\boldsymbol{q}}\times\delta\boldsymbol{j}(-\boldsymbol{q})]\right\},\end{aligned} \tag{D.4}$$

where $\hat{\boldsymbol{q}} = \boldsymbol{q}/|\boldsymbol{q}|$. At this point, δT_s^{LR} and δE_{xc}^{LR} are given as functionals of n_0 (inside $\Pi_{L/T}$) and δj^ν but not yet as functionals of the complete current

$j^\nu = g^{\nu 0} n_0 + \delta j^\nu$. Two paths can be followed towards the construction of actual current density functionals: On the one hand, one can rewrite the complete linear response energies (D.3, D.4) as highly nonlocal current density functionals utilising either that $j^\nu(\boldsymbol{x}) - j^\nu(\boldsymbol{y}) = \delta j^\nu(\boldsymbol{x}) - \delta j^\nu(\boldsymbol{y})$ [5, 6] or that $\nabla j^\nu(\boldsymbol{x}) = \nabla \delta j^\nu(\boldsymbol{x})$ [121]. On the other hand, one can restrict oneself to a long-wavelength expansion of (D.3, D.4) assuming $\delta j^\nu(\boldsymbol{q})$ to be strongly localised, i.e. $\delta j^\nu(\boldsymbol{x})$ to be rather delocalised. This approach leads to (semilocal) gradient corrections and has been persued extensively in the case of the nonrelativistic $E_{xc}[j^\nu]$. However, due to the limited information available for the relativistic polarisation functions $\Pi_{L/T}$ (and even on their small-$\boldsymbol{q}$ expansions) no applications of (D.4) have been reported so far.

In order to illustrate the derivation of gradient corrections we thus consider $T_s[j^\nu]$. Extracting the long-wavelength limit of (D.3),

$$\delta T_s^{[0]} = -\frac{1}{2}\int \frac{d^3q}{(2\pi)^3} \frac{\delta j^0(\boldsymbol{q})\delta j^0(-\boldsymbol{q})}{\Pi_L^{(0)}(0,\boldsymbol{0})}, \tag{D.5}$$

which is absorbed in the RLDA in accordance with the compressibility sum rule, one finds by insertion of (B.36–B.38) into (D.3) and subsequent Fourier transformation,

$$\begin{aligned}
\delta T_s^{[2]} &= \frac{1}{72m}\int d^3x \frac{1}{n_0\eta}\left[1 + 2\frac{\beta}{\eta}\operatorname{arsinh}(\beta)\right][\nabla \delta j^0(\boldsymbol{x})]^2 \\
&\quad + \frac{3\pi}{4}\int d^3x \int d^3y \frac{1}{\operatorname{arsinh}(\beta)} \frac{\delta \boldsymbol{j}(\boldsymbol{x})\cdot \delta \boldsymbol{j}(\boldsymbol{y})}{|\boldsymbol{x}-\boldsymbol{y}|} \\
\delta T_s^{[4]} &= \frac{1}{4320m}\int d^3x \frac{1}{n_0 k_F^2 \eta}[\nabla^2 \delta j^0(\boldsymbol{x})]^2 \\
&\quad \times \left[5\left(1 + 2\frac{\beta}{\eta}\operatorname{arsinh}(\beta)\right)^2 + 3(1-4\beta^2) + 12\frac{\beta^3}{\eta}\right] \\
&\quad + \frac{3\pi^2}{10m^2}\int d^3x \frac{1}{\operatorname{arsinh}(\beta)^2}\left(\frac{\eta}{\beta} - 1\right)\delta \boldsymbol{j}(\boldsymbol{x})\cdot \delta \boldsymbol{j}(\boldsymbol{x}),
\end{aligned}$$

where current conservation has been used and $[k]$ denotes the order of the small $\boldsymbol{q}$ expansion. For the density components of $\delta T_s^{[2/4]}$ (i.e. the first and third line) one now can simply replace $\nabla \delta j^0(\boldsymbol{x}) = \nabla j^0(\boldsymbol{x})$ and, correct to second order, $k_F = [3\pi^2 j^0(\boldsymbol{x})]^{\frac{1}{3}}$. While one can also utilise $\delta \boldsymbol{j}(\boldsymbol{x}) = \boldsymbol{j}(\boldsymbol{x})$, the density dependent prefactor $1/\operatorname{arsinh}(k_F/m)$ of the first current component cannot be expressed umambigously in terms of $j^0(\boldsymbol{x})$ as now two spatial variables are available. Similar to the situation for the complete linear response corrections (D.3, D.4) one is left with a choice for this substitution[27]. Abbreviating this (symmetric)

[27] In contrast to the linear response approach the direct gradient expansion of $T_s[j^\mu]$ determines the current contribution to the second order gradient correction completely [122].

function of x and y by $\bar{\beta}(x, y)$ one thus obtains

$$\delta T_s^{[2]}[j^\nu] = \frac{m^2}{24\pi^2}\int d^3x (\nabla\beta)^2 \frac{\beta}{\eta}\left[1 + 2\frac{\beta}{\eta}\operatorname{arsinh}(\beta)\right] + \frac{3\pi}{4}\int d^3x \int d^3y \frac{1}{\operatorname{arsinh}(\bar{\beta}(x, y))}\frac{j(x)\cdot j(y)}{|x - y|} \tag{D.6}$$

$$\delta T_s^{[4]}[j^\nu] = \frac{1}{1440\pi^2}\int d^3x \frac{(\nabla^2\beta)^2}{\beta\eta} \times\left[5\left(1 + 2\frac{\beta}{\eta}\operatorname{arsinh}(\beta)\right)^2 + 3(1 - 4\beta^2) + 12\frac{\beta^3}{\eta}\right] + \frac{3\pi^2}{10m^2}\int d^3x \frac{1}{\operatorname{arsinh}(\beta)^2}\left(\frac{\eta}{\beta} - 1\right) j(x)^2, \tag{D.7}$$

where β is understood as $\beta = [3\pi^2 j^0(x)]^{1/3}/m$ ($\eta = \sqrt{1 + \beta^2}$). One should note that the vacuum parts of $\Pi^{(0)}_{L/T}$ are responsible for the rightmost terms in both the density and the current component of $\delta T_s^{[4]}$. We also remark that the response approach can be extended to quadratic and cubic response which allows the derivation of the complete gradient expansion to fourth order [60].

E Direct Gradient Expansion of $T_s[j^\nu]$

A density (or current density) functional representation of the relativistic noninteracting kinetic energy T_s can either be obtained by the (linear) response technique discussed in Appendix D or by a direct gradient expansion (GE) on the basis of (2.38), whose kinetic contribution is given by

$$T_s = -i\int d^3x \lim_{y\to x} {}_s\, tr[(-i\gamma\cdot\nabla + m)G(x, y)] - \langle 0|\hat{H}_e|0\rangle + \Delta T_s^{inhom}. \tag{E.1}$$

The Greens function involved in (E.1) is the propagator for a given external potential V^μ (here to be understood as the KS-potential), which has already been discussed in Section 2. It satisfies the differential equation

$$(i\partial\!\!\!/_x - m - V\!\!\!/(x))G(x, y) = \delta^{(4)}(x - y), \tag{E.2}$$

(a factor of e has been absorbed into V^μ for brevity) and can be expanded in powers of V^μ in the form (2.36) without requiring renormalisation.

In order to set up the GE for $T_s[j^\nu]$ one proceeds in four steps [123]. In the first step the differential equation (E.2) is solved by an expansion in powers of gradients of V^μ. The starting point of this expansion is the known solution of

(E.2) for a constant external potential,

$$G^{[0]}(x,y) = e^{-i(x-y)\cdot V}\int\frac{d^4p}{(2\pi)^4}e^{-ip\cdot(x-y)}G^{[0]}(p_\mu,V_\nu) \tag{E.3}$$

$$G^{[0]}(p_\mu,V_\nu) = G_V^0(p) - 2\pi i(\not p + m)\frac{\delta(p^0-E)}{2E}\Theta(\varepsilon_F - V^0 - p^0). \tag{E.4}$$

The complete expansion is then written in the form

$$G(x,y) = e^{-i(x-y)\cdot V(x)}\int\frac{d^4p}{(2\pi)^4}e^{-ip\cdot(x-y)}\sum_{k=0}^{\infty}G^{[k]}(p_\mu,V_\nu(\boldsymbol{x})), \tag{E.5}$$

where $[k]$ denotes the order of the gradients and an overall phase factor has been extracted in analogy to (E.3). For $x^0 = y^0$ (E.5) coincides with an expansion in powers of $\hbar$ and is therefore called a semiclassical expansion. Upon insertion into (E.2) the form (E.5) leads to the recursion relation

$$(\not p - m)G^{[k]}(p_\mu,V_\nu(\boldsymbol{x})) = \left[(i\not\partial V_\nu(\boldsymbol{x}))\frac{\partial}{\partial p_\nu} - i\not\partial\right]G^{[k-1]}(p_\mu,V_\nu(\boldsymbol{x})), \tag{E.6}$$

whose solution with appropriate boundary conditions is given by

$$G^{[k]}(p_\mu,V_\nu(\boldsymbol{x})) = \left\{G^{[0]}(p_\mu,V_\nu)\left[(i\not\partial V_\nu(\mathbf{x}))\frac{\partial}{\partial p_\nu} - i\not\partial\right]\right\}^k G^{[0]}(p_\mu,V_\nu). \tag{E.7}$$

(E.7) is then evaluated explicitly to the (finite) order of gradients one is interested in.

In a second step the semiclassical expansion (E.7) is used to evaluate the semiclassical expansions of the four current j^ν, Eq. (2.39), and of T_s, Eq. (E.1). The symmetric limits in (2.39) and (E.1) introduce UV-divergencies which are eliminated by the corresponding counterterms $\Delta j^{\nu,(0)}$ and ΔT_s^{inhom} (as discussed in Section 2).

In order to illustrate this point explicitly, we consider the expansion of $j^0(\boldsymbol{x}) = n(\boldsymbol{x})$ and $T_s[n]$ for the case of a purely electrostatic potential $V^\mu = (V,\boldsymbol{0})$ to second order in the gradients of V. Using dimensional regularisation for the evaluation of the integrals in question one finds for the semiclassical expansions without counterterms,

$$\begin{aligned}\tilde{n}[V]_{reg} &= \frac{p^3}{3\pi^2} + \frac{1}{12\pi^2}\Gamma\left(2-\frac{d}{2}\right)(\nabla^2V) \\ &\quad - \frac{1}{12\pi^2}\left\{\left[\frac{E}{p} + 2\operatorname{arsinh}\left(\frac{p}{m}\right)\right](\nabla^2V) + \left[\frac{E^2}{p^2} - 3\right]\frac{(\nabla V)^2}{2p}\right\}\end{aligned} \tag{E.8}$$

$$\begin{aligned}\tilde{t}_s[V]_{reg} &= \frac{m^4}{16\pi^2}\Gamma\left(2-\frac{d}{2}\right) + \frac{1}{8\pi^2}\left\{pE^3 + p^3E - \operatorname{arsinh}\left(\frac{p}{m}\right)\right\} \\ &\quad + \frac{1}{24\pi^2}\Gamma\left(2-\frac{d}{2}\right)(\nabla V)^2 - \frac{1}{12\pi^2}\left[\frac{E^2}{p} + p\right](\nabla^2V) \\ &\quad - \frac{1}{12\pi^2}\left[\frac{E^3}{2p^3} - \frac{E}{p} + \operatorname{arsinh}\left(\frac{p}{m}\right)\right](\nabla V)^2,\end{aligned} \tag{E.9}$$

where

$$E = \varepsilon_F - V(x); \quad p = \sqrt{E^2 - m^2}\,\Theta(E^2 - m^2). \tag{E.10}$$

One recognises divergent contributions (in the limit $d \to 4$) , namely the divergent contribution of the kinetic energy of the Dirac sea

$$\begin{aligned}\langle 0|\hat{\mathscr{H}}_e|0\rangle &= -i \lim_{y\to x} tr[(-i\gamma\cdot\nabla + m)G_V^0(x,y)] \\ &= \frac{m^4}{16\pi^2}\Gamma\left(2-\frac{d}{2}\right),\end{aligned} \tag{E.11}$$

and the UV-divergencies arising from taking the symmetric limit, which are proportional to $\nabla^2 V$. While the former is eliminated by the subtraction of the vacuum energy of noninteracting fermions according to (E.1), the UV-divergencies are cancelled by the counterterms $\Delta j^{\nu,(0)}$ and ΔT_s^{inhom}, given in Eqs. (C.5, C.13). One thus explicitly verifies that from (E.8, E.9) finite semiclassical expansions $\tilde{n}[V]$ and $\tilde{t}_s[V]$ are obtained,

$$\tilde{n}[V] = \tilde{n}[V]_{reg} + \Delta j^{0,(0)} \tag{E.12}$$

$$\tilde{T}_s[V] = \int d^3x\,\tilde{t}_s[V]_{reg} - \langle 0|\hat{H}_e|0\rangle + \Delta T_s^{inhom}. \tag{E.13}$$

In the third step the semiclassical expansion of j^ν is inverted order by order to the order required. This yields V^ν as a function of j^ν and its gradients, so that by insertion into the semiclassical expansion for T_s (the fourth step) one obtains the desired relativistic gradient expansion (RGE). For the example considered here ($V = 0$) this leads to [21]

$$T_s^{RGE0}[n] = \int d^3x\, t_s^{NRHEG}(n)\,\Phi_{s,0}(\beta) \tag{E.14}$$

$$T_s^{RGE2}[n] = \frac{1}{72m}\int d^3x \frac{(\nabla n)^2}{n}\Phi_{s,2}(\beta) \tag{E.15}$$

$$\Phi_{s,2}(\beta) = \frac{1}{\eta}\left[1 + 2\frac{\beta}{\eta}\operatorname{arsinh}(\beta)\right], \tag{E.16}$$

where the electron rest mass has been subtracted and t_s^{NRHEG} and $\Phi_{s,0}$ are given by (B.43, B.44). The relativistic correction factors $\Phi_{s,0/2}$ are plotted in Fig. E.1. In contrast to $\Phi_{x,c}$ both $\Phi_{s,0}$ and $\Phi_{s,2}$ decrease in the ultrarelativistic limit.

The fourth order correction corresponding to a purely electrostatic potential has also been evaluated [124],

$$T_s^{RGE4}[n] = \int d^3x[t_{s,V}^{RGE4}[n] + t_{s,D}^{REG4}[n]]$$

$$t_{s,V}^{RGE4} = \frac{1}{360\pi^2}\left\{3\frac{\beta^2}{\eta^2}(\nabla^2\beta)^2 + 6\frac{\beta}{\eta^4}(\nabla^2\beta)(\nabla\beta)^2 + \frac{3-\beta^4\eta^2}{\eta^6}(\nabla\beta)^4\right\}$$

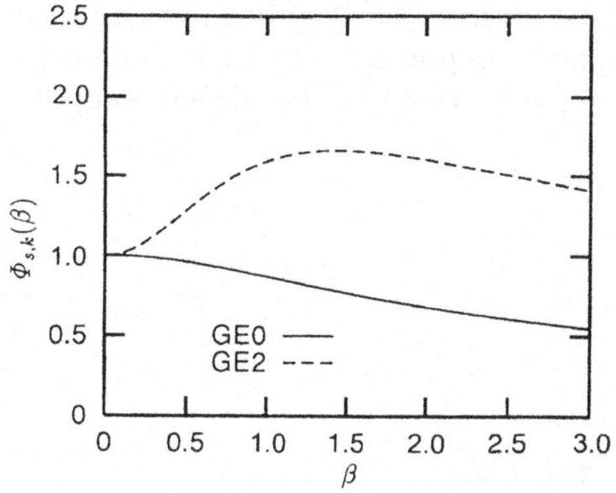

Fig. E.1. Relativistic corrections to kinetic energy densities. Eq. (B.44)—solid line (GE0), Eq. (E.16)—dashed line (GE2)

$$\begin{aligned} t_{s,D}^{RGE4} = \frac{1}{5760\pi^2} \Bigg\{ & 4\frac{(\nabla^2\beta)^2}{\beta\eta}\left[3(1-4\beta^2) + 5\left(1 + 2\frac{\beta}{\eta}\operatorname{arsinh}(\beta)\right)^2\right] \\ & + 2\frac{(\nabla^2\beta)(\nabla\beta)^2}{\beta^2\eta^3}\left[-41\beta^2 + 20\left(1 + 2\frac{\beta}{\eta}\operatorname{arsinh}(\beta)\right)\right. \\ & \left.\times\left(\frac{\beta^2+\eta^2}{2} + 2\frac{\beta}{\eta}\operatorname{arsinh}(\beta)\right)\right] \\ & + \frac{(\nabla\beta)^4}{\beta^3\eta^5}\left[3 - 19\beta^2 - 8\beta^4 + 8\beta^6 + 16\beta^8\right. \\ & \left. + 20\left(\frac{\beta^2+\eta^2}{2} + 2\frac{\beta}{\eta}\operatorname{arsinh}(\beta)\right)^2\right]\Bigg\}. \end{aligned}$$

The lowest order term $T_s^{RGE0}[n]$, the relativistic kinetic energy in the Thomas–Fermi limit, has first been calculated by Vallarta and Rosen [12]. In the second order contribution (which is given in a form simplified by partial integration) explicit vacuum corrections do not occur after renormalisation. Finite radiative corrections, originating from the vacuum part of the propagator (E.5), first show up in fourth order, where the term in $T_{s,V}^{RGE4}$ proportional to $(\nabla\beta)^4$ can be identified with the electrostatic part of the Euler–Heisenberg energy of QED [26]. One also recognises that both $T_s^{RGE2}[n]$ and the linear response component of $T_s^{RGE4}[n]$ agree with the results (D.6, D.7). In the nonrelativistic limit the individual contributions reduce to the appropriate results of the nonrelativistic gradient expansion [125, 126, 127].

The direct gradient expansion has also been applied to evaluate the current dependence of T_s to second order [122],

$$\begin{aligned} T_s^{REG2}[\boldsymbol{j}] = \frac{3}{16}\int d^3x \frac{1}{\operatorname{arsinh}(\beta(\boldsymbol{x}))} \\ \times \int d^3y \int d^3z \sum_{k,l=1}^{3} \frac{\partial_k^y j^l(\boldsymbol{y})\partial_k^z j^l(\boldsymbol{z}) - \partial_k^y j^l(\boldsymbol{y})\partial_l^z j^k(\boldsymbol{z})}{|\boldsymbol{x}-\boldsymbol{y}||\boldsymbol{x}-\boldsymbol{z}|}. \end{aligned} \tag{E.17}$$

The direct GE provides an unambigous result for $T_s^{RGE2}[j]$ in contrast to the linear response approach as here no expansion in powers of V^{μ} itself is involved. Neglecting the x-dependence of β one can reduce (E.17) to the current component of (D.6).

8 References

1. Jones RO, Gunnarsson O (1989) Rev Mod Phys 61: 689
2. Becke AD, (1992) J Chem Phys 96: 2155
3. Johnson BG, Gill PMW, Pople JA (1993) J Chem Phys 98: 5612
4. Gross EKU, Dreizler RM (eds) (1995) Density Functional Theory. NATO ASI Series B, Vol 337, Plenum, New York
5. Hohenberg P, Kohn W (1964) Phys Rev 136 B: 864
6. Kohn W, Sham LJ (1965) Phys Rev 140 A: 1133
7. Eliav E, Kaldor U, Ishikawa Y (1994) Phys Rev A 49: 1724
8. Pyykkö P (1988) Chem Rev 88: 563
9. Pepper M, Bursten BE (1991) Chem Rev 91: 719
10. Becke AD (1988) Phys Rev A 38: 3098
11. Perdew JP (1991) In: Ziesche P, Eschrig H (eds) Electronic Structure of Solids 1991. Akademie Verlag, Berlin
12. Vallarta MS, Rosen N (1932) Phys Rev 41: 708
13. Jensen H (1933) Z Phys 82: 794
14. Thomas LH (1927) Proc Cambridge Phil Soc 23: 542
15. Fermi E (1928) Z Phys 48: 73
16. Gross EKU, Dreizler RM (1981) Phys Lett 81A: 447
17. Rajagopal AK, Callaway J (1973) Phys Rev B 7: 1912
18. Rajagopal AK (1978) J Phys C 11: L943
19. MacDonald AH, Vosko SH (1979) J Phys C 12: 2977
20. Malzacher P, Dreizler RM (1986) Z Phys D 2: 37
21. Engel E, Dreizler RM (1987) Phys Rev A 35: 3607
22. Engel E, Müller H, Speicher C, Dreizler RM (1995) In: Ref. [4], p 65
23. Liberman D, Waber JT, Cromer DT (1965) Phys Rev 137 A: 27
24. Andersen OK (1970) Phys Rev B 2: 883
25. Ramana MV, Rajagopal AK (1983) Adv Chem Phys 54: 231
26. Itzykson C, Zuber J-B (1980) Quantum Field Theory. McGraw-Hill, New York
27. Källén AOG (1958) In: Handbuch der Physik, Band V, Teil 1. Springer, Berlin
28. Schmutzer E (1972) Symmetrien und Erhaltungssätze der Physik. Akademie, Berlin
29. Bjorken JD, Drell SD (1965) Relativistic Quantum Fields. McGraw-Hill, New York, chap 14.6
30. Ramana MV, Rajagopal AK (1981) J Phys C 14: 4291
31. Plunien G, Müller B, Greiner W (1986) Phys Rep 134: 87
32. Rafelski J, Fulcher LP, Klein A (1978) Phys Rep 38C: 227
33. Greiner W, Müller B, Rafelski J (1985) Quantum Electrodynamics of Strong Fields. Springer, Berlin
34. Schweber SS (1966) Relativistic Quantum Field Theory. Harper, New York
35. Shadwick BA, Talman JD, Norman MR (1989) Comp Phys Commun 54: 95
36. Engel E, Keller S, Facco Bonetti A, Müller H, Dreizler RM (1995) Phys Rev A 52: 2750
37. Vignale G, Rasolt M (1987) Phys Rev Lett 59: 2360
38. Vignale G, Rasolt M (1988) Phys Rev B 37: 10685
39. For recent comparisons of QED results for atomic systems with experimental data see eg Chen MH, Cheng KT, Johnson WR (1993) Phys Rev A 47: 3692; Riis E, Sinclair AG, Poulsen O, Drake GWF, Rowley WRC, Levick AP (1994) Phys Rev A 49: 207
40. Ramana MV, Rajagopal AK (1979) J Phys C 12: L845

41. MacDonald AH (1983) J Phys C 16: 3869
42. Xu BX, Rajagopal AK, Ramana MV (1984) J Phys C 17: 1339
43. Eschrig H, Seifert G, Ziesche P (1985) Solid State Commun 56: 777
44. Diener G, Gräfenstein J (1989) J Phys Condens Matter 1: 8445
45. Nagy A (1994) Phys Rev A 49: 3074
46. Engel E, Dreizler RM (1988) Phys Rev A 38: 3909
47. Dreizler RM, Gross EKU (1990) Density Functional Theory. Springer, Berlin
48. Mann JB, Johnson WR (1971) Phys Rev A 4: 41
49. Parpia FA, Johnson WR (1984) J Phys B 17: 531
50. Parpia FA, Johnson WR (1983) J Phys B 16: L375
51. Parpia FA, Johnson WR (1983) Phys Lett 99 A: 172
52. Talman JD, Shadwick WF (1976) Phys Rev A 14: 36
53. Sharp RT, Horton GK (1953) Phys Rev 90: 317
54. Sahni V, Gruenebaum J, Perdew JP (1982) Phys Rev B 26: 4371
55. Langreth DC, Mehl MJ (1983) Phys Rev B 28: 1809
56. Sham LJ (1985) Phys Rev B 32: 3876
57. Engel E, Chevary JA, Macdonald LD, Vosko SH (1992) Z Phys D 23: 7
58. Krieger JB, Li Y, Iafrate GJ (1992) Phys Rev B 45: 101
59. Engel E, Vosko SH (1993) Phys Rev A 47: 2800
60. Engel E, Vosko SH (1994) Phys Rev A 50: 10498
61. Kotani T (1994) Phys Rev B 50: 14816
62. Kotani T (1995) Phys Rev Lett 74: 2989
63. Bylander DM, Kleinman L (1995) Phys Rev Lett 74: 3660
64. Grabo T, Gross EKU (1995) 240: 141 Chem Phys Lett
65. Krieger JB, Li Y, Iafrate GJ (1990a) Phys Lett 146 A: 256
66. Krieger JB, Li Y, Iafrate GJ (1990b) Phys Lett 148 A: 470
67. Li Y, Krieger JB, Norman MR, Iafrate GJ (1991) Phys Rev B 44: 10437
68. Dyall KG, Grant IP, Johnson CT, Parpia FA, Plummer EP (1989) Comp Phys Commun 55: 425
69. In all our calculations the nuclei were represented by uniformly charged spheres with nuclear radii given by $R_{nucl} = 1.0793 A^{1/3} + 0.73587 fm$, A being the atomic mass (weighted by isotopic abundances) taken from Table III.7 of Ref. [128] unless explicitly stated otherwise. The speed of light has been set to $c = 137.0359895$.
70. Vosko SH, Wilk L, Nusair M (1980) Can J Phys 58: 1200
71. Perdew JP, Zunger A (1981) Phys Rev B 23: 5048
72. Perdew JP, Wang Y (1991) Phys Rev B 45: 13244
73. Ortiz G, Ballone P (1994) Phys Rev B 50: 1391
74. Ceperley DM, Alder BJ (1980) Phys Rev Lett 45: 566
75. Ma S-K, Brueckner KA (1968) Phys Rev 165: 18
76. Sham LJ (1971) In: Marcus PM, Janak JF, Williams AR (eds) Computational Methods in Band Theory. Plenum, New York, p 458
77. Geldart DJW, Rasolt M (1976) Phys Rev B 13: 1477
78. Antoniewicz PR, Kleinman L (1985) Phys Rev B 31: 6779
79. Kleinman L, Lee S (1988) Phys Rev B 37: 4634
80. Chevary JA, Vosko SH (1990) Phys Rev B 42: 5320
81. Engel E, Vosko SH (1990) Phys Rev B 42: 4940; Phys Rev B 44: 1446(E)
82. Bagno P, Jepsen O, Gunnarsson O (1989) Phys Rev B 40: 1997
83. Langreth DC, Mehl MJ (1981) Phys Rev Lett 47: 446
84. Becke AD (1986) J Chem Phys 84: 4524
85. Perdew JP, Wang Y (1986) Phys Rev B 33: 8800
86. Perdew JP (1986) Phys Rev B 33: 8822; Phys Rev B 34: 7406(E)
87. Garcia A, Elsässer Ch, Zhu J, Louie SG, Cohen ML (1992) Phys Rev B 46: 9829; Phys Rev B 47: 4150(E)
88. Singh DJ, Ashkenazi J (1992) Phys Rev B 46: 11570
89. Singh DJ (1993) Phys Rev B 48: 14099
90. Dufek P, Blaha P, Sliwko V, Schwarz K (1994) Phys Rev B 49: 10170
91. Dufek P, Blaha P, Schwarz K (1994) Phys Rev B 50: 7279
92. Keller S (1988) Diploma thesis, Frankfurt/Main
93. Akhiezer IA, Peletminskii SV (1960) Sov Phys JETP 11: 1316

94. Zapolsky HS (1960) Cornell University LNS Report. unpublished
95. Jancovici B (1962) Nuovo Cim XXV: 428
96. Freedman BA, McLerran LD (1977) Phys Rev D 16: 1130, 1147, 1169
97. Ramana MV, Rajagopal AK, Johnson WR (1982) Phys Rev A 25: 96
98. Ramana MV, Rajagopal AK (1981) Phys Rev A 24: 1689
99. Müller H, Serot BD (1995) unpublished
100. Gunnarsson O, Jonson M, Lundqvist BI (1977) Solid State Commun 24: 765
101. Alonso JA, Girifalco LA (1978) Phys Rev B 17: 3735
102. Ellis DE (1977) J Phys B 10: 1
103. A simple scheme to overcome this problem has been suggested in Przybylski H, Borstel G (1984) Solid State Commun 49: 317; Solid State Commun 52: 713
104. Das MP, Ramana MV, Rajagopal AK (1980) Phys Rev A 22: 9
105. MacDonald AH, Daams JM, Vosko SH, Koelling DD (1981) Phys Rev B 23: 6377
106. Engel E, Vosko SH (1993) Phys Rev B 47: 13164
107. Davidson ER, Hagstrom SA, Chakravorty SJ, Meiser Umar V, Froese Fischer C (1991) Phys Rev A 44: 7071
108. Ishikawa Y, Koc K (1994) Phys Rev A 50: 4733
109. Chakravorty SJ, Gwaltney SR, Davidson ER, Parpia FA, Froese Fischer C (1993) Phys Rev A 47: 3649
110. Stroucken HJAC, Dreizler RM (1991) Phys Rev A 43: 3401
111. Rieger MM, Vogl P (1995) Phys Rev A 52: 282
112. t'Hooft G, Veltman M (1972) Nucl Phys B 44: 189
113. Landsman NP, van Weert ChG (1987) Phys Rep 145: 141
114. Lim K, Horowitz CJ (1989) Nucl Phys A 501: 729
115. Källén G, Sabry A (1955) Dan Mat Fys Medd 29: No.17
116. Barbieri R, Remiddi E (1973) Nuovo Cim A 13: 99
117. Kapusta JI (1989) Finite-temperature Field Theory. Cambridge Univ. Press, Cambridge
118. Braaten E, Pisarski RD (1990) Nucl Phys B 337: 569
119. Chin SA (1977) Ann Phys (NY) 108: 301
120. Brovman EG, Kagan Y (1969) Zh Eksp Teor Fiz 57: 1329 [Sov Phys JETP 30: 721 (1970)]
121. Engel E (1995) Phys Rev A 51: 1159
122. Müller H, Engel E, Dreizler RM (1989) Phys Rev A 40: 5542
123. Engel E, Müller H, Dreizler RM (1989) Phys Rev A 39: 4873
124. Pohlner WF, Dreizler RM (1991) Phys Rev A 44: 7165
125. Kirzhnits DA (1967) Field Theoretical Methods in Many-Body Systems. Pergamon, London
126. Hodges CH (1973) Can J Phys 51: 1428
127. Jennings BK (1973) PhD Thesis, McMaster University; Brack M, Jennings BK, Chu YH (1976) Phys Lett 65B: 1
128. Hisaka K, et al. (Particle Data Group) (1992) Phys Rev D 45: Number 11, Part II

Density Functional Theory of Time-Dependent Phenomena

E.K.U. Gross[1], J.F. Dobson[2], and M. Petersilka[1]

[1] Institut für Theoretische Physik, Universität Würzburg, Am Hubland, D-97074 Würzburg, Germany
[2] School of Science, Griffith University, Nathan, Queensland 4111, Australia

Table of Contents

A density-functional formalism comparable to the theory of Hohenberg, Kohn and Sham is developed for electronic systems subject to time-dependent external fields. The formalism leads to a set of time-dependent Kohn-Sham equations which, in addition to the external potential, contain a time-dependent Hartree term and a local time-dependent exchange-correlation potential. Rigorous properties and explicit approximations of the latter are discussed in detail. Generalizations of the basic formalism to incorporate the nuclear motion and to deal with magnetic effects are described. Within the regime of linear-response theory, the time-dependent Kohn-Sham equations lead to a formally exact representation of the frequency-dependent linear density response. Applications within the linear-response regime include the computation of photoabsorbtion cross sections, the determination of van der Waals forces and the calculation of excitation energies. The latter is based on the fact that the frequency-dependent linear density response has poles at the true excitation energies of the interacting many-body system. The time-dependent Kohn-Sham formalism then leads to a simple additive correction of the Kohn-Sham single-particle excitation energies. Beyond the linear-response regime, the time-dependent Kohn-Sham scheme is applied to atoms in strong femto-second laser pulses to describe multi-photon ionization and harmonic generation in a non-perturbative way.

1 Introduction

The basic idea of density functional theory (DFT) is to describe an interacting many-particle system exclusively and completely in terms of its density. The formalism rests on two basic theorems:

I. Every observable quantity can be calculated, at least in principle, from the density alone, i.e. each quantum mechanical observable can be written as a functional of the density.
II. The density of the interacting system of interest can be obtained as the density of an auxiliary system of non-interacting particles moving in an effective *local* single-particle potential, the so-called Kohn Sham potential.

In the original work of Hohenberg and Kohn (HK) [1] and Kohn and Sham (KS) [2] these theorems were proven for the ground-state density of *static* many-body systems. On the basis of these theorems, DFT has provided an extremely successful description of ground-state properties of atoms, molecules and solids [3–5]. The quality of approximations for the Kohn-Sham potential has steadily improved over the years and the currently best functionals yield ground-state properties in very close agreement with configuration interaction results [6]. Excited-state properties, however, are notoriously difficult to calculate within the traditional density functional framework and time-dependent phenomena are not accessible at all.

Time-dependent density functional theory (TDDFT) as a complete formalism [7] is a more recent development, although the historical roots date back to the time-dependent Thomas-Fermi model proposed by Bloch [8] as early as 1933. The first and rather successful steps towards a time-dependent Kohn-Sham (TDKS) scheme were taken by Peuckert [9] and by Zangwill and Soven [10]. These authors treated the linear density response of rare-gas atoms to a time-dependent external potential as the response of non-interacting electrons to an effective time-dependent potential. In analogy to stationary KS theory, this effective potential was assumed to contain an exchange-correlation (xc) part, $v_{xc}(\mathbf{r}, t)$, in addition to the time-dependent external and Hartree terms:

$$v_s(\mathbf{r}, t) = v(\mathbf{r}, t) + \int d^3 r' \frac{n(\mathbf{r}', t)}{|\mathbf{r} - \mathbf{r}'|} + v_{xc}(\mathbf{r}, t). \qquad (1)$$

Peuckert suggested an iterative scheme for the calculation of v_{xc}, while Zangwill and Soven adopted the functional form of the static exchange-correlation potential in local density approximation. Significant steps towards a rigorous foundation of time-dependent density functional theory were taken by Deb and Ghosh [11–14] and by Bartolotti [15–18] who formulated and explored HK and KS type theorems for the time-dependent density. Each of these derivations, however, was restricted to a rather narrow set of allowable time-dependent potentials (to potentials periodic in time in the theorems of Deb and Ghosh, and to adiabatic processes in the work of Bartolotti). A general proof of statements I and II above for the time-dependent density was given by Runge and Gross [7]. A novel feature of this formalism, not present in ground-state density functional theory, is the dependence of the respective density functionals on the initial (many-particle) state. A detailed description of the time-dependent density functional formalism will be presented in Sect. 2. The central result is a set of TDKS equations which are structurally similar to the time-dependent Hartree equations but include (in principle exactly) all many-body effects through a *local* time-dependent exchange-correlation potential. In Sect. 2, we focus on the motion of electrons only. In many experimental situations, however, the nuclear motion and its coupling to the electronic motion is important as well. If, for example, a molecule is placed in the focus of a strong laser, the electric field of the laser can either couple directly to the nuclei (in the infrared frequency regime) or the coupling to the electrons can lead to photoionization with subsequent Coulomb explosion (dissociation) of the molecule. To deal with situations of this type a DF formalism for the coupled system of electrons and nuclei is developed in Sect. 3. The central result is a set of coupled TDKS equations for the electrons and for each nuclear species. While Sects. 2 and 3 exclusively deal with time-dependent electric fields, magnetic effects will be considered in Sect. 4. Both the ordinary Zeeman coupling and the coupling of magnetic fields to the orbital currents will be included.

To date, most applications of TDDFT fall in the regime of linear response. The linear response limit of time-dependent density functional theory will be discussed in Sect. 5.1. After that, in Sect. 5.2, we shall describe the density-functional calculation of higher orders of the density response. For practical applications, approximations of the time-dependent xc potential are needed. In Sect. 6 we shall describe in detail the construction of such approximate functionals. Some exact constraints, which serve as guidelines in the construction, will also be derived in this section. Finally, in Sects. 7 and 8, we will discuss applications of TDDFT within and beyond the perturbative regime. Apart from linear response calculations of the photoabsorbtion spectrum (Sect. 7.1) which, by now, is a mature and widely applied subject, we also describe some very recent developments such as the density functional calculation of excitation energies (Sect. 7.2), van der Waals forces (Sect. 7.3) and atoms in superintense laser pulses (Sect. 8).

2 Basic Formalism for Electrons in Time-Dependent Electric Fields

2.1 One-to-One Mapping Between Time-Dependent Potentials and Time-Dependent Densities

Density functional theory is based on the existence of an exact mapping between densities and external potentials. In the ground-state formalism [1], the existence proof relies on the Rayleigh-Ritz minimum principle for the energy. Straightforward extension to the time-dependent domain is not possible since a minimum principle is not available in this case. The proof given by Runge and Gross [7] for time-dependent systems is based directly on the Schrödinger equation (atomic units are used throughout):

$$i\frac{\partial}{\partial t}\Psi(t) = \hat{H}(t)\Psi(t). \tag{2}$$

We shall investigate the densities $n(\mathbf{r}, t)$ of electronic systems evolving from a *fixed* initial (many-particle) state

$$\Psi(t_0) = \Psi_0 \tag{3}$$

under the influence of different external potentials of the form

$$\hat{V}(t) = \sum_{i=1}^{N} v(\mathbf{r}_i, t) \ . \tag{4}$$

In the following discussion, the initial time t_0 is assumed to be finite and the potentials are required to be expandable in a Taylor series about t_0. No further assumptions concerning the size of the radius of convergence are made. It is sufficient that the radius of convergence is greater than zero. The initial state Ψ_0 is not required to be the ground state or some other eigenstate of the initial potential $v(\mathbf{r}, t_0) = v_0(\mathbf{r})$. This means that the case of sudden switching is included in the formalism. On the other hand, potentials that are switched-on adiabatically from $t_0 = -\infty$ are excluded by the Taylor-expandability condition because adiabatic switching involves an *essential* singularity at $t_0 = -\infty$.

Besides an external potential of the form (4), the Hamiltonian in Eq. (2) contains the kinetic energy of the electrons and their mutual Coulomb repulsion:

$$\hat{H}(t) = \hat{T} + \hat{U} + \hat{V}(t) \tag{5}$$

with

$$\hat{T} = \sum_{i=1}^{N} -\frac{1}{2}\nabla_i^2 \tag{6}$$

and

$$\hat{U} = \frac{1}{2} \sum_{\substack{i,j=1 \\ i \neq j}}^{N} \frac{1}{|\mathbf{r}_i - \mathbf{r}_j|} \ . \tag{7}$$

With these preliminaries, we can formulate the following Hohenberg-Kohn-type theorem: The densities $n(\mathbf{r}, t)$ and $n'(\mathbf{r}, t)$ evolving from a common initial state $\Psi_0 = \Psi(t_0)$ under the influence of two potentials $v(\mathbf{r}, t)$ and $v'(\mathbf{r}, t)$ (both Taylor expandable about the initial time t_0) are always different provided that the potentials differ by more than a purely time-dependent ($\mathbf{r}$-independent) function:[1]

$$v(\mathbf{r}, t) \neq v'(\mathbf{r}, t) + c(t) \ . \tag{8}$$

To prove this theorem, we use the condition that the potentials v and v' can be

[1] If v and v' differ by a purely time-dependent function, the resulting wave functions $\Psi(t)$ and $\Psi'(t)$ differ by a purely time-dependent phase factor and, consequently, the resulting densities ρ and ρ' are identical. This trivial case is excluded by the condition (8), in analogy to the ground-state formalism where the potentials are required to differ by more than a constant.

expanded in Taylor series:

$$v(\mathbf{r}, t) = \sum_{k=0}^{\infty} \frac{1}{k!} v_k(\mathbf{r}) (t - t_0)^k , \tag{9}$$

$$v'(\mathbf{r}, t) = \sum_{k=0}^{\infty} \frac{1}{k!} v'_k(\mathbf{r}) (t - t_0)^k . \tag{10}$$

Equation (8) is equivalent to the statement that for the expansion coefficients $v_k(\mathbf{r})$ and $v'_k(\mathbf{r})$ there exists a smallest integer $k \geq 0$ such that

$$v_k(\mathbf{r}) - v'_k(\mathbf{r}) = \frac{\partial^k}{\partial t^k} (v(\mathbf{r}, t) - v'(\mathbf{r}, t)) \bigg|_{t=t_0} \neq \text{const.} \tag{11}$$

From this inequality we prove in a first step that the current densities

$$\mathbf{j}(\mathbf{r}, t) = \langle \Psi(t) | \hat{\mathbf{j}}_p(\mathbf{r}) | \Psi(t) \rangle \tag{12}$$

and

$$\mathbf{j}'(\mathbf{r}, t) = \langle \Psi'(t) | \hat{\mathbf{j}}_p(\mathbf{r}) | \Psi'(t) \rangle \tag{13}$$

are different for different potentials v and v'. Here,

$$\hat{\mathbf{j}}_p(\mathbf{r}) = \frac{1}{2i} \sum_{j=1}^{N} (\nabla_{\mathbf{r}_j} \delta(\mathbf{r} - \mathbf{r}_j) + \delta(\mathbf{r} - \mathbf{r}_j) \nabla_{\mathbf{r}_j}) \tag{14}$$

is the usual paramagnetic current density operator. In a second step we shall show that the densities n and n' are different.

Using the quantum mechanical equation of motion for the expectation value of an operator $\hat{Q}(t)$,

$$\frac{\partial}{\partial t} \langle \Psi(t) | \hat{Q}(t) | \Psi(t) \rangle = \langle \Psi(t) | \left(\frac{\partial \hat{Q}}{\partial t} - i [\hat{Q}(t), \hat{H}(t)] \right) | \Psi(t) \rangle , \tag{15}$$

we obtain for the current densities:

$$\frac{\partial}{\partial t} \mathbf{j}(\mathbf{r}, t) = \frac{\partial}{\partial t} \langle \Psi(t) | \hat{\mathbf{j}}_p(\mathbf{r}) | \Psi(t) \rangle = -i \langle \Psi(t) | [\hat{\mathbf{j}}_p(\mathbf{r}), \hat{H}(t)] | \Psi(t) \rangle , \tag{16}$$

$$\begin{aligned} \frac{\partial}{\partial t} \mathbf{j}'(\mathbf{r}, t) &= \frac{\partial}{\partial t} \langle \Psi'(t) | \hat{\mathbf{j}}_p(\mathbf{r}) | \Psi'(t) \rangle \\ &= -i \langle \Psi'(t) | [\hat{\mathbf{j}}_p(\mathbf{r}), \hat{H}'(t)] | \Psi'(t) \rangle . \end{aligned} \tag{17}$$

Since Ψ and Ψ' evolve from the same initial state

$$\Psi(t_0) = \Psi'(t_0) = \Psi_0 \tag{18}$$

we can write

$$\frac{\partial}{\partial t}(\mathbf{j}(\mathbf{r},t)-\mathbf{j}'(\mathbf{r},t))\bigg|_{t=t_0} = -i\langle\Psi_0|[\hat{\mathbf{j}}_p(\mathbf{r}),\hat{H}(t_0)-\hat{H}'(t_0)]|\Psi_0\rangle$$

$$= -n_0(\mathbf{r})\,\nabla(v(\mathbf{r},t_0)-v'(\mathbf{r},t_0)) \tag{19}$$

with the initial density

$$n_0(\mathbf{r}) = n(\mathbf{r},t_0). \tag{20}$$

If the condition (11) is satisfied for $k=0$ the right-hand side of (19) cannot vanish identically and **j** and **j**′ will become different infinitesimally later than t_0. If (11) holds for some finite $k>0$ we use Eq. (15) $(k+1)$ times and obtain after some algebra:

$$\left(\frac{\partial}{\partial t}\right)^{k+1}(\mathbf{j}(\mathbf{r},t)-\mathbf{j}'(\mathbf{r},t))\bigg|_{t=t_0} = -n_0(\mathbf{r})\,\nabla w_k(\mathbf{r}) \neq 0 \tag{21}$$

with

$$w_k(\mathbf{r}) = \left(\frac{\partial}{\partial t}\right)^{k}(v(\mathbf{r},t)-v'(\mathbf{r},t))\bigg|_{t=t_0} \tag{22}$$

Once again, we conclude that

$$\mathbf{j}(\mathbf{r},t) \neq \mathbf{j}'(\mathbf{r},t) \tag{23}$$

provided that (11) holds for v and v'. To prove the corresponding statement for the densities we use the continuity equation

$$\frac{\partial}{\partial t}(n(\mathbf{r},t)-n'(\mathbf{r},t)) = -\nabla\cdot(\mathbf{j}(\mathbf{r},t)-\mathbf{j}'(\mathbf{r},t)) \tag{24}$$

and calculate the $(k+1)$th time derivative of Eq. (24) at $t=t_0$:

$$\left(\frac{\partial}{\partial t}\right)^{k+2}(n(\mathbf{r},t)-n'(\mathbf{r}t))\bigg|_{t=t_0} = \nabla\cdot(n_0(\mathbf{r})\,\nabla w_k(\mathbf{r})). \tag{25}$$

In order to prove that the densities $n(\mathbf{r},t)$ and $n'(\mathbf{r},t)$ will become different infinitesimally later than t_0, we have to demonstrate that the right-hand side of Eq. (25) cannot vanish identically. This is done by reductio ad absurdum: Assume

$$\nabla\cdot(n_0(\mathbf{r})\,\nabla w_k(\mathbf{r})) \equiv 0 \tag{26}$$

and evaluate the integral

$$\int d^3r\, n_0(\mathbf{r})\, [\nabla w_k(\mathbf{r})]^2$$
$$= -\int d^3r\, w_k(\mathbf{r})\, \nabla \cdot (n_0(\mathbf{r})\, \nabla w_k(\mathbf{r})) + \oint d\mathbf{S} \cdot (n_0(\mathbf{r})\, w_k(\mathbf{r})\, \nabla w_k(\mathbf{r})), \quad (27)$$

where we have used Green's theorem. The first integral on the right-hand side of (27) vanishes by assumption. For physically realistic potentials (i.e. potentials arising from normalizable external charge densities), the surface integral vanishes as well, because for such potentials the quantities $w_k(\mathbf{r})$ fall off at least as $1/r$. Since the integrand on the left-hand side is non-negative one concludes that

$$n_0(\mathbf{r})\, [\nabla w_k(\mathbf{r})]^2 \equiv 0 \quad (28)$$

in contradiction to $w_k(\mathbf{r}) \neq const.$ This completes the proof of the theorem. We mention that more general potentials may also be considered. The precise conditions have been formulated in [19].

We note in passing that the right-hand side of Eq. (25) is linear in w_k. Consequently, the difference between $n(\mathbf{r}, t)$ and $n'(\mathbf{r}, t)$ is non-vanishing already in first order of $v(\mathbf{r}, t) - v'(\mathbf{r}, t)$. This result will be of importance in Sect. 5 because it ensures the invertibility of linear response operators.

By virtue of the 1–1 correspondence established above (for a given Ψ_0), the time-dependent density determines the external potential uniquely up to within an additive purely time-dependent function. The potential, on the other hand, determines the time-dependent wave function, which can therefore be considered as a functional of the time-dependent density, unique up to within a purely time-dependent phase $\alpha(t)$:

$$\Psi(t) = e^{-i\alpha(t)}\, \tilde{\Psi}[n](t)\,. \quad (29)$$

As a consequence, the expectation value of any quantum mechanical operator $\hat{Q}(t)$ is a *unique* functional of the density:

$$Q[n](t) = \langle \tilde{\Psi}[n](t) \,|\, \hat{Q}(t) \,|\, \tilde{\Psi}[n](t) \rangle\,. \quad (30)$$

The ambiguity in the phase cancels out. As a particular example, the right-hand side of Eq. (16) can be considered as a density functional which depends parametrically on $\mathbf{r}$ and t:

$$\mathbf{P}[n](\mathbf{r}, t) \equiv -i \langle \Psi[n](t) \,|\, [\hat{\mathbf{j}}_p(\mathbf{r}), \hat{H}(t)] \,|\, \Psi[n](t) \rangle\,. \quad (31)$$

This implies that the time-dependent particle and current densities can always be calculated (in principle exactly) from the following set of "hydrodynamical" equations:

$$\frac{\partial}{\partial t} n(\mathbf{r}, t) = -\nabla \cdot \mathbf{j}(\mathbf{r}, t) \quad (32)$$

$$\frac{\partial}{\partial t}\mathbf{j}(\mathbf{r},t) = \mathbf{P}[n](\mathbf{r},t)\,. \tag{33}$$

In practice, of course, the functional $\mathbf{P}[n]$ is only known approximately.

2.2 Stationary-Action Principle

The solution of the time-dependent Schrödinger equation (2) with initial condition (3) corresponds to a stationary point of the quantum mechanical action integral

$$\mathscr{A} = \int_{t_0}^{t_1} dt\,\langle\Psi(t)|\,i\frac{\partial}{\partial t} - \hat{H}(t)\,|\Psi(t)\rangle\,. \tag{34}$$

Since there is a 1–1 mapping between time-dependent wave functions, $\Psi(t)$, and time-dependent densities, $n(\mathbf{r}, t)$, the corresponding density functional

$$\mathscr{A} = \int_{t_0}^{t_1} dt\,\langle\Psi[n](t)|\,i\frac{\partial}{\partial t} - \hat{H}(t)\,|\Psi[n](t)\rangle \tag{35}$$

must have a stationary point at the correct time-dependent density (corresponding to the Hamiltonian $\hat{H}(t)$ and the initial state Ψ_0). Thus the correct density can be obtained by solving the Euler equation

$$\frac{\delta\mathscr{A}[n]}{\delta n(\mathbf{r},t)} = 0 \tag{36}$$

with appropriate boundary conditions. The functional $\mathscr{A}[n]$ can be written as

$$\mathscr{A}[n] = \mathscr{B}[n] - \int_{t_0}^{t_1} dt \int d^3r\, n(\mathbf{r},t)\, v(\mathbf{r},t) \tag{37}$$

with a *universal* (Ψ_0-dependent) functional $\mathscr{B}[n]$, formally defined as

$$\mathscr{B}[n] = \int_{t_0}^{t_1} dt\,\langle\Psi[n](t)|\,i\frac{\partial}{\partial t} - \hat{T} - \hat{U}\,|\Psi[n](t)\rangle\,. \tag{38}$$

On the exact level, the hydrodynamical equations (32, 33) and the variational equation (36) are of course equivalent. The functionals $\mathbf{P}[n]$, $\mathscr{A}[n]$, $\mathscr{B}[n]$ are well-defined only for v-representable densities, i.e. for densities that come from some time-dependent potential satisfying Eq. (9). In view of this, a Levy-Lieb-type [20–22] extension of the respective functionals to arbitrary (non-negative, normalizable) functions $n(\mathbf{r}, t)$ appears desirable. Two different proposals of this type have been put forward so far [23, 24].

2.3 Time-Dependent Kohn-Sham Scheme

The 1–1 correspondence between time-dependent densities and time-dependent potentials can be established for any *given* interaction $\hat{U}$, in particular also for $\hat{U} \equiv 0$, i.e. for non-interacting particles. Therefore the external potential $v_s[n](\mathbf{r}, t)$ of a non-interacting system reproducing a given density $n(\mathbf{r}, t)$ is uniquely determined. However, the 1–1 correspondence only ensures the uniqueness of $v_s[n]$ for all v-representable densities but not its existence for an *arbitrary* $n(\mathbf{r}, t)$. In order to derive a time-dependent KS scheme we have to assume, similar to the static case, non-interacting v-representability, i.e., we have to assume that a potential v_s exists that reproduces the time-dependent density of the interacting system of interest. Under this assumption, the density of the interacting system can be obtained from

$$n(\mathbf{r}, t) = \sum_{j=1}^{N} |\varphi_j(\mathbf{r}, t)|^2 \tag{39}$$

with orbitals $\varphi_j(\mathbf{r}, t)$ satisfying the time-dependent KS equation

$$i\frac{\partial}{\partial t}\varphi_j(\mathbf{r}, t) = \left(-\frac{\nabla^2}{2} + v_s[n](\mathbf{r}, t)\right)\varphi_j(\mathbf{r}, t). \tag{40}$$

Usually, the single-particle potential v_s is written as

$$v_s[n](\mathbf{r}, t) = v(\mathbf{r}, t) + \int d^3r' \frac{n(\mathbf{r}', t)}{|\mathbf{r} - \mathbf{r}'|} + v_{xc}[n](\mathbf{r}, t), \tag{41}$$

where $v(\mathbf{r}, t)$ is the external time-dependent field. Equation (41) *defines* the time-dependent xc potential. In practice, this quantitiy has to be approximated. As in the static case, the great advantage of the time-dependent KS scheme lies in its computational simplicity compared to other methods such as time-dependent Hartree-Fock or time-dependent configuration interaction [25–32]. One has to emphasize that, in contrast to time-dependent Hartree-Fock, the effective single-particle potential v_s is a *local* potential, i.e., a multiplicative operator in configuration space.

A few remarks are in order at this point:

(i) An important difference between the ordinary ground state density functional theory and the time-dependent formalism developed above is that in the time-dependent case the 1–1 correspondence between potentials and densities can be established only for a *fixed* initial many-body state Ψ_0. Consequently, the functionals $\mathbf{P}[n]$, $\mathcal{A}[n]$ and $\mathcal{B}[n]$ implicitly depend on Ψ_0. In the same way, $v_s[n]$ and $v_{xc}[n]$ implicitly depend on the initial KS Slater determinant. The formalism provides no guideline of how to choose the initial KS orbitals $\varphi_j(\mathbf{r}, t_0)$ as long as they reproduce the initial interacting density n_0 corresponding to Ψ_0. In general, there exist infinitely many Slater determinants reproducing a given density [33, 34]. From a formal point of view there is no problem with that; any choice of initial orbitals $\varphi_j(\mathbf{r}, t_0)$ reproducing the initial interacting density n_0 will do the job because the dependence of $v_s[n]$ on the initial state is such that the interacting density will be reproduced in each case. In practice, however, the dependence on the initial state is a nuisance. Of course one would prefer to have functionals of the density alone rather than functionals of $n(\mathbf{r}, t)$ and Ψ_0. One has to emphasize, however, that for a large class of systems, namely those where both Ψ_0 and the initial KS Slater determinant are non-degenerate ground states, $\mathbf{P}[n]$ and $v_s[n]$ are indeed functionals of the density alone. This is because any non-degenerate ground state Ψ_0 is a unique functional of its density $n_0(\mathbf{r})$ by virtue of the traditional HK theorem. In particular, the initial KS orbitals are *uniquely* determined as well in this case.

(ii) We emphasize that the KS scheme does not follow from the variational principle. Incidentally, the same statement holds true in the static case as well. The KS scheme follows from the basic 1-1 mapping (applied to non-interacting particles) and the assumption of non-interacting v-representability. The variational principle yields an additional piece of information, namely the equation

$$v_{xc}[n](\mathbf{r}, t) = \frac{\delta \mathcal{A}_{xc}[n]}{\delta n(\mathbf{r}, t)}, \tag{42}$$

where $\mathcal{A}_{xc}$ is the xc part of the action functional, formally defined by

$$\mathcal{A}_{xc}[n] = \mathcal{B}_s[n] - \mathcal{B}[n] - \frac{1}{2}\int_{t_0}^{t_1} dt \int d^3r \int d^3r' \frac{n(\mathbf{r}, t)\, n(\mathbf{r}', t)}{|\mathbf{r} - \mathbf{r}'|}. \tag{43}$$

Here $\mathcal{B}_s[n]$ is the non-interacting analogue of the functional $\mathcal{B}[n]$, i.e.,

$$\mathcal{B}_s[n] = \int_{t_0}^{t_1} dt \, \langle \Phi[n](t) | \, i\frac{\partial}{\partial t} - \hat{T} \, | \Phi[n](t) \rangle \tag{44}$$

where $\Phi[n](t)$ is the unique time-dependent Slater determinant corresponding to the density n.

(iii) The current density

$$\mathbf{j}(\mathbf{r}, t) = \frac{1}{2i} \sum_{k=1}^{N} (\varphi_k^*(\mathbf{r}, t)\, \nabla \varphi_k(\mathbf{r}, t) - \varphi_k(\mathbf{r}, t)\, \nabla \varphi_k^*(\mathbf{r}, t)) \tag{45}$$

following from the TDKS orbitals is identical with the true current density of the interacting system at hand. In order to prove this statement we recall that the first part of the Runge-Gross proof described above establishes a 1–1 mapping, $v(\mathbf{r}, t) \leftrightarrow \mathbf{j}(\mathbf{r}, t)$, between external potentials and current densities of interacting particles and likewise, for $\hat{U} \equiv 0$, a 1–1 mapping, $v_s(\mathbf{r}, t) \leftrightarrow \mathbf{j}_s(\mathbf{r}, t)$, between external potentials and current densities of noninteracting particles. Making once again the assumption of non-interacting v-representability of the interacting current density $\mathbf{j}(\mathbf{r}, t)$, one can establish an alternative "current-density version" of the KS scheme,

$$i \frac{\partial}{\partial t} \tilde{\varphi}_j(\mathbf{r}, t) = (-\tfrac{1}{2}\nabla^2 + \tilde{v}_s[\mathbf{j}](\mathbf{r}, t))\, \tilde{\varphi}_j(\mathbf{r}, t) \tag{46}$$

$$\mathbf{j}(\mathbf{r}, t) = \frac{1}{2i} \sum_{k=1}^{N} (\tilde{\varphi}_k^*(\mathbf{r}, t)\, \nabla \tilde{\varphi}_k(\mathbf{r}, t) - \tilde{\varphi}_k(\mathbf{r}, t)\, \nabla \tilde{\varphi}_k^*(\mathbf{r}, t)) \tag{47}$$

whose solution reproduces the current density, $\mathbf{j}$, of the *interacting* system of interest. To prove Eq. (45), we show that the solutions $\tilde{\varphi}_j(\mathbf{r}, t)$ of (46,47) are in fact identical with the solutions φ_j of the ordinary TDKS scheme (39–41). To this end we prove that the density

$$\tilde{n}(\mathbf{r}, t) = \sum_{k=1}^{N} |\tilde{\varphi}_k(\mathbf{r}, t)|^2 \tag{48}$$

is identical with the density resulting from (39–41):

$$\tilde{n}(\mathbf{r}, t) = n(\mathbf{r}, t)\,. \tag{49}$$

Then the uniqueness of the potential $v_s[n]$ reproducing $n(\mathbf{r}, t)$ implies that $\tilde{v}_s(\mathbf{r}, t) = v_s(\mathbf{r}, t)$ so that the solutions of (46) and (40) are identical. In order to prove Eq. (49) we observe that the full many-body Schrödinger Eq. (2) implies the continuity equation

$$\frac{\partial n(\mathbf{r}, t)}{\partial t} = -\nabla \cdot \mathbf{j}(\mathbf{r}, t) \tag{50}$$

while the Schrödinger equation (46) implies the continuity equation

$$\frac{\partial \tilde{n}(\mathbf{r}, t)}{\partial t} = -\nabla \cdot \mathbf{j}(\mathbf{r}, t)\,. \tag{51}$$

Comparing (50) and (51) we find that $n(\mathbf{r}, t)$ and $\tilde{n}(\mathbf{r}, t)$ can differ at most by a *time-independent* function $\eta(\mathbf{r})$ so that, at the initial time t_0,

$$n(\mathbf{r}, t_0) = \tilde{n}(\mathbf{r}, t_0) + \eta(\mathbf{r})\,. \tag{52}$$

Hence, if the initial orbitals are chosen to be identical,

$$\varphi_k(\mathbf{r}, t_0) = \tilde{\varphi}_k(\mathbf{r}, t_0) \quad k = 1, \ldots, N\,, \tag{53}$$

it follows that $\eta(\mathbf{r}) \equiv 0$ and Eq. (49) will be satisfied for all times. It remains to be shown that the choice (53) is always possible. This is not obvious a priori because, by construction, the orbitals $\varphi_k(\mathbf{r}, t_0)$ must reproduce the initial density $n(\mathbf{r}, t_0)$ while the orbitals $\tilde{\varphi}_k(\mathbf{r}, t_0)$ must yield the initial current density $\mathbf{j}(\mathbf{r}, t_0)$. In order to prove that the choice (53) is possible we show that a given density $n_0(\mathbf{r})$ and a given current density $j_0(\mathbf{r})$ can always be simultaneously reproduced by a single Slater determinant

$$\Phi(\mathbf{r}_1, \ldots, \mathbf{r}_N) = \frac{1}{\sqrt{N!}} \det\{\phi_j(\mathbf{r}_i)\} . \tag{54}$$

This can be shown with a current-density generalization [24] of the so-called Harriman construction [34]. Here we reproduce the construction for one spatial dimension. The three dimensional case can be treated in analogy to Ref. [35]. Given the densities $n_0(x)$ and $j_0(x)$ we define the following functions

$$q(x) := \frac{2\pi}{N} \int_a^x dx' \, n_0(x') \tag{55}$$

$$s(x) := \int_a^x dx' \, \frac{j_0(x')}{n_0(x')} \tag{56}$$

so that

$$\frac{dq(x)}{dx} = \frac{2\pi}{N} n_0(x) \tag{57}$$

and

$$\frac{ds(x)}{dx} = \frac{j_0(x)}{n_0(x)} . \tag{58}$$

In terms of these quantities and the particle number N we define the single-particle orbitals

$$\phi_k(x) := \sqrt{\frac{n_0(x)}{N}} \, e^{i(kq(x) + s(x) - (M/N)\, q(x))} \tag{59}$$

where k is an arbitrary integer while M is a fixed integer to be determined below. The functions $\{\phi_k\colon k \text{ integer}\}$ form a complete and orthonormal set (see e.g. [4]). Constructing the Slater determinant (54) from these orbitals it is readily verified that

$$\sum_{j=1}^{N} |\phi_{k_j}(x)|^2 = n_0(x) \tag{60}$$

and

$$\frac{1}{2i} \sum_{j=1}^{N} \left(\phi_{k_j}^*(x) \frac{d}{dx} \phi_{k_j}(x) - \phi_{k_j}(x) \frac{d}{dx} \phi_{k_j}^*(x) \right)$$
$$= n_0(x) \frac{ds}{dx} + \frac{n_0(x)}{N} \frac{dq}{dx} \left[\left(\sum_{j=1}^{N} k_j \right) - M \right] . \tag{61}$$

Hence, by virtue of Eq. (58), Eq. (61) reproduces the given current density $j_0(x)$ if M is chosen equal to $(\sum_{j=1}^{N} k_j)$. This completes the proof.

3 Motion of the Nuclei

3.1 Quantum Mechanical Treatment of Nuclear Motion

The formalism developed so far is adequate whenever the motion of the atomic nuclei can be neglected. Then the electron-nucleus interaction only enters as a *static* contribution to the potential $v(\mathbf{r}, t)$ in Eq. (41). This is a good approximation for atoms in strong laser fields above the infrared frequency regime. When the nuclei are allowed to move, the nuclear motion couples dynamically to the electronic motion and the situation becomes more complicated.

In this section we shall describe a TDDFT for systems consisting of N electrons and N_A nuclei of charge Z_A and mass M_A (in a.u.), $A = 1, \ldots, K$. K is the number of different nuclear species. Let $\mathbf{R}_{A\alpha}$ be the configuration space vector of the αth nucleus of species A. Then the complete system of electrons and nuclei is described by the Schrödinger equation

$$i\frac{\partial}{\partial t}\Psi(\mathbf{r}_1 \ldots \mathbf{r}_N, \{\mathbf{R}_{A\alpha}\}, t) = [\hat{H}_e(\mathbf{r}_1 \ldots \mathbf{r}_N, t) + \hat{H}_n(\{\mathbf{R}_{A\alpha}\}, t) + \hat{H}_{en}(\mathbf{r}_1 \ldots \mathbf{r}_N, \{\mathbf{R}_{A\alpha}\})]\, \Psi(\mathbf{r}_1 \ldots \mathbf{r}_N, \{\mathbf{R}_{A\alpha}\}, t) \tag{62}$$

with the electronic Hamiltonian

$$\hat{H}_e = \sum_{j=1}^{N}\left(-\frac{1}{2}\nabla_{\mathbf{r}_j}^2 + v_{\text{ext}}(\mathbf{r}_j, t)\right) + \frac{1}{2}\sum_{\substack{j,k=1\\ j\neq k}}^{N}\frac{1}{|\mathbf{r}_j - \mathbf{r}_k|}, \tag{63}$$

the nuclear Hamiltonian

$$\hat{H}_n = \sum_{A=1}^{K}\sum_{\alpha=1}^{N_A}\left(-\frac{1}{2M_A}\nabla_{\mathbf{R}_{A\alpha}}^2 + V_{\text{ext}}^A(\mathbf{R}_{A\alpha}, t)\right) + \frac{1}{2}\underbrace{\sum_{A=1}^{K}\sum_{\alpha=1}^{N_A}\sum_{B=1}^{K}\sum_{\beta=1}^{N_B}}_{(A\alpha)\neq(B\beta)}\frac{Z_A Z_B}{|\mathbf{R}_{A\alpha} - \mathbf{R}_{B\beta}|} \tag{64}$$

and the electron-nucleus interaction

$$\hat{H}_{en} = -\sum_{j=1}^{N}\sum_{A=1}^{K}\sum_{\alpha=1}^{N_A}\frac{Z_A}{|\mathbf{r}_j - \mathbf{R}_{A\alpha}|}. \tag{65}$$

Based on an extension [35] of the Runge-Gross theorems descried in Sect. 2 to arbitrary multi-component systems one can develop [36] a TDDFT for the coupled system of electrons and nuclei described above. In analogy to Sects. 2.1–2.3, one can establish three basic statements: First of all, there exists a rigorous 1–1 mapping between the vector of external potentials and the vector of electronic and nuclear densities,

$$(v_{\text{ext}}(\mathbf{r},t);\, V^1_{\text{ext}}(\mathbf{R},t), \ldots, V^K_{\text{ext}}(\mathbf{R},t)) \xleftrightarrow{1-1} (n(\mathbf{r},t);\, n_1(\mathbf{R},t), \ldots, n_K(\mathbf{R},t)). \tag{66}$$

Once again, this 1–1 correspondence is valid for a fixed initial many-body state $\Psi(\mathbf{r}_1, \ldots \mathbf{r}_N, \{\mathbf{R}_{A\alpha}\}; t_0)$. Besides this HK-type statement, one can derive a stationary-action principle and a set of coupled TDKS equations for electrons and nuclei. The latter read as follows:

$$i\frac{\partial}{\partial t}\varphi_j(\mathbf{r},t) = \left(-\tfrac{1}{2}\nabla^2_{\mathbf{r}} + v_s[n,\{n_B\}](\mathbf{r},t)\right)\varphi_j(\mathbf{r},t), \quad j = 1, \ldots, N \tag{67}$$

$$\mathrm{i}\frac{\partial}{\partial t}\psi_{A\alpha}(\mathbf{R},t) = \left(-\frac{1}{2M_A}\nabla^2_{\mathbf{R}} + V^A_s[n,\{n_B\}](\mathbf{R},t)\right)\psi_{A\alpha}(\mathbf{R},t), \qquad A = 1, \ldots, K; \quad \alpha = 1, \ldots, N_A \tag{68}$$

with the nuclear densities

$$n_A(\mathbf{R},t) = \sum_{\alpha=1}^{N_A} n_{A\alpha}(\mathbf{R},t), \quad n_{A\alpha}(\mathbf{R},t) = |\psi_{A\alpha}(\mathbf{R},t)|^2, \tag{69}$$

the electronic density

$$n(\mathbf{r},t) = \sum_{j=1}^{N} |\varphi_j(\mathbf{r},t)|^2, \tag{70}$$

and the KS potentials

$$v_s[n,\{n_B\}](\mathbf{r},t) = v_{\text{ext}}(\mathbf{r},t) + \int d^3r'\,\frac{n(\mathbf{r}',t)}{|\mathbf{r}-\mathbf{r}'|} - \sum_{B=1}^{K}\int d^3R\,\frac{Z_B n_B(\mathbf{R},t)}{|\mathbf{r}-\mathbf{R}|} + v_{\text{xc}}[n,\{n_B\}](\mathbf{r},t), \tag{71}$$

$$V^A_s[n,\{n_B\}](\mathbf{R},t) = V^A_{\text{ext}}(\mathbf{R},t) - Z_A\int d^3r'\,\frac{n(\mathbf{r}',t)}{|\mathbf{R}-\mathbf{r}'|} + Z_A\sum_{B=1}^{K}\int d^3R'\,\frac{Z_B n_B(\mathbf{R}',t)}{|\mathbf{R}-\mathbf{R}'|} + V^A_{\text{xc}}[n,\{n_B\}](\mathbf{R},t). \tag{72}$$

The xc potentials in (71) and (72) are formally given by functional derivatives of the xc part of the quantum mechanical action functionals:

$$v_{\mathrm{xc}}(\mathbf{r}, t) = \frac{\delta \mathscr{A}_{\mathrm{xc}}[n, \{n_B\}]}{\delta n(\mathbf{r}, t)} \tag{73}$$

$$V_{\mathrm{xc}}^{A}(\mathbf{R}, t) = \frac{\delta \mathscr{A}_{\mathrm{xc}}[n, \{n_B\}]}{\delta n_{\mathrm{A}}(\mathbf{R}, t)} . \tag{74}$$

In molecules and clusters, *genuine* exchange (as well as correlation) among identical nuclei is very small because, at typical internuclear separations, the overlap of nuclear wave functions is rather small.[2] However, the exact xc functional also contains self-exchange contributions which are not small and which cancel the self-interaction terms contained in the Hartree potentials in Eqs. (71) and (72). Hence it will be a very good approximation to represent V_{xc}^{A} by the self-exchange terms alone. This leads to

$$\begin{aligned} V_{\mathrm{s}}^{A\alpha}[n, \{n_B\}](\mathbf{R}, t) = V_{\mathrm{ext}}^{A}(\mathbf{R}, t) &- Z_A \int d^3 r' \frac{n(\mathbf{r}', t)}{|\mathbf{R} - \mathbf{r}'|} \\ &+ Z_A \sum_{\substack{B=1 \\ B \neq A}}^{K} \int d^3 \mathrm{R}' \frac{Z_B n_B(\mathbf{R}', t)}{|\mathbf{R} - \mathbf{R}'|} \\ &+ Z_A \sum_{\substack{\beta=1 \\ \beta \neq \alpha}}^{N_A} \int d^3 R' \frac{Z_A n_{A\beta}(\mathbf{R}', t)}{|\mathbf{R} - \mathbf{R}'|} . \end{aligned} \tag{75}$$

Note that, within this approximation, the nuclear KS potential depends on the state $\psi_{A\alpha}$ it acts on. This is analogous to the SIC scheme of Perdew and Zunger [37].

Clearly, a complete numerical solution of the coupled KS equations (67, 68) for electrons and nuclei will be rather involved. Usually only the valence electrons need to be treated dynamically. The core electrons can be taken into account approximately by replacing the electron-nucleus interaction (65) by suitable pseudopotentials and by replacing the nuclear Coulomb potential in Eq. (64) by the appropriate *ionic* Coulomb potential [38]. This procedure reduces the number of electronic KS equations and hence the numerical effort considerably.

3.2 Classical Treatment of Nuclear Motion

Further simplification is achieved by treating the nuclear motion classically. Numerical schemes of this type have been derived in various ways [38–43]. In

[2] In atomic scattering processes, this is not necessarily the case.

this section we shall use the multicomponent formalism developed in Sect. 3.1 as a starting point to derive classical equations of motion for the nuclei.

Applying Ehrenfest's theorem to the nuclear KS equation (68), the classical trajectory

$$\mathbf{R}_{A\alpha}^{\text{class}}(t) = \langle \psi_{A\alpha}(t) | \hat{\mathbf{R}} | \psi_{A\alpha}(t) \rangle = \int d^3R\, \mathbf{R}\, n_{A\alpha}(\mathbf{R}, t) \tag{76}$$

of the αth nucleus of species A satisfies the equation of motion

$$M_A \frac{d^2}{dt^2} \mathbf{R}_{A\alpha}^{\text{class}}(t) = \mathbf{F}_{A\alpha}(t) \tag{77}$$

where the force is given by

$$\mathbf{F}_{A\alpha}(t) = -\langle \psi_{A\alpha}(t) | \nabla_{\mathbf{R}} V_s^A | \psi_{A\alpha}(t) \rangle . \tag{78}$$

Since the TDKS equations (67–72) reproduce the exact nuclear densities, Eq. (76) yields the *exact* classical trajectory whenever species A contains only one nucleus. When species A contains more than one nucleus we have a system of *indistinguishable* particles and then, strictly speaking, the trajectories of *single* nuclei cannot be told apart: Only the total density $n_A(\mathbf{R}, t)$ and hence the center-of-mass trajectory of species A can be measured. In this case, trajectories of single nuclei can be *defined* by Eq. (76) within some effective single-particle theory. TDKS theory is particularly suitable for this purpose since the TDKS partial densities $n_{A\alpha}$ lead to the *exact* total density n_A.

Employing the approximation (75) for the nuclear KS potential the force (78) on the nucleus $(A\alpha)$ simplifies to

$$\begin{aligned} \mathbf{F}_{A\alpha}(t) = -\int d^3R\, n_{A\alpha}(\mathbf{R}, t) \Bigg[& \nabla_{\mathbf{R}} V_{\text{ext}}^A(\mathbf{R}, t) - Z_A \int d^3r\, \nabla_{\mathbf{R}} \frac{n(\mathbf{r}, t)}{|\mathbf{R} - \mathbf{r}|} \\ & + Z_A \sum_{\substack{B=1 \\ B \neq A}}^{K} \int d^3R'\, \nabla_{\mathbf{R}} \frac{Z_B n_B(\mathbf{R}', t)}{|\mathbf{R} - \mathbf{R}'|} \\ & + Z_A \sum_{\substack{\beta=1 \\ \beta \neq \alpha}}^{N_A} \int d^3R'\, \nabla_{\mathbf{R}} \frac{Z_A n_{A\beta}(\mathbf{R}', t)}{|\mathbf{R} - \mathbf{R}'|} \Bigg] . \end{aligned} \tag{79}$$

In many cases, the nuclear densities $n_{A\alpha}(\mathbf{R}, t)$ will be rather narrow functions, with a strong peak at the classical trajectory $\mathbf{R}_{A\alpha}^{\text{class}}(t)$. In such a situation, integrals of the form $\int d^3R\, n_{A\alpha}(\mathbf{R}, t) G(\mathbf{R})$ are well represented by Taylor-expanding $G(\mathbf{R})$ around the classical trajectory. This leads to

$$\int d^3R\, n_{A\alpha}(\mathbf{R}, t) G(\mathbf{R}) = G(\mathbf{R}_{A\alpha}^{\text{class}}(t)) + O(\mathbf{R} - \mathbf{R}_{A\alpha}^{\text{class}})^2 . \tag{80}$$

The first-order term vanishes due to the definition (76) of the classical trajectory. Neglecting terms of second and higher order is equivalent to replacing the nuclear densities by δ-functions:

$$n_{A\alpha}(\mathbf{R}, t) = \delta(\mathbf{R} - \mathbf{R}^{\text{class}}_{A\alpha}(t))\,. \tag{81}$$

In this way the Newton equations (77, 79) reduce to

$$M_A \frac{d^2}{dt^2} \mathbf{R}^{\text{class}}_{A\alpha}(t) = -\nabla_{\mathbf{R}^{\text{class}}_{A\alpha}} \Bigg[V^A_{\text{ext}}(\mathbf{R}^{\text{class}}_{A\alpha}, t) - \int d^3r \, \frac{Z_A n(\mathbf{r}, t)}{|\mathbf{R}^{\text{class}}_{A\alpha} - \mathbf{r}|} + \underbrace{\sum_{B=1}^{K} \sum_{\beta=1}^{N_B}}_{(B\beta) \neq (A\alpha)} \frac{Z_A Z_B}{|\mathbf{R}^{\text{class}}_{A\alpha} - \mathbf{R}^{\text{class}}_{B\beta}|} \Bigg] \tag{82}$$

and the electronic KS equations simplify to

$$i \frac{\partial}{\partial t} \varphi_j(\mathbf{r}, t) = \Bigg(-\frac{1}{2}\nabla^2 + v_{\text{ext}}(\mathbf{r}, t) + \int d^3r' \, \frac{n(\mathbf{r}', t)}{|\mathbf{r} - \mathbf{r}'|} - \sum_{B=1}^{K} \sum_{\beta=1}^{N_B} \frac{Z_B}{|\mathbf{r} - \mathbf{R}^{\text{class}}_{B\beta}(t)|} + v_{\text{xc}}[n, \{\mathbf{R}^{\text{class}}_{B\beta}(t)\}](\mathbf{r}, t) \Bigg) \varphi_j(\mathbf{r}, t)\,. \tag{83}$$

Equations (82) and (83) are coupled and have to be solved simultaneously. This scheme has been applied rather successfully to describe the melting of bulk sodium [38]. Compared to the Car-Parrinello method [44–46] the scheme has the advantage of not requiring the imposition of orthonormality constraints in the electronic equations of motion.

One has to emphasize that Eqs. (82) and (83) do not involve the Born-Oppenheimer approximation although the nuclear motion is treated classically. This is an important advantage over the quantum molecular dynamics approach [47–54] where the nuclear Newton equations (82) are solved simultaneously with a set of *ground-state* KS equations at the instantaneous nuclear positions. In spite of the obvious numerical advantages one has to keep in mind that the classical treatment of nuclear motion is justified only if the probability densities $n_{A\alpha}(\mathbf{R}, t)$ remain narrow distributions during the whole process considered. The splitting of the nuclear wave packet found, e.g., in pump-probe experiments [55–58] cannot be properly accounted for by treating the nuclear motion classically. In this case, one has to face the complete system (67–72) of coupled TDKS equations for electrons and nuclei.

4 Electrons in Time-Dependent Electromagnetic Fields

Up to this point we have exclusively dealt with time-dependent electric fields. The objective of the present chapter is to incorporate magnetic effects. For simplicity, only the electronic degrees of freedom are being discussed, i.e., the nuclear motion is not considered. Magnetic fields couple both to the spin and to the electronic orbital currents. Hence, the most general TDDFT should encompass both of these couplings at the same time. However, to keep matters as simple as possible, we shall treat the two couplings separately in the following sections.

4.1 Coupling to Spin

In order to account for the coupling of a magnetic field $\mathbf{B}(\mathbf{r}, t)$ to the electronic spin, the external potential

$$\hat{V}(t) = \sum_{j=1}^{N} v(\mathbf{r}_j, t) = \int d^3 r\, \hat{n}(\mathbf{r})\, v(\mathbf{r}, t) \tag{84}$$

represented in terms of the density operator

$$\hat{n}(\mathbf{r}) = \sum_{j=1}^{N} \delta(\mathbf{r} - \mathbf{r}_j) \tag{85}$$

has to be complemented by a Zeeman term, i.e., $\hat{V}(t)$ has to be replaced by

$$\hat{V}_B(t) = \int d^3 r\, \hat{n}(\mathbf{r})\, v(\mathbf{r}, t) - \int d^3 r\, \hat{\mathbf{m}}(\mathbf{r}) \cdot \mathbf{B}(\mathbf{r}, t) \tag{86}$$

where $\hat{\mathbf{m}}(\mathbf{r})$ represents the operator of the spin magnetization. For simplicity we assume that the vector $\mathbf{B}$ has only one non-vanishing component, the z-component, so that

$$\hat{V}_B(t) = \int d^3 r\, \hat{n}(\mathbf{r})\, v(\mathbf{r}, t) - \int d^3 r\, \hat{m}_z(\mathbf{r})\, B_z(\mathbf{r}, t)\,. \tag{87}$$

If the system contains $N_\uparrow$ spin-up electrons and $N_\downarrow = N - N_\uparrow$ spin-down electrons we can define spin-up and spin-down density operators by

$$\hat{n}_\uparrow(\mathbf{r}) := \sum_{j=1}^{N_\uparrow} \delta(\mathbf{r} - \mathbf{r}_j) \tag{88}$$

$$\hat{n}_\downarrow(\mathbf{r}) := \sum_{j=N_\uparrow+1}^{N} \delta(\mathbf{r} - \mathbf{r}_j)\,. \tag{89}$$

In terms of these operators the total density $\hat{n}(\mathbf{r})$ and the magnetization $\hat{m}_z(\mathbf{r})$ can be expressed as

$$\hat{n}(\mathbf{r}) = \hat{n}_\uparrow(\mathbf{r}) + \hat{n}_\downarrow(\mathbf{r}) \tag{90}$$

$$\hat{m}_z(\mathbf{r}) = -\mu_0\,[\hat{n}_\uparrow(\mathbf{r}) - \hat{n}_\downarrow(\mathbf{r})] \tag{91}$$

where μ_0 is the Bohr magneton.

Defining furthermore the spin-up and spin-down potentials

$$v_\uparrow(\mathbf{r}, t) = v(\mathbf{r}, t) + \mu_0 B_z(\mathbf{r}, t) \tag{92}$$

and

$$v_\downarrow(\mathbf{r}, t) = v(\mathbf{r}, t) - \mu_0 B_z(\mathbf{r}, t), \tag{93}$$

Eq. (87) simplifies to

$$\hat{V}_B(t) = \int d^3 r\, \hat{n}_\uparrow(\mathbf{r})\, v_\uparrow(\mathbf{r}, t) + \int d^3 r\, \hat{n}_\downarrow(\mathbf{r})\, v_\downarrow(\mathbf{r}, t). \tag{94}$$

Starting from the time-dependent many-body Schrödinger equation

$$i \frac{\partial}{\partial t} \Psi(t) = (\hat{T} + \hat{U} + \hat{V}_B(t))\, \Psi(t), \tag{95}$$

a time-dependent HKS formalism can be established [59] in analogy to Sect 2: The time-dependent spin densities

$$n_\uparrow(\mathbf{r}, t) = \langle \Psi(t) | \hat{n}_\uparrow(\mathbf{r}) | \Psi(t) \rangle \tag{96}$$

$$n_\downarrow(\mathbf{r}, t) = \langle \Psi(t) | \hat{n}_\downarrow(\mathbf{r}) | \Psi(t) \rangle \tag{97}$$

evolving from a fixed initial many-body state $\Psi(t_0)$ are in 1–1 correspondence with the potentials $(v_\uparrow(\mathbf{r}, t), v_\downarrow(\mathbf{r}, t))$ provided that the latter can be expanded in Taylor series around the initial time t_0. The spin densities thus determine the potentials $v_\uparrow = v_\uparrow[n_\uparrow, n_\downarrow]$, $v_\downarrow = v_\downarrow[n_\uparrow, n_\downarrow]$ uniquely up to within purely time-dependent ($\mathbf{r}$-independent) additive functions. Consequently the many-body wave function can be considered as a functional $\Psi(t) = \Psi[n_\uparrow, n_\downarrow](t)$ of the spin densities which is unique up to within a purely time-dependent phase factor. Furthermore, following the arguments in Sect. 2.2, the spin densities of a given interacting system can be determined variationally by solving the Euler-Lagrange equations

$$\frac{\delta \mathscr{A}_B[n_\uparrow, n_\downarrow]}{\delta n_\uparrow(\mathbf{r}, t)} = 0 \tag{98}$$

$$\frac{\delta \mathscr{A}_B[n_\uparrow, n_\downarrow]}{\delta n_\downarrow(\mathbf{r}, t)} = 0, \tag{99}$$

where the action functional is formally defined as

$$\mathscr{A}_B[n_\uparrow, n_\downarrow] := \int_{t_0}^{t_1} dt\, \langle \Psi[n_\uparrow, n_\downarrow](t) | i \frac{\partial}{\partial t} - \hat{T} - \hat{U} - \hat{V}_B | \Psi[n_\uparrow, n_\downarrow](t) \rangle . \tag{100}$$

Finally, assuming non-interacting v-representability, the spin densities of the interacting system of interest can be obtained from time-dependent spin orbitals

$$n_\sigma(\mathbf{r}, t) = \sum_{j=1}^{N_\sigma} |\varphi_{j\sigma}(\mathbf{r}, t)|^2 \tag{101}$$

coming from the time-dependent KS equations

$$i \frac{\partial}{\partial t} \varphi_{j\sigma}(\mathbf{r}, t) = \left(-\tfrac{1}{2}\nabla^2 + v_{s\sigma}[n_\uparrow, n_\downarrow](\mathbf{r}, t)\right) \varphi_{j\sigma}(\mathbf{r}, t) \tag{102}$$

$$j = 1 \dots N_\sigma, \quad \sigma = \uparrow\downarrow$$

where the spin-dependent effective single-particle potential for electrons with spin $\sigma = \uparrow, \downarrow$ is given by

$$v_{s\sigma}[n_\uparrow, n_\downarrow](\mathbf{r}, t) = v_\sigma(\mathbf{r}, t) + \int d^3 r' \frac{n(\mathbf{r}', t)}{|\mathbf{r} - \mathbf{r}'|} + v_{xc\sigma}[n_\uparrow, n_\downarrow](\mathbf{r}, t). \tag{103}$$

Eqs. (101)–(103) constitute the KS scheme of time-dependent spin-density functional theory. With the xc action functional $\mathcal{A}_{xc}[n_\uparrow, n_\downarrow]$ defined in analogy to Eq. (43), the spin-dependent xc potentials can be represented as functional derivatives:

$$v_{xc\sigma}[n_\uparrow, n_\downarrow](\mathbf{r}, t) = \frac{\delta \mathcal{A}_{xc}[n_\uparrow, n_\downarrow]}{\delta n_\sigma(\mathbf{r}, t)}. \tag{104}$$

In the limit of vanishing magnetic field the external potentials in Eq. (103) become identical

$$v_\uparrow(\mathbf{r}, t) = v_\downarrow(\mathbf{r}, t) = v(\mathbf{r}, t) \quad \text{for} \quad \mathbf{B}(\mathbf{r}, t) \equiv 0. \tag{105}$$

Nevertheless, Eqs. (102) and (103) do not necessarily reduce to the ordinary TDKS equations (40) and (41) in this limit. This is because the spin-dependent xc potentials $v_{xc_\uparrow}$ and $v_{xc_\downarrow}$ are not identical except for the case of spin-saturated systems ($n_\uparrow \equiv n_\downarrow$).

4.2 Coupling to Orbital Currents

In order to describe the coupling of time-dependent magnetic fields to the electronic orbital currents, the kinetic energy $\hat{T}$ has to be replaced by

$$\hat{T}_{\mathbf{A}}(t) = \sum_{j=1}^{N} \frac{1}{2}\left(-i\nabla_{\mathbf{r}_j} + \frac{1}{c}\mathbf{A}(\mathbf{r}_j, t)\right)^2 \tag{106}$$

where $\mathbf{A}(\mathbf{r}, t)$ is the time-dependent vector potential related to the magnetic field by

$$\mathbf{B}(\mathbf{r}, t) = \nabla \times \mathbf{A}(\mathbf{r}, t). \tag{107}$$

Since the vector potential is not a gauge-invariant quantity, particular attention has to be paid to gauge transformations: If $\Psi(\mathbf{r}_1, \ldots, \mathbf{r}_N, t)$ is a solution of the time-dependent Schrödinger equation

$$i \frac{\partial}{\partial t} \Psi(t) = (\hat{T}_{\mathbf{A}}(t) + U + V(t))\, \Psi(t), \tag{108}$$

then the transformed wave function

$$\tilde{\Psi}(\mathbf{r}_1, \ldots, \mathbf{r}_N, t) = \exp\left\{ -\frac{i}{c} \sum_{j=1}^{N} \Lambda(\mathbf{r}_j, t) \right\} \Psi(\mathbf{r}_1, \ldots, \mathbf{r}_N, t) \tag{109}$$

is a solution of (108) with the gauge-transformed potentials[3]

$$\tilde{v}(\mathbf{r}, t) = v(\mathbf{r}, t) + \frac{1}{c} \frac{\partial}{\partial t} \Lambda(\mathbf{r}, t) \tag{110}$$

$$\tilde{\mathbf{A}}(\mathbf{r}, t) = \mathbf{A}(\mathbf{r}, t) + \nabla \Lambda(\mathbf{r}, t). \tag{111}$$

The physical (i.e. gauge-invariant) current density is given by

$$\mathbf{j}(\mathbf{r}, t) = \langle \Psi(t) | \hat{\mathbf{j}}_{\mathrm{p}}(\mathbf{r}) | \Psi(t) \rangle + \frac{1}{c} n(\mathbf{r}, t)\, \mathbf{A}(\mathbf{r}, t) \tag{112}$$

where $\hat{\mathbf{j}}_{\mathrm{p}}(\mathbf{r})$ is the paramagnetic current density operator defined in Eq. (14). With these preliminaries, the central Hohenberg-Kohn-like theorem [24, 60] to be proven subsequently can be formulated as follows:

The current densities $\mathbf{j}(\mathbf{r}, t)$ and $\mathbf{j}'(\mathbf{r}, t)$ evolving from a common initial state $\Psi_0 = \Psi(\mathbf{r}_1, \ldots, \mathbf{r}_N, t_0)$ under the influence of two four-potentials $(v(\mathbf{r}, t), \mathbf{A}(\mathbf{r}, t))$ and $(v'(\mathbf{r}, t), \mathbf{A}'(\mathbf{r}, t))$ which differ by more than a gauge transformation with $\Lambda(\mathbf{r}, t_0) = 0$ are always different provided that the potentials can be expanded in Taylor series around the initial time t_0.

Since the current density is gauge invariant the proof of the theorem can be carried out with an arbitrary representative of the gauge class of $(v, \mathbf{A})$ and an arbitrary representative of the gauge class of $(v', \mathbf{A}')$. As representatives we choose those four-potentials having a vanishing electric potential, i.e., for $v(\mathbf{r}, t)$ we make a gauge transformation (110) satisfying

$$\frac{\partial}{\partial t} \Lambda(\mathbf{r}, t) = -c\, v(\mathbf{r}, t), \quad \Lambda(\mathbf{r}, t_0) = 0 \tag{113}$$

---

[3] In the context of electrodynamics, the gauge transformation (111) is usually complemented by the transformation $\tilde{\varphi}(\mathbf{r}, t) = \varphi(\mathbf{r}, t) - 1/c\, \partial\Lambda(\mathbf{r}, t)/\partial t$ where $\varphi(\mathbf{r}, t)$ is the electric *potential*. In quantum mechanics, on the other hand, one works with the *potential energy* $v(\mathbf{r}, t) = q\varphi(\mathbf{r}, t)$. For electrons $(q = -e)$ the gauge transformation of ϕ then leads to (110) in atomic units.

and for $v'(\mathbf{r}, t)$ we make a gauge transformation satisfying

$$\frac{\partial}{\partial t}\Lambda'(\mathbf{r}, t) = -c\,v'(\mathbf{r}, t), \quad \Lambda'(\mathbf{r}, t_0) = 0. \tag{114}$$

The corresponding gauge-transformed vector potentials are denoted by $\tilde{\mathbf{A}}(\mathbf{r}, t)$ and $\tilde{\mathbf{A}}'(\mathbf{r}, t)$. Thus we have to show that

$$(0, \tilde{\mathbf{A}}(\mathbf{r}, t)) \neq (0, \tilde{\mathbf{A}}'(\mathbf{r}, t)) \tag{115}$$

implies

$$\mathbf{j}(\mathbf{r}, t) \neq \mathbf{j}'(\mathbf{r}, t). \tag{116}$$

If $\tilde{\mathbf{A}}(\mathbf{r}, t_0) \neq \tilde{\mathbf{A}}'(\mathbf{r}, t_0)$, then the statement of the theorem is trivially true because the initial paramagnetic currents and the initial densities are identical so that

$$\mathbf{j}(\mathbf{r}, t_0) - \mathbf{j}'(\mathbf{r}, t_0) = \frac{1}{c}\,n(\mathbf{r}, t_0)\,(\tilde{\mathbf{A}}(\mathbf{r}, t_0) - \tilde{\mathbf{A}}'(\mathbf{r}, t_0)) \neq 0. \tag{117}$$

If $\tilde{\mathbf{A}}(\mathbf{r}, t_0) = \tilde{\mathbf{A}}'(\mathbf{r}, t_0)$ the potentials must differ in some higher Taylor coefficient, i.e.,

$$\frac{\partial^k}{\partial t^k}(\tilde{\mathbf{A}}(\mathbf{r}, t) - \tilde{\mathbf{A}}'(\mathbf{r}, t))\bigg|_{t=t_0} \begin{cases} = 0\,; & k < l \\ \neq 0\,; & k = l \end{cases} \tag{118}$$

must be satisfied with an integer $l > 0$. Calculating the l-th time derivative of the densities $\mathbf{j}(\mathbf{r}, t)$ and $\mathbf{j}'(\mathbf{r}, t)$ by applying the Heisenberg equation of motion l times and taking the difference at the initial time t_0 we obtain

$$\left(i\frac{\partial}{\partial t}\right)^l (\mathbf{j}(\mathbf{r}, t) - \mathbf{j}'(\mathbf{r}, t))\bigg|_{t=t_0} = \frac{1}{c}\,n(\mathbf{r}, t_0)\left(i\frac{\partial}{\partial t}\right)^l (\tilde{\mathbf{A}}(\mathbf{r}, t) - \tilde{\mathbf{A}}'(\mathbf{r}, t))\bigg|_{t=t_0} \neq 0, \tag{119}$$

where $n(\mathbf{r}, t_0)$ is the particle density at t_0. By virtue of (119) the current densities $\mathbf{j}(\mathbf{r}, t)$ and $\mathbf{j}'(\mathbf{r}, t)$ will become different at times infinitesimally later than t_0. This completes the proof. As a consequence of this theorem the physical current density $\mathbf{j}(\mathbf{r}, t)$ determines the potentials $v[\mathbf{j}]$, $\mathbf{A}[\mathbf{j}]$ uniquely up to within a gauge transformation (110), (111). Hence, by virtue of the Schrödinger equation (108), the many-body wave function is a current-density functional $\Psi[\mathbf{j}](\mathbf{r}_1, \ldots, \mathbf{r}_N, t)$, unique up to within a gauge transformation (109). In a *fixed* gauge, of course, v, $\mathbf{A}$ and Ψ are determined uniquely by the current density. Applying the theorem to noninteracting particles then, once again, the potentials $v_s[\mathbf{j}]$, $A_s[\mathbf{j}]$ and the Slater determinant $\Phi[\mathbf{j}]$ leading to the current density $\mathbf{j}(\mathbf{r}, t)$ are uniquely determined in a fixed gauge.

In order to derive a TDKS scheme we consider a particular interacting system with current density $\mathbf{j}_0(\mathbf{r}, t)$, produced by the external potentials $v_0(\mathbf{r}, t)$,

$\mathbf{A}_0(\mathbf{r}, t)$ (in a given gauge). Assuming noninteracting v-representability, i.e., assuming the existence of potentials $v_{s,0}$, $\mathbf{A}_{s,0}$ leading to $\mathbf{j}_0$, we can calculate $\mathbf{j}_0$ from the equations

$$i\frac{\partial}{\partial t}\varphi_j(\mathbf{r}, t) = \left(\frac{1}{2}\left(-i\nabla + \frac{1}{c}\mathbf{A}_{s,0}[\mathbf{j}_0](\mathbf{r}, t)\right)^2 + v_{s,0}[\mathbf{j}_0](\mathbf{r}, t)\right)\varphi_j(\mathbf{r}, t) \tag{120}$$

$$\mathbf{j}_0(\mathbf{r}, t) = \frac{1}{2i}\sum_{k=1}^{N}(\varphi_k^*(\mathbf{r}, t)\nabla\varphi_k(\mathbf{r}, t) - \varphi_k(\mathbf{r}, t)\nabla\varphi_k^*(\mathbf{r}, t)) + \frac{1}{c}\left(\sum_{k=1}^{N}|\varphi_k(\mathbf{r}, t)|^2\right)\mathbf{A}_{s,0}(\mathbf{r}, t). \tag{121}$$

Once the existence of $v_{s,0}$ and $\mathbf{A}_{s,0}$ is assumed, uniqueness follows from the above theorem. Up to this point the time-dependent HKS formalism is quite similar to the case without magnetic fields developed in Sect. 2. The variational representation of $v_{s,0}$ and $\mathbf{A}_{s,0}$, however, turns out to be much more complicated. Following Wacker, Kümmel and Gross [60] the quantum mechanical action functional

$$\mathscr{A}_{v_0, \mathbf{A}_0}[\mathbf{j}] = \int_{t_0}^{t_1} dt \langle\Psi[\mathbf{j}](t)|i\frac{\partial}{\partial t} - \hat{T}_{\mathbf{A}_0}(t) - \hat{U} - \hat{V}_0(t)|\Psi[\mathbf{j}](t)\rangle \tag{122}$$

has a stationary point at the true current density $\mathbf{j}_0$, i.e., the latter can be determined from the variational equation

$$\left.\frac{\delta\mathscr{A}_{v_0, \mathbf{A}_0}[\mathbf{j}]}{\delta\mathbf{j}(\mathbf{r}, t)}\right|_{\mathbf{j}_0} = 0. \tag{123}$$

Correspondingly, the action functional

$$\mathscr{A}^{s}_{v_{s0}, \mathbf{A}_{s0}}[\mathbf{j}] = \int_{t_0}^{t_1} dt \langle\Phi[\mathbf{j}](t)|i\frac{\partial}{\partial t} - \hat{T}_{\mathbf{A}_{s0}}(t) - \hat{V}_{s0}(t)|\Phi[\mathbf{j}](t)\rangle \tag{124}$$

of noninteracting particles moving in the external potentials v_{s0}, $\mathbf{A}_{s0}$ has a stationary point at $\mathbf{j}_0$ as well, i.e.,

$$\left.\frac{\delta\mathscr{A}^{s}_{v_{s0}, \mathbf{A}_{s0}}[\mathbf{j}]}{\delta\mathbf{j}(\mathbf{r}, t)}\right|_{\mathbf{j}_0} = 0. \tag{125}$$

In order to deduce an integral equation determining the potentials v_{s0}, $\mathbf{A}_{s0}$, we decompose $\mathscr{A}_{v_0, \mathbf{A}_0}[\mathbf{j}]$ into a universal part, $\mathscr{B}[\mathbf{j}]$, and a functional $\mathscr{Q}_{v_0, \mathbf{A}_0}[\mathbf{j}]$ that depends on the external potentials $v_0(\mathbf{r}, t)$, $\mathbf{A}_0(\mathbf{r}, t)$:

$$\mathscr{A}_{v_0, \mathbf{A}_0}[\mathbf{j}] = \mathscr{B}[\mathbf{j}] - \mathscr{Q}_{v_0, \mathbf{A}_0}[\mathbf{j}]. \tag{126}$$

The universal part is given by

$$\mathscr{B}[\mathbf{j}] = \int_{t_0}^{t_1} dt \, \langle \Psi[\mathbf{j}](t) | \, i \frac{\partial}{\partial t} - \hat{T} - \hat{U} \, | \Psi[\mathbf{j}](t) \rangle , \tag{127}$$

where $\hat{T}$ is the kinetic energy (6). In terms of the functionals

$$n[\mathbf{j}](\mathbf{r}, t) = -\int_{t_0}^{t} dt' \operatorname{div} \mathbf{j}(\mathbf{r}, t') \tag{128}$$

and

$$\mathbf{j}_p[\mathbf{j}](\mathbf{r}, t) = \mathbf{j}(\mathbf{r}, t) - \frac{1}{c} n[\mathbf{j}](\mathbf{r}, t) \mathbf{A}[\mathbf{j}](\mathbf{r}, t) \tag{129}$$

the non-universal contribution to $\mathscr{A}[\mathbf{j}]$ can be expressed as

$$\begin{aligned} \mathscr{Q}_{v_0, \mathbf{A}_0}[\mathbf{j}] = \int_{t_0}^{t_1} dt \int d^3 r \Bigg\{ \left(v_0(\mathbf{r}, t) + \frac{1}{2c^2} \mathbf{A}_0(\mathbf{r}, t)^2 \right) n[\mathbf{j}](\mathbf{r}, t) \\ + \frac{1}{c} \mathbf{A}_0(\mathbf{r}, t) \cdot \mathbf{j}_p[\mathbf{j}](\mathbf{r}, t) \Bigg\} . \end{aligned} \tag{130}$$

Similarly, the action functional of noninteracting particles can be written as

$$\mathscr{A}^s_{v_{s0}, \mathbf{A}_{s0}}[\mathbf{j}] = \mathscr{B}^s[\mathbf{j}] - \mathscr{Q}^s_{v_{s0}, \mathbf{A}_{s0}}[\mathbf{j}] . \tag{131}$$

where

$$\mathscr{B}^s[\mathbf{j}] = \int_{t_0}^{t_1} dt \, \langle \Phi[\mathbf{j}](t) | \, i \frac{\partial}{\partial t} - \hat{T} \, | \Phi[\mathbf{j}](t) \rangle \tag{132}$$

$$\begin{aligned} \mathscr{Q}^s_{v_{s0}, \mathbf{A}_{s0}}[\mathbf{j}] = \int_{t_0}^{t_1} dt \int d^3 r \Bigg\{ \left(v_{s0}(\mathbf{r}, t) + \frac{1}{2c^2} \mathbf{A}_{s0}(\mathbf{r}, t)^2 \right) n[\mathbf{j}](\mathbf{r}, t) \\ + \frac{1}{c} \mathbf{A}_{s0}(\mathbf{r}, t) \cdot \mathbf{j}_{ps}[\mathbf{j}](\mathbf{r}, t) \Bigg\} . \end{aligned} \tag{133}$$

Note that the functional

$$\mathbf{j}_{ps}[\mathbf{j}] = \mathbf{j}(\mathbf{r}, t) - \frac{1}{c} n[\mathbf{j}](\mathbf{r}, t) \, \mathbf{A}_s[\mathbf{j}](\mathbf{r}, t) \tag{134}$$

is different, in general, from the functional $\mathbf{j}_p[\mathbf{j}]$ given by Eq. (129). Defining the universal xc functional as

$$\mathscr{A}_{xc}[\mathbf{j}] = \mathscr{B}^s[\mathbf{j}] - \mathscr{B}[\mathbf{j}] \tag{135}$$

the total action functional of the interacting system can be expressed as

$$\mathscr{A}_{v_0, \mathbf{A}_0}[\mathbf{j}] = \mathscr{B}^{\mathrm{s}}[\mathbf{j}] - \mathscr{Q}_{v_0, \mathbf{A}_0}[\mathbf{j}] - \mathscr{A}_{\mathrm{xc}}[\mathbf{j}]. \tag{136}$$

Equating the functional derivatives in (123) and (125) and inserting the expressions (131) and (136) we obtain

$$\left.\frac{\delta \mathscr{Q}^{\mathrm{s}}_{v_{\mathrm{s}0}, \mathbf{A}_{\mathrm{s}0}}[\mathbf{j}]}{\delta \mathbf{j}(\mathbf{r}, t)}\right|_{\mathbf{j}_0} = \left.\frac{\delta \mathscr{Q}_{v_0, \mathbf{A}_0}[\mathbf{j}]}{\delta \mathbf{j}(\mathbf{r}, t)}\right|_{\mathbf{j}_0} + \left.\frac{\delta \mathscr{A}_{\mathrm{xc}}[\mathbf{j}]}{\delta \mathbf{j}(\mathbf{r}, t)}\right|_{\mathbf{j}_0}. \tag{137}$$

This equation defines the TDKS potentials v_{s0}, $\mathbf{A}_{s0}$ implicitly in terms of the functionals $\mathbf{A}[\mathbf{j}]$ and $\mathbf{A}_s[\mathbf{j}]$. Clearly, Eq. (137) is rather complicated. The external-potential terms $\mathscr{Q}$ and $\mathscr{Q}^s$ are *simple* functionals of the *density* and the *paramagnetic current density*. The complexity of Eq. (137) arises from the fact that the density, Eq. (128), and the paramagnetic currents, Eqs. (129), (134), are complicated functionals of $\mathbf{j}$. Hence a formulation directly in terms of the density and the *paramagnetic* current density would be desirable. For electrons in *static* electromagnetic fields, Vignale and Rasolt [61–63] have formulated a current-density functional theory in terms of the density and the paramagnetic current density which has been successfully applied to a variety of systems [63]. A time-dependent HKS formalism in terms of the density and the paramagnetic current density, however, has not been achieved so far.

Several extensions of the formalism presented here have been proposed to deal with more general situations. Those include superconductors in time-dependent electromagnetic fields [60, 64] and time-dependent ensembles either for the electrons alone [65, 66] or for the coupled system of electrons and ions [42, 67]. As long as the number of photons is large, i.e., $\gg 1$ in a volume given by the wave length cubed, the electromagnetic fields can be treated as *classical* fields. For smaller photon densities the quantum nature of electromagnetic radiation becomes important. In this case, a time-dependent functional theory can be formulated [68] on the basis of quantum electrodynamics. In this formulation the electromagnetic field is treated as a quantum field to be determined self-consistently with the four-current vector of the Dirac matter field.

5 Perturbative Regime, Basic Equations

5.1 Time-Dependent Linear Density Response

In this section we shall derive [69] a formally exact representation of the linear density response $n_1(\mathbf{r}, \omega)$ of an interacting many-electron system in terms of the response function of the corresponding (non-interacting) Kohn-Sham system and a frequency-dependent xc kernel.

We consider electronic systems subject to external potentials of the form

$$v_{ext}(\mathbf{r}, t) = \begin{cases} v_0(\mathbf{r}) & ; \quad t \leq t_0 \\ v_0(\mathbf{r}) + v_1(\mathbf{r}, t); & t > t_0 \end{cases} \tag{138}$$

where $v_0(\mathbf{r})$ denotes the static external potential of the unperturbed system (typically the nuclear Coulomb attraction) and $v_1(\mathbf{r}, t)$ is a time-dependent perturbation. We assume, that at times $t \leq t_0$ the system is in the *ground state* corresponding to $v_0(\mathbf{r})$. In this case, the initial density $n_0(\mathbf{r})$ can be obtained from the self-consistent solution of the ordinary ground-state Kohn-Sham equations:

$$\left(-\frac{1}{2}\nabla^2 + v_0(\mathbf{r}) + \int d^3r' \frac{n_0(\mathbf{r}')}{|\mathbf{r} - \mathbf{r}'|} + v_{xc}[n_0](\mathbf{r}) \right) \phi_j(\mathbf{r}) = \varepsilon_j \phi_j(\mathbf{r}), \tag{139}$$

$$n_0(\mathbf{r}) = \sum_{j=1}^{N} |\phi_j(\mathbf{r})|^2. \tag{140}$$

By virtue of the static HK-theorem, the initial many-body ground state is uniquely determined by the initial ground-state density n_0. Hence, in this case, the time-dependent density $n(\mathbf{r}, t)$ is a functional of the external potential alone,

$$n(\mathbf{r}, t) = n[v_{ext}](\mathbf{r}, t), \tag{141}$$

i.e., there is no additional dependence on the initial many-body state. By virtue of the fundamental 1–1 correspondence between time-dependent densities and time-dependent potentials, proven by Runge and Gross [7], the functional $n[v_{ext}]$ can be inverted, i.e.,

$$v_{ext}(\mathbf{r}, t) = v_{ext}[n](\mathbf{r}, t). \tag{142}$$

Within the realm of perturbation theory, i.e., for sufficiently small $v_1(\mathbf{r}, t)$ the functional $n[v_{ext}]$ can be expanded into a functional Taylor series with respect to the perturbation $v_1(\mathbf{r}, t)$,

$$n(\mathbf{r}, t) - n_0(\mathbf{r}) = n_1(\mathbf{r}, t) + n_2(\mathbf{r}, t) + n_3(\mathbf{r}, t) + \ldots, \tag{143}$$

where the lower indices denote the orders in v_1. The *first* order density response n_1 is given by

$$n_1(\mathbf{r}, t) = \int dt' \int d^3r' \chi(\mathbf{r}, t, \mathbf{r}', t') v_1(\mathbf{r}', t') \tag{144}$$

with the density-density response function

$$\chi(\mathbf{r}, t, \mathbf{r}', t') = \left. \frac{\delta n[v_{ext}](\mathbf{r}, t)}{\delta v_{ext}(\mathbf{r}', t')} \right|_{v_0}. \tag{145}$$

Owing to the static HK theorem, the initial potential $v_0 = v_{ext}[n_0]$ is a functional of the unperturbed ground-state density n_0, so that the response function χ, by Eq. (145), is a functional of n_0 as well.

For *non-interacting* particles moving in external potentials $v_s(\mathbf{r}, t)$, the Runge-Gross theorem holds as well. Therefore the functional

$$n(\mathbf{r}, t) = n[v_s](\mathbf{r}, t) \tag{146}$$

can be inverted,

$$v_s(\mathbf{r}, t) = v_s[n](\mathbf{r}, t), \tag{147}$$

and the Kohn-Sham response function, i.e., the density-density response function of non-interacting particles with unperturbed density n_0 is given by

$$\chi_s(\mathbf{r}, t, \mathbf{r}', t') = \left.\frac{\delta n[v_s](\mathbf{r}, t)}{\delta v_s(\mathbf{r}', t')}\right|_{v_s[n_0]}. \tag{148}$$

By inserting the functional (141) into the right-hand side of Eq. (147) one has formally constructed a unique functional $v_s[v_{\text{ext}}]$ such that the time-dependent density of *non*interacting particles moving in $v_s(\mathbf{r}, t)$ is identical with the density of Coulomb-interacting particles moving in $v_{\text{ext}}(\mathbf{r}, t)$. The potential $v_s(\mathbf{r}, t)$ corresponding to a *given* $v_{\text{ext}}(\mathbf{r}, t)$, is the time-dependent Kohn-Sham potential (41):

$$v_s(\mathbf{r}, t) = v_{\text{ext}}(\mathbf{r}, t) + \int d^3r' \frac{n(\mathbf{r}', t)}{|\mathbf{r} - \mathbf{r}'|} + v_{\text{xc}}(\mathbf{r}, t). \tag{149}$$

By virtue of the functional chain rule, the functional derivative of v_s with respect to v_{ext} provides a link of the interacting response function (145) to its noninteracting counterpart:

$$\chi(\mathbf{r}, t, \mathbf{r}', t') = \int d^3x \int d\tau \left.\frac{\delta n(\mathbf{r}, t)}{\delta v_s(\mathbf{x}, \tau)} \frac{\delta v_s(\mathbf{x}, \tau)}{\delta v_{\text{ext}}(\mathbf{r}', t')}\right|_{n_0}. \tag{150}$$

Making use of the functional chain rule once more to calculate the functional derivative of v_s with respect to v_{ext} one gets

$$\left.\frac{\delta v_s(\mathbf{r}, t)}{\delta v_{\text{ext}}(\mathbf{r}', t')}\right|_{n_0} = \delta(\mathbf{r} - \mathbf{r}')\,\delta(t - t') + \int d^3x \int d\tau \left(\frac{\delta(t - \tau)}{|\mathbf{r} - \mathbf{x}|} + \frac{\delta v_{\text{xc}}(\mathbf{r}, t)}{\delta n(\mathbf{x}, \tau)}\right) \frac{\delta n(\mathbf{x}, t)}{\delta v_{\text{ext}}(\mathbf{r}', t')}. \tag{151}$$

By inserting (151) into (150) and using the defintions (145) and (148) we end up with a Dyson-type equation relating the interacting and the noninteracting response functions to each other:

$$\chi(\mathbf{r}, t, \mathbf{r}', t') = \chi_s(\mathbf{r}, t, \mathbf{r}', t') + \int d^3x \int d\tau \int d^3x' \int d\tau'\, \chi_s(\mathbf{r}, t, \mathbf{x}, \tau) \times \left(\frac{\delta(\tau - \tau')}{|\mathbf{x} - \mathbf{x}'|} + f_{\text{xc}}[n_0](\mathbf{x}, \tau, \mathbf{x}', \tau')\right) \chi(\mathbf{x}', \tau', \mathbf{r}', t'), \tag{152}$$

where the so-called time-dependent xc kernel

$$f_{xc}[n_0](\mathbf{r}, t, \mathbf{r}', t') := \left.\frac{\delta v_{xc}[n](\mathbf{r}, t)}{\delta n(\mathbf{r}', t')}\right|_{n_0} \tag{153}$$

is a functional of the initial ground-state density n_0. Equations (152)–(153) are the central result of our analysis. In previous work, see e.g. Ref. [70], it has been common practice to define f_{xc} by Eq. (152). The present derivation of Eq. (152) from TDDFT shows that f_{xc}, apart from its relation to the response functions χ and χ_s, can also be represented as the functional derivative of the TD xc potential. Multiplying Eq. (152) by the perturbing potential $v_1(\mathbf{r}', t')$ and integrating over $\mathbf{r}'$ and t' leads to the time-dependent Kohn-Sham equations for the linear density response:

$$n_1(\mathbf{r}, t) = \int dt' \int d^3r' \, \chi_s(\mathbf{r}, t, \mathbf{r}', t') \, v_{s,1}(\mathbf{r}', t') \tag{154}$$

where the effective potential

$$v_{s,1}(\mathbf{r}, t) = v_1(\mathbf{r}, t) + \int d^3r' \frac{n_1(\mathbf{r}', t)}{|\mathbf{r} - \mathbf{r}'|} + \int d^3r' \int dt' \, f_{xc}[n_0](\mathbf{r}, t, \mathbf{r}', t') \, n_1(\mathbf{r}', t') \tag{155}$$

consists of the external perturbation v_1 and the Hartree- and exchange-correlation contributions to first order in the perturbing potential v_1. We emphasize that Eqs. (154) and (155), postulated in previous work [10, 71, 72], constitute an exact representation of the linear density response. In other words, the *exact* linear density response $n_1(\mathbf{r}, t)$ of an interacting system can be written as the linear density response of a *noninteracting* system to the effective perturbation $v_{s,1}(\mathbf{r}, t)$. Combining Eqs. (154) and (155) and taking the Fourier transform with respect to time, the *exact* frequency-dependent linear density response is seen to be

$$n_1(\mathbf{r}, \omega) = \int d^3y \, \chi_s(\mathbf{r}, \mathbf{y}; \omega) \, v_1(\mathbf{y}, \omega) + \int d^3y \int d^3y' \, \chi_s(\mathbf{r}, \mathbf{y}; \omega) \left(\frac{1}{|\mathbf{y} - \mathbf{y}'|} + f_{xc}[n_0](\mathbf{y}, \mathbf{y}'; \omega) \right) n_1(\mathbf{y}', \omega). \tag{156}$$

The Kohn-Sham response function χ_s is readily expressed in terms of the static unperturbed Kohn-Sham orbitals ϕ_k:

$$\chi_s(\mathbf{r}, \mathbf{r}'; \omega) = \sum_{j,k} (f_k - f_j) \frac{\phi_j(\mathbf{r}) \, \phi_k^*(\mathbf{r}) \, \phi_j^*(\mathbf{r}') \, \phi_k(\mathbf{r}')}{\omega - (\varepsilon_j - \varepsilon_k) + i\eta}. \tag{157}$$

Here, (f_k, f_j) are the occupation numbers (0 or 1) of the KS orbitals. The summation in (157) ranges over both occupied and unoccupied orbitals, including the continuum states.

In this section we only dealt with the linear response to time-dependent electric fields of systems at zero temperature. The corresponding formalism for systems at finite temperature in thermal equilibrium was developed in [73, 74]. The current-density-functional response theory for arbitrary time-dependent electromagnetic fields has been worked out by Ng [75]. The exchange-correlation kernel f_{xc}, given by Eq. (153), comprises all dynamic exchange and correlation effects to linear order in the perturbing potential. Depending on physical context, f_{xc} has different names: In the theory of the homogeneous electron gas [70, 76–78] the Fourier transform, $f_{xc}(q, \omega)$, of $f_{xc}(\mathbf{r}, t, \mathbf{r}', t')$ with respect to $(\mathbf{r} - \mathbf{r}')$ and $(t - t')$ is proportional to the so-called *local field correction*

$$\mathscr{G}(q, \omega) = -\frac{q^2}{4\pi} f_{xc}(q, \omega) . \tag{158}$$

In the theory of classical liquids [79], f_{xc} plus the particle-particle interaction is known as *Ornstein-Zernike function*. In practice, of course, this quantity is only approximately known. Suitable approximations of f_{xc} will be discussed in section 6. In order to construct such approximate functionals, it is useful to express f_{xc} in terms of the full response function χ. An exact representation of f_{xc} is readily obtained by solving Eq. (144) for v_1 and inserting the result in Eq. (155). Equation (154) then yields

$$f_{xc}[n_0](\mathbf{r}, t, \mathbf{r}', t') = \chi_s^{-1}[n_0](\mathbf{r}, t, \mathbf{r}', t') - \chi^{-1}[n_0](\mathbf{r}, t, \mathbf{r}', t') - \frac{\delta(t - t')}{|\mathbf{r} - \mathbf{r}'|} , \tag{159}$$

where χ_s^{-1} and χ^{-1} stand for the kernels of the corresponding inverse integral operators whose existence on the set of densities specified by Eqs. (138) and (144) follows from Eq. (253), as mentioned in section 2.1. The frequency-dependent response operators $\chi(\mathbf{r}, \mathbf{r}'; \omega)$ and $\chi_s(\mathbf{r}, \mathbf{r}'; \omega)$, on the other hand, can be non-invertible at isolated frequencies [80, 81]. Ng and Singwi [73, 82] have argued, however, that these examples are typical of *finite* systems while for large systems in the thermodynamic limit invertibility of the frequency-dependent response operators is guaranteed by the second law of thermodynamics.

As a consequence of causality, the response functions $\chi(\mathbf{r}, t, \mathbf{r}', t')$ and $\chi_s(\mathbf{r}, t, \mathbf{r}', t')$ vanish for $t' > t$. The same statement holds true for the kernels $\chi^{-1}(\mathbf{r}, t, \mathbf{r}', t')$ and $\chi_s^{-1}(\mathbf{r}, t, \mathbf{r}', t')$ of the inverse response operators and hence, by Eq. (159), the xc kernel must satisfy

$$f_{xc}(\mathbf{r}, t, \mathbf{r}', t') = 0 \quad \text{for} \quad t' > t . \tag{160}$$

In particular, $f_{xc}(\mathbf{r}, t, \mathbf{r}', t')$ is not symmetric under exchange of $(\mathbf{r}, t)$ and $(\mathbf{r}', t')$. Hence, by virtue of Schwarz' lemma for functionals [83], $f_{xc}(\mathbf{r}, t, \mathbf{r}', t')$ cannot be a second functional derivative $\delta^2 \mathscr{A}_{xc} / \delta n(\mathbf{r}, t)\, \delta n(\mathbf{r}', t')$. Since, on the other hand, f_{xc} is the functional derivative of v_{xc} one concludes that the exact $v_{xc}[n](\mathbf{r}, t)$ cannot be a functional derivative. This is in contradiction to the stationary action principle described in Sect. 2.2. which leads to the representation (42) of v_{xc} as a functional derivative. This contradiction is currently an unresolved problem. It appears that causality somehow has to be taken into account explicitly in the variational principle (36). We emphasize once more that these considerations do not affect the validity of the TDKS equations (39)–(41) nor do they affect the validity of the response equations (152)–(156). Only the variational representation (42) of v_{xc} appears doubtful.

We finally mention that the chain of arguments leading to Eq. (152) can be repeated within *static* HKS theory as well. This yields

$$\chi^{stat}(\mathbf{r},\mathbf{r}') = \chi_s^{stat}(\mathbf{r},\mathbf{r}') + \int d^3x \int d^3x' \, \chi_s^{stat}(\mathbf{r},\mathbf{x}) \times \left(\frac{1}{|\mathbf{x}-\mathbf{x}'|} + f_{xc}^{stat}[n_0](\mathbf{x},\mathbf{x}')\right) \chi^{stat}(\mathbf{x}',\mathbf{r}'), \tag{161}$$

where χ^{stat} and χ_s^{stat} are the full and the KS response functions to static perturbations and

$$f_{xc}[n_0](\mathbf{x},\mathbf{x}') := \left.\frac{\delta v_{xc}[n](\mathbf{x})}{\delta n(\mathbf{x}')}\right|_{n_0} = \left.\frac{\delta^2 E_{xc}[n]}{\delta n(\mathbf{x})\delta n(\mathbf{x}')}\right|_{n_0}. \tag{162}$$

On the other hand, taking the Fourier transform of Eq. (152) with respect to $(t - t')$ one obtains

$$\chi(\mathbf{r},\mathbf{r}';\omega) = \chi_s(\mathbf{r},\mathbf{r}';\omega) + \int d^3x \int d^3x' \, \chi_s(\mathbf{r},\mathbf{x};\omega) \times \left(\frac{1}{|\mathbf{x}-\mathbf{x}'|} + f_{xc}[n_0](\mathbf{x},\mathbf{x}';\omega)\right) \chi(\mathbf{x}',\mathbf{r}';\omega). \tag{163}$$

Subtracting the zero-frequency limit of this equation from Eq. (161) and using the fact that

$$\chi_s^{stat}(\mathbf{r},\mathbf{r}') = \chi_s(\mathbf{r},\mathbf{r}';\omega = 0) \tag{164}$$

$$\chi^{stat}(\mathbf{r},\mathbf{r}') = \chi(\mathbf{r},\mathbf{r}';\omega = 0) \tag{165}$$

one concludes that

$$\int d^3x \int d^3x' \, \chi_s(\mathbf{r},\mathbf{x};\omega = 0) \left(f_{xc}[n_0](\mathbf{x},\mathbf{x}';\omega = 0) - f_{xc}^{stat}[n_0](\mathbf{x},\mathbf{x}')\right) \chi(\mathbf{x}',\mathbf{r}';\omega = 0) = 0. \tag{166}$$

Acting on this equation with $\hat{\chi}_s^{-1}$ from the left and with $\hat{\chi}^{-1}$ from the right, one obtains the rigorous identity [84]

$$\lim_{\omega \to 0} f_{xc}[n_0](\mathbf{x},\mathbf{x}';\omega) = \left.\frac{\delta^2 E_{xc}[n]}{\delta n(\mathbf{x})\,\delta n(\mathbf{x}')}\right|_{n_0}. \tag{167}$$

5.2 Time-Dependent Higher-Order Response

Recently there has been a great deal of interest in nonlinear phenomena, both from a fundamental point of view, and for the development of new nonlinear optical and optoelectronic devices. Even in the optical case, the nonlinearity is usually engendered by a solid or molecular medium whose properties are typically determined by nonlinear response of an interacting many-electron system. To be able to predict these response properties we need an efficient description of exchange and correlation phenomena in many-electron systems which are not necessarily near to equilibrium. The objective of this chapter is to develop the basic formalism of time-dependent nonlinear response within density functional theory, i.e., the calculation of the higher-order terms of the functional Taylor expansion Eq. (143). In the following this will be done explicitly for the second- and third-order terms

$$n_2(x) = \frac{1}{2!}\int dy \int dy'\, \chi^{(2)}(x, y, y')\, v_1(y)\, v_1(y') \tag{168}$$

$$n_3(x) = \frac{1}{3!}\int dy \int dy' \int dy''\, \chi^{(3)}(x, y, y', y'')\, v_1(y)\, v_1(y')\, v_1(y'') . \tag{169}$$

The extension to terms of arbitrary order is straightforward. For convenience, we use the four-vector notation

$$x \equiv (\mathbf{r}, t) \quad \text{and} \quad \int dx \equiv \int d^3 r \int dt . \tag{170}$$

The second- and third-order response functions of the interacting system are formally given by the functional derivatives

$$\chi^{(2)}(x, y, y') = \left. \frac{\delta^2 n(x)}{\delta v_{\mathrm{ext}}(y)\, \delta v_{\mathrm{ext}}(y')} \right|_{n_0} \tag{171}$$

$$\chi^{(3)}(x, y, y', y'') = \left. \frac{\delta^3 n(x)}{\delta v_{\mathrm{ext}}(y)\, \delta v_{\mathrm{ext}}(y')\, \delta v_{\mathrm{ext}}(y'')} \right|_{n_0} \tag{172}$$

of the time-dependent density with respect to the time-dependent external potential v_{ext} evaluated at the initial ground-state density n_0. From ordinary time-dependent perturbation theory, these quantities are given by [85]

$$\chi^{(2)}(x, y, y') = (-i)^2 \sum_{\mathscr{P}} \theta(t-\tau)\, \theta(t-\tau') \times \langle \Psi_0 | [[\hat{n}_H(x), \hat{n}_H(y)], \hat{n}_H(y')] | \Psi_0 \rangle \tag{173}$$

$$\chi^{(3)}(x, y, y', y'') = (-i)^3 \sum_{\mathscr{P}} \theta(t-\tau)\, \theta(t-\tau')\, \theta(t'-\tau'') \times \langle \Psi_0 | [[[\hat{n}_H(x), \hat{n}_H(y)], \hat{n}_H(y')], \hat{n}_H(y'')] | \Psi_0 \rangle \tag{174}$$

where the sum has to be taken over all permutations $\mathscr{P}$ of y, y', y'' and the index H denotes the Heisenberg picture corresponding to the unperturbed Hamiltonian. From the time-translational invariance of the unperturbed system it follows that the response functions (145), (171) and (172) only depend on the differences of the time-arguments. Obviously, the full response functions (171) and (172) are very hard to calculate.

The response functions of systems of noninteracting particles, on the other hand, are functional derivatives of the density with respect to the time-dependent single-particle potential v_s:

$$\chi_s^{(2)}(x, y, y') = \left.\frac{\delta^2 n(x)}{\delta v_s(y)\,\delta v_s(y')}\right|_{n_0} \tag{175}$$

$$\chi_s^{(3)}(x, y, y', y'') = \left.\frac{\delta^3 n(x)}{\delta v_s(y)\,\delta v_s(y')\,\delta v_s(y'')}\right|_{n_0} . \tag{176}$$

These functions can be expressed in terms of single-particle orbitals, similar to the linear response function (157).

To obtain the higher-order expressions of the density response, we use the functional chain rule in Eq. (171):

$$\begin{aligned}\frac{\delta^2 n(x)}{\delta v_{\text{ext}}(y)\,\delta v_{\text{ext}}(y')} &= \frac{\delta}{\delta v_{\text{ext}}(y)} \int dz\, \frac{\delta n(x)}{\delta v_s(z)} \frac{\delta v_s(z)}{\delta v_{\text{ext}}(y')} \\ &= \int dz \int dz'\, \frac{\delta^2 n(x)}{\delta v_s(z')\,\delta v_s(z)} \frac{\delta v_s(z')}{\delta v_{\text{ext}}(y)} \frac{\delta v_s(z)}{\delta v_{\text{ext}}(y')} \\ &\quad + \int dz\, \frac{\delta n(x)}{\delta v_s(z)} \frac{\delta^2 v_s(z)}{\delta v_{\text{ext}}(y)\, v_{\text{ext}}(y')} .\end{aligned} \tag{177}$$

As has been outlined in Sect. 2, the full time-dependent Kohn-Sham potential

$$v_s(x) = v_{\text{ext}}(x) + v_H(x) + v_{\text{xc}}(x) \tag{178}$$

is a unique functional of the external potential v_{ext}. Hence, we get

$$\begin{aligned}\frac{\delta^2 v_s(z)}{\delta v_{\text{ext}}(y)\,\delta v_{\text{ext}}(y')} &= \frac{\delta}{\delta v_{\text{ext}}(y)} \int dz'\, \frac{\delta (v_H(z) + v_{\text{xc}}(z))}{\delta n(z')} \frac{\delta n(z')}{\delta v_{\text{ext}}(y')} \\ &= \int dz' \int dz''\, \frac{\delta^2 (v_H(z) + v_{\text{xc}}(z))}{\delta n(z'')\,\delta n(z')} \frac{\delta n(z'')}{\delta v_{\text{ext}}(y)} \frac{\delta n(z')}{\delta v_{\text{ext}}(y')} \\ &\quad + \int dz'\, \frac{\delta (v_H(z) + v_{\text{xc}}(z))}{\delta n(z')} \frac{\delta^2 n(z')}{\delta v_{\text{ext}}(y)\,\delta v_{\text{ext}}(y')} .\end{aligned} \tag{179}$$

Combining Eqs. (177) and (179) and evaluating all functionals at the ground-state density n_0, we obtain a Dyson-type relation for the second-order response function (171):

$$\chi^{(2)}(x, y, y') = \int dz \int dz' \, \chi_s^{(2)}(x, z, z') \left.\frac{\delta v_s(z)}{\delta v_{ext}(y)}\right|_{n_0} \left.\frac{\delta v_s(z')}{\delta v_{ext}(y')}\right|_{n_0}$$

$$+ \int dz \, \chi_s(x, z) \int dz' \int dz'' \, g_{xc}(z, z', z'') \, \chi(z', y) \, \chi(z'', y')$$

$$+ \int dz \, \chi_s(x, z) \int dz' \, (w(z, z') + f_{xc}(z, z')) \, \chi^{(2)}(z', y, y') \,, \quad (180)$$

where the time-dependent second-order xc kernel g_{xc} is defined as:

$$g_{xc}(z, z', z'') \equiv \left.\frac{\delta^2 v_{xc}(z)}{\delta n(z') \delta n(z'')}\right|_{n_0} . \quad (181)$$

To arrive at Eq. (180) we have used the definitions (145), (148), (171) and (175) of the density response functions. Furthermore, we have abbreviated the kernel of the (instantaneous) Coulomb interaction by $w(x, x') \equiv \delta(t - t')/|\mathbf{r} - \mathbf{r}'|$. Finally, by inserting Eq. (180) into (168) one arrives at the time-dependent Kohn-Sham equations for the second-order density response:

$$n_2(x) = \tfrac{1}{2} \int dz \int dz' \, \chi_s^{(2)}(x, z, z') \, v_{s,1}(z) v_{s,1}(z')$$

$$+ \tfrac{1}{2} \int dz \int dz' \int dz'' \, \chi_s(x, z) \, g_{xc}(z, z', z'') \, n_1(z') \, n_1(z'')$$

$$+ \int dz \int dz' \, \chi_s(x, z) \, (w(z, z') + f_{xc}(z, z')) n_2(z') \, . \quad (182)$$

Solving Eqs. (154) and (155) first, allows for the subsequent solution of the selfconsistent Eq. (182) which is quadratic in the (effective) perturbing potential (155).

In similar fashion, one can set up the equation for the third-order density response (169):

$$n_3(x) = \tfrac{1}{6} \int dy \int dy' \int dy'' \, \chi_s^{(3)}(x, y, y', y'') \, v_{s,1}(y) \, v_{s,1}(y') \, v_{s,1}(y'')$$

$$+ \tfrac{1}{2} \int dy \int dy' \int dz \int dz' \, \chi_s^{(2)}(x, y, y') \, v_{s,1}(y) \, g_{xc}(y', z, z') \, n_1(z) \, n_1(z')$$

$$+ \int dy \int dy' \int dy'' \, \chi_s^{(2)}(x, y, y') \, v_{s,1}(y) \, (w(y', y'') + f_{xc}(y', y'')) \, n_2(y'')$$

$$+ \tfrac{1}{6} \int dy \int dz \int dz' \int dz'' \, \chi_s(x, y) \, h_{xc}(y, z, z', z'') \, n_1(z) \, n_1(z') \, n_1(z'')$$

$$+ \int dy \int dy' \int dy'' \, \chi_s(x, y) \, g_{xc}(y, y', y'') \, n_1(y') \, n_2(y'')$$

$$+ \int dy \int dy' \, \chi_s(x, y) \, (w(y, y') + f_{xc}(y, y')) \, n_3(y') \, . \quad (183)$$

The quantity h_{xc} occurring in this equation is the third-order functional derivative of the time-dependent xc potential with respect to the time-dependent densities:

$$h_{xc}(y, z, z', z'') \equiv \left.\frac{\delta^3 v_{xc}(y)}{\delta n(z) \delta n(z') \delta n(z'')}\right|_{n_0} . \quad (184)$$

Interestingly, the Eqs. (154), (155), (182) and (183) for the i-th order density responses all exhibit the same structure:

$$n_i(x) = \mathcal{M}_i(x) + \int dy \int dy' \, \chi_s(y, y') \, (w(y, y') + f_{xc}(x, y)) \, n_i(y') \quad i = 1, 2, 3, \tag{185}$$

where the functionals $\mathcal{M}_i(x)$ are known after the solution of the $(i-1)$th order. This establishes a hierarchy of Kohn-Sham equations for the time-dependent density response.

The frequency-dependent nonlinear density responses are given by the Fourier transforms of Eqs. (185). For monochromatic perturbations, the expressions for the higher-order frequency dependent density shifts decouple in the frequency variable. The corresponding formulae and explicit expressions for the Kohn-Sham response functions up to third order are given in work of Senatore and Subbaswamy [86]. The corresponding *static* higher-order response has been worked out and applied to solids by Gonze and Vigneron [87].

6 The Time-Dependent Exchange-Correlation Potential: Rigorous Properties and Approximate Functionals

6.1 Approximations Based on the Homogeneous Electron Gas

The simplest possible approximation of the time-dependent xc potential is the so-called *time-dependent* or "*adiabatic*" *local density approximation* (ALDA). It employs the functional form of the static LDA with a time-dependent density:

$$v_{xc}^{ALDA}[n](\mathbf{r}, t) = v_{xc}^{hom}(n(\mathbf{r}, t)) = \frac{d}{d\rho}(\rho \varepsilon_{xc}^{hom}(\rho))\bigg|_{\rho = n(\mathbf{r}, t)} . \tag{186}$$

Here ε_{xc}^{hom} is the xc energy per particle of the homogeneous electron gas. By its very definition, the ALDA can be expected to be a good approximation only for nearly homogeneous densities, i.e., for functions $n(\mathbf{r}, t)$ that are slowly varying both spatially and temporally. It will turn out, however, that the ALDA gives rather accurate results even for rapidly varying densities (see Sects. 7 and 8). For the time-dependent xc kernel (153), Eq. (186) leads to

$$f_{xc}^{ALDA}[n_0](\mathbf{r}, t, \mathbf{r}', t') = \delta(t - t') \, \delta(\mathbf{r} - \mathbf{r}') \frac{d^2}{d\rho^2} (\rho \varepsilon_{xc}^{hom}(\rho))\bigg|_{\rho = n_0(\mathbf{r})} . \tag{187}$$

The Fourier-transformed quantity

$$f_{xc}^{ALDA}[n_0](\mathbf{r}, \mathbf{r}'; \omega) = \delta(\mathbf{r} - \mathbf{r}') \frac{d^2}{d\rho^2} (\rho \varepsilon_{xc}^{hom}(\rho))\bigg|_{\rho = n_0(\mathbf{r})} . \tag{188}$$

has no frequency-dependence at all.

In order to incorporate the frequency-dependence of f_{xc} in some approximate fashion, Gross and Kohn [71] suggested to use the frequency-dependent xc kernel f_{xc}^{hom} of the homogeneous electron gas in the sense of an LDA:

$$f_{xc}^{LDA}[n_0](\mathbf{r}, \mathbf{r}'; \omega) := f_{xc}^{hom}(n_0(\mathbf{r}), |\mathbf{r} - \mathbf{r}'|; \omega). \tag{189}$$

The LDA of non-local quantities, such as response functions, always involves some ambiguity [1, 2] as to whether the inhomogeneous n_0 is to be evaluated at $\mathbf{r}$, at $\mathbf{r}'$, or at some suitably chosen mean value of $\mathbf{r}$ and $\mathbf{r}'$. Of course, in the limit of slowly varying $n_0(\mathbf{r})$ (i. e. in the limit where the LDA should be a good approximation) the choice does not matter. In addition to the LDA replacement $f_{xc} \to f_{xc}^{hom}$, Gross and Kohn [71] made the assumption that $n_1(\mathbf{r}, \omega)$ is slowly varying on the length scale given by the range of $f_{xc}^{hom}(n_0(\mathbf{r}), |\mathbf{r} - \mathbf{r}'|; \omega)$. Under this assumption, the change in the xc potential can be calculated as

$$v_{xc}^{(1)}(\mathbf{r}, \omega) = n_1(\mathbf{r}, \omega) \int d^3r' \, f_{xc}^{hom}(n_0(\mathbf{r}), |\mathbf{r} - \mathbf{r}'|; \omega). \tag{190}$$

In terms of the Fourier transform of f_{xc}^{hom} with respect to $(\mathbf{r} - \mathbf{r}')$, Eq. (190) amounts to the approximation

$$f_{xc}^{GK}[n_0](\mathbf{r}, \mathbf{r}'; \omega) = \delta(\mathbf{r} - \mathbf{r}') \, f_{xc}^{hom}(n_0(\mathbf{r}), q = 0; \omega). \tag{191}$$

This approximation requires the xc kernel of the homogeneous electron gas as input. In order to investigate this quantity we consider Eq. (159) in the homogeneous case. Fourier transformation with respect to $(\mathbf{r} - \mathbf{r}')$ and $(t - t')$ leads to

$$f_{xc}^{hom}(n_0, q; \omega) = \frac{1}{\chi_s^{hom}(n_0, q; \omega)} - \frac{1}{\chi^{hom}(n_0, q; \omega)} - \frac{4\pi}{q^2}. \tag{192}$$

The response function χ_s^{hom} of a non-interacting homogeneous system is the well-known Lindhard function. The full response function χ^{hom}, on the other hand, is not known analytically. However, some exact features of χ^{hom} are known. From these, the following exact properties of f_{xc}^{hom} can be deduced:

1. As a consequence of the compressibility sum rule one finds [76]

$$\lim_{q \to 0} f_{xc}^{hom}(q, \omega = 0) = \frac{d^2}{dn^2}(n\varepsilon_{xc}^{hom}(n)) \equiv f_0(n). \tag{193}$$

   This shows that f_{xc}^{ALDA}, as given by Eq. (188), is identical with the zero-frequency limit of f_{xc}^{GK}.
2. The third-frequency-moment sum rule leads to [88]

$$\lim_{q \to 0} f_{xc}^{hom}(q, \omega = \infty)$$
$$= -\frac{4}{5} n^{2/3} \frac{d}{dn}\left(\frac{\varepsilon_{xc}^{hom}(n)}{n^{2/3}}\right) + 6n^{1/3} \frac{d}{dn}\left(\frac{\varepsilon_{xc}^{hom}(n)}{n^{1/3}}\right) \equiv f_\infty(n). \tag{194}$$

3. According to the best estimates [89, 90] *of* ε_{xc}^{hom}, the following relation holds for all densities:

$$f_0(n) < f_\infty(n) < 0\,. \tag{195}$$

4. The large-q behavior at zero frequency is given by [91]

$$\lim_{q\to\infty} f_{xc}^{hom}(q, \omega = 0) = -2n^{1/3}\frac{d}{dn}(n^{-1/3}\,\varepsilon_{xc}^{hom}(n)) - \frac{4\pi}{q^2}B(n)\,. \tag{196}$$

The function $B(n)$ has been fitted [92] to Monte-Carlo results. The resulting parametrization

$$B(n) = \frac{1 + 2.15x + 0.435x^3}{3 + 1.57x + 0.409x^3}\,, \quad x = \sqrt{r_s} = \left(\frac{3}{4\pi n}\right)^{1/6} \tag{197}$$

reproduces the Monte-Carlo results with a precision of about 1% in the density range $0 \le r_s \le 10$.

5. The short-wavelength behavior in the high-frequency limit is given by [93, 94]

$$\lim_{q\to\infty} f_{xc}^{hom}(q, \omega = \infty) = -\frac{2}{3}\cdot\frac{4\pi}{q^2}(1 - g(0)) + 6n^{1/3}\frac{d}{dn}(n^{-1/3}\,\varepsilon_{xc}^{hom}(n)) \tag{198}$$

where $g(r)$ denotes the pair correlation function.

6. $f_{xc}^{hom}(q, \omega)$ is a complex-valued function satisfying the symmetry relations

$$\Re f_{xc}^{hom}(q, \omega) = \Re f_{xc}^{hom}(q, -\omega) \tag{199}$$

$$\Im f_{xc}^{hom}(q, \omega) = -\Im f_{xc}^{hom}(q, -\omega)\,. \tag{200}$$

7. $f_{xc}^{hom}(q, \omega)$ is an analytic function of ω in the upper half of the complex ω-plane and approaches a real function $f_\infty(q)$ for $\omega \to \infty$ [70]. Therefore, the function $(f_{xc}^{hom}(q, \omega) - f_\infty(q))$ satisfies standard Kramers-Kronig relations:

$$\Re f_{xc}^{hom}(q, \omega) - f_\infty(q) = \mathbf{P}\int \frac{d\omega'}{\pi}\,\frac{\Im f_{xc}^{hom}(q, \omega')}{\omega' - \omega} \tag{201}$$

$$\Im f_{xc}^{hom}(q, \omega) = -\mathbf{P}\int \frac{d\omega'}{\pi}\,\frac{\Re f_{xc}^{hom}(q, \omega') - f_\infty(q)}{\omega' - \omega}\,. \tag{202}$$

8. The imaginary part of f_{xc}^{hom} exhibits the high-frequency behavior

$$\lim_{\omega\to\infty} \Im f_{xc}^{hom}(q, \omega) = -\frac{c}{\omega^{3/2}} \tag{203}$$

for any $q < \infty$ [95]. A second-order perturbation expansion [95, 96] of the irreducible polarization propagator leads to the high-density limit

$$c = \frac{23\pi}{15}\,. \qquad (204)$$

Other authors [97, 98] find $c = 46\pi/15$; see also Ref. [94].

9. In the same limit, the real part of f_{xc}^{hom} behaves like [71]

$$\lim_{\omega \to \infty} \Re f_{xc}^{hom}(q, \omega) = f_\infty(q) + \frac{c}{\omega^{3/2}}\,. \qquad (205)$$

Since $c > 0$, the infinite-frequency value f_∞ is approached from above. This implies, in view of the relation (195), that $\Re f_{xc}^{hom}(q = 0, \omega)$ cannot grow monotonically from f_0 to f_∞.

The above features of f_{xc}^{hom} are valid for a three-dimensional electron gas. Analogous results have been obtained for the two-dimensional case [95, 99, 100].

Taking into account the exact high- and low-frequency limits, Gross and Kohn [71] proposed the following parametrization for the imaginary part of f_{xc}^{hom}:

$$\Im f_{xc}^{hom}(q = 0, \omega) = \frac{a(n)\,\omega}{(1 + b(n)\,\omega^2)^{5/4}}\,, \qquad (206)$$

where

$$a(n) = -c(\gamma/c)^{5/3}\,(f_\infty(n) - f_0(n))^{5/3} \qquad (207)$$

$$b(n) = (\gamma/c)^{4/3}\,(f_\infty(n) - f_0(n))^{4/3} \qquad (208)$$

$$\gamma = \frac{(\Gamma(1/4))^2}{4\sqrt{2\pi}}\,. \qquad (209)$$

f_0, f_∞, and c are given by Eqs. (193), (194), and (204), respectively. Using the Kramers-Kronig relation (201), the real part can be expressed as

$$\Re f_{xc}^{hom}(q = 0, \omega) = f_\infty + \frac{a}{\pi s^2}\sqrt{\frac{8}{b}}\left[2E\left(\frac{1}{\sqrt{2}}\right) - \frac{1+s}{2}\Pi\left(\frac{1-s}{2}, \frac{1}{\sqrt{2}}\right) - \frac{1-s}{2}\Pi\left(\frac{1+s}{2}, \frac{1}{\sqrt{2}}\right)\right], \qquad s^2 = 1 + b\omega^2\,. \qquad (210)$$

E and Π are complete elliptic integrals of the second and third kind in the standard notation of Byrd and Friedman [101]. This completes the explicit form of the Gross-Kohn approximation (191).

Figs. 1 and 2 show the real and imaginary part of f_{xc}^{hom} as calculated from (206) and (210). The functions are plotted for the two density values corresponding to $r_s = 2$ and $r_s = 4$. For the lower density value ($r_s = 4$), a considerable frequency dependence is found. The dependence on ω becomes less pronounced for higher densities. In the extreme high-density limit, the difference between f_0 and f_∞ tends to zero. One finds the exact result

$$f_\infty - f_0 \sim r_s^2 \qquad \text{for } r_s \to 0 \ . \tag{211}$$

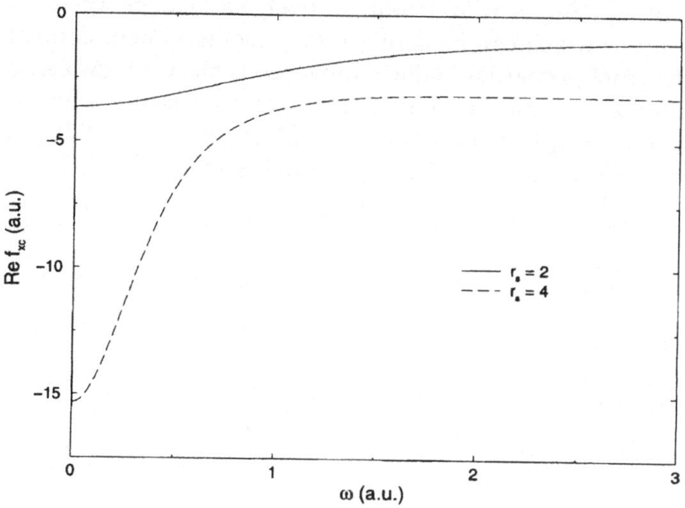

Fig. 1. Real part of the parametrization for f_{xc}^{hom} $(q = 0, \omega)$, from Ref. [88]

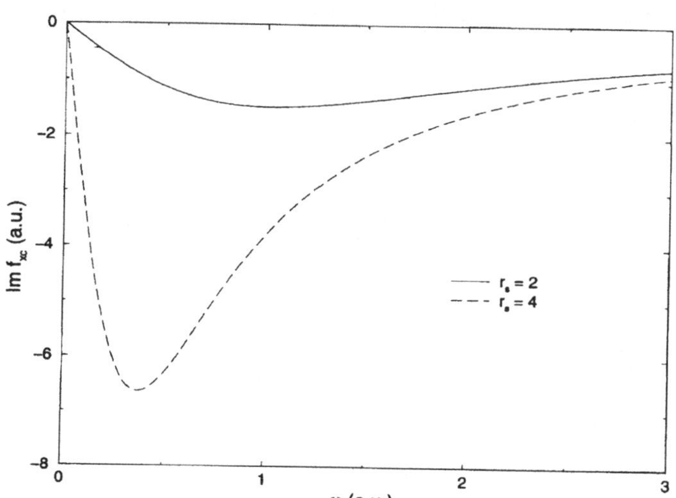

Fig. 2. Imaginary part of the parametrization for f_{xc}^{hom} $(q = 0, \omega)$, from Ref. [88]

At the same time, the depth of the minimum of $\Im f_{\mathrm{xc}}^{\mathrm{hom}}$ decreases, within the parametrization (206) proportional to r_s^2.

We finally mention that an extension of the parametrization (206) to non-vanishing q was given by Dabrowski [102]. The spin-dependent case was treated by Liu [103]. A similar interpolation for the exchange-correlation kernel of the 2-dimensional electron gas has been derived by Holas and Singwi [95].

In the construction and improvement of static ground-state density functionals, various exact constraints such as xc hole normalization [104] and scaling relations [105] have been extremely useful. While the development of explicit time-dependent functionals is at a comparatively early stage, there are some constraint conditions which can be useful in the time-dependent context. First of all, some of the exact properties of the homogeneous-electron-gas kernel $f_{\mathrm{xc}}^{\mathrm{hom}}$ are readily generalized to the inhomogeneous case: Causality leads to Kramers-Kronig relations for $f_{\mathrm{xc}}(\mathbf{r}, \mathbf{r}'; \omega)$ analogous to Eqs. (201) and (202), and the fact that $f_{\mathrm{xc}}(\mathbf{r}, t, \mathbf{r}', t')$ is a real-valued quantity implies that

$$f_{\mathrm{xc}}(\mathbf{r}, \mathbf{r}'; \omega) = f_{\mathrm{xc}}(\mathbf{r}, \mathbf{r}'; -\omega)^* . \tag{212}$$

Besides that, the response functions χ_s and χ satisfy the symmetry relations [106]

$$\chi(\mathbf{r}, \mathbf{r}'; \omega) = \chi(\mathbf{r}', \mathbf{r}; \omega) \tag{213}$$

$$\chi_s(\mathbf{r}, \mathbf{r}'; \omega) = \chi_s(\mathbf{r}', \mathbf{r}; \omega) \tag{214}$$

provided that the unperturbed system has time-reversal symmetry. Equation (163) then implies that

$$f_{\mathrm{xc}}(\mathbf{r}, \mathbf{r}'; \omega) = f_{\mathrm{xc}}(\mathbf{r}', \mathbf{r}; \omega) . \tag{215}$$

Further exact constraints can be deduced from the quantum mechanical equation of motion (15). For the operator

$$\hat{\mathbf{r}} = \int d^3r\, \mathbf{r}\, \hat{n}(\mathbf{r}) \tag{216}$$

Eq. (15) leads to

$$\frac{d}{dt}\langle\Psi(t)|\,\mathbf{r}\,|\Psi(t)\rangle = \frac{d}{dt}\int d^3r\, \mathbf{r}\, n(\mathbf{r}, t) = i\langle\Psi(t)|\,[\hat{H}(t), \hat{\mathbf{r}}]\,|\Psi(t)\rangle \tag{217}$$

where

$$\hat{H}(t) = \hat{T} + \hat{U} + \hat{V}_{\mathrm{ext}}(t) . \tag{218}$$

Taking the time derivative of Eq. (217) and employing the equation of motion (15) once more one obtains

$$\frac{d^2}{dt^2}\int d^3r\, \mathbf{r}\, n(\mathbf{r}, t) = -\langle\Psi(t)|\,[\hat{H}(t), [\hat{H}(t), \hat{\mathbf{r}}]]\,|\Psi(t)\rangle \tag{219}$$

because $\frac{\partial}{\partial t}[\hat{H}(t), \hat{\mathbf{r}}] \equiv 0$. Using the translational invariance of the Coulomb interaction $\hat{U}$ the double commutator in (219) is readily calculated, leading to the traditional Ehrenfest theorem:

$$\frac{d^2}{dt^2}\int d^3r\,\mathbf{r}\,n(\mathbf{r},t) = -\int d^3r\,n(\mathbf{r},t)\,\nabla v_{\mathrm{ext}}(\mathbf{r},t). \tag{220}$$

Likewise, for noninteracting systems described by Hamiltonians of the form

$$\hat{H}_s(t) = \hat{T} + \hat{V}_s(t) \tag{221}$$

one obtains

$$\frac{d^2}{dt^2}\int d^3r\,\mathbf{r}\,n_s(\mathbf{r},t) = -\int d^3r\,n_s(\mathbf{r},t)\,\nabla v_s(\mathbf{r},t). \tag{222}$$

For the unique KS potential

$$v_s[n](\mathbf{r},t) = v_{\mathrm{ext}}(\mathbf{r},t) + v_H[n](\mathbf{r},t) + v_{\mathrm{xc}}[n](\mathbf{r},t) \tag{223}$$

which reproduces the density $n(\mathbf{r}, t)$ of the interacting system, Eq. (222) leads to

$$\frac{d^2}{dt^2}\int d^3r\,\mathbf{r}\,n(\mathbf{r},t) = -\int d^3r\,n(\mathbf{r},t)\,\nabla\big(v_{\mathrm{ext}}(\mathbf{r},t) + v_H[n](\mathbf{r},t) + v_{\mathrm{xc}}[n](\mathbf{r},t)\big). \tag{224}$$

Subtracting Eq. (220) from Eq. (224) one obtains the rigorous result

$$\int d^3r\,n(\mathbf{r},t)\,\nabla v_{\mathrm{xc}}[n](\mathbf{r},t) = 0. \tag{225}$$

To arrive at Eq. (225) we have used the fact that the Hartree potential

$$v_H[n](\mathbf{r},t) = \int d^3r'\,\frac{n(\mathbf{r}',t)}{|\mathbf{r}-\mathbf{r}'|} \tag{226}$$

satisfies the equation

$$\int d^3r\,n(\mathbf{r},t)\,\nabla v_H[n](\mathbf{r},t) = 0. \tag{227}$$

Equation (225) was first obtained by Vignale [107] from invariance properties of the xc action functional $\mathcal{A}_{\mathrm{xc}}$ defined in Eq. (43). The derivation given here [108, 109] has the advantage of being independent of the stationary action principle.

Applying the equation of motion (15) to the angle operator $\hat{\varphi}$ and using the rotational invariance of the Coulomb interaction $\hat{U}$, one obtains

$$\frac{d^2}{dt^2}\langle\Psi(t)|\hat{\varphi}|\Psi(t)\rangle = -\int d^3r\,n(\mathbf{r},t)\,\mathbf{r}\times\nabla v_{\mathrm{ext}}(\mathbf{r},t). \tag{228}$$

Subtraction of the corresponding equation for the KS potential (223) then leads to the exact constraint

$$\int d^3 r n(\mathbf{r}, t)\, \mathbf{r} \times \nabla v_{xc}[n](\mathbf{r}, t) = 0 . \tag{229}$$

Corresponding properties of the exact xc kernel are obtained by evaluating the left-hand sides of Eqs. (225) and (229) at the density

$$n(\mathbf{r}, t) = n_0(\mathbf{r}) + \delta n(\mathbf{r}, t), \tag{230}$$

where $\delta n(\mathbf{r}, t)$ is an arbitrary deviation from the ground-state density $n_0(\mathbf{r})$. To first order in δn one obtains from Eq. (225)

$$0 = \int d^3 r\, n_0(\mathbf{r}) \nabla v_{xc}[n_0](\mathbf{r}) + \int d^3 r' \int dt'\, \delta n(\mathbf{r}', t') \Big[\delta(t - t') \nabla_{\mathbf{r}'} v_{xc}[n_0](\mathbf{r}') + \int d^3 r\, n_0(\mathbf{r}) \nabla_{\mathbf{r}} f_{xc}[n_0](\mathbf{r}, t, \mathbf{r}', t') \Big] . \tag{231}$$

The first integral on the right-hand side of this equation must vanish. (This is the static limit [110] of Eq. (225).) Since δn is arbitrary, the second integral leads to the identity

$$\int d^3 r\, n_0(\mathbf{r}) \nabla_{\mathbf{r}} f_{xc}[n_0](\mathbf{r}, t, \mathbf{r}', t') = -\delta(t - t') \nabla_{\mathbf{r}'} v_{xc}[n_0](\mathbf{r}') . \tag{232}$$

Taking the Fourier-Transform of this equation with respect to $(t - t')$ one obtains the constraint

$$\int d^3 r\, n_0(\mathbf{r}) \nabla_{\mathbf{r}} f_{xc}[n_0](\mathbf{r}\, \mathbf{r}'; \omega) = -\nabla_{\mathbf{r}'} v_{xc}[n_0](\mathbf{r}') . \tag{233}$$

Applying the same procedure to Eq. (229) one arrives at [108, 109]

$$\int d^3 r\, n_0(\mathbf{r})\, \mathbf{r} \times \nabla_{\mathbf{r}} f_{xc}[n_0](\mathbf{r}, \mathbf{r}'; \omega) = -\mathbf{r}' \times \nabla_{\mathbf{r}'} v_{xc}[n_0](\mathbf{r}') . \tag{234}$$

Finally, multiplying Eqs. (233) and (234) by $n_0(\mathbf{r}')$ and integrating over $\mathbf{r}'$ leads to

$$\int d^3 r \int d^3 r' n_0(\mathbf{r}) n_0(\mathbf{r}') \nabla_{\mathbf{r}} f_{xc}[n_0](\mathbf{r}, \mathbf{r}'; \omega) = 0 \tag{235}$$

and

$$\int d^3 r \int d^3 r' n_0(\mathbf{r}) n_0(\mathbf{r}')\, \mathbf{r} \times \nabla_{\mathbf{r}} f_{xc}[n_0](\mathbf{r}, \mathbf{r}'; \omega) = 0 . \tag{236}$$

Equation (233) was first obtained by Vignale [111] from a new sum rule for the response function. The ALDA satisfies the constraints (223) and (234) while the Gross-Kohn approximation (191) is easily seen to violate them. This fact is closely related to the violation of the Harmonic Potential Theorem which will be discussed in detail below.

Another type of constraint on theories of time-dependent phenomena in interacting inhomogeneous systems is obtained by taking expectation values of repeated commutators of current operators with the Hamiltonian. In this way one obtains exact relations for frequency moments of response functions. Very recently Sturm [94] has given a detailed study of the odd frequency moments of the dielectric function in *inhomogeneous* systems, and has explored odd moments up to the seventh. The ALDA satisfies the first, third and fifth frequency moment sum rules but violates the seventh as was demonstrated by Sturm for metals in the nearly free electron approximation.

Finally, another rigorous constraint [112] is known as the Harmonic Potential Theorem (HPT), as it relates to the motion of interacting many-electron systems in an externally-imposed harmonic oscillator potential $v(\mathbf{r}) = \frac{1}{2}\mathbf{r}\cdot\mathbf{K}\cdot\mathbf{r}$ plus the potential $-\mathbf{F}(t)\cdot\mathbf{r}$ describing a spatially uniform, time-dependent external force $\mathbf{F}(t)$. Here $\mathbf{K}$ is a spring-constant matrix which can be assumed symmetric without loss of generality. (Suitable choices of $\mathbf{K}$ yield various physical situations: for example, the choice $\mathbf{K} = \mathbf{diag}(k, k, k)$ corresponds to a spherical quantum dot or "Hooke's atom", while the choice $\mathbf{K} = \mathbf{diag}(0, 0, k)$ yields a parabolic quantum well such as may be grown in the $Ga_{1-x}Al_xAs$ system by molecular beam epitaxy.) The harmonic external potential is special, being the only confining potential which retains its form when one transforms to a homogeneously accelerated reference frame. To see that it does so, consider [113] a moving frame whose origin has the space coordinate $\mathbf{X}(t)$ relative to the rest frame. The observer in this frame sees a total external potential

$$\bar{v}(\bar{\mathbf{r}}, t) = \tfrac{1}{2}\mathbf{r}\cdot\mathbf{K}\cdot\mathbf{r} + m\ddot{\mathbf{X}}\cdot\bar{\mathbf{r}} - \mathbf{F}(t)\cdot\mathbf{r} \tag{237}$$

where $\bar{\mathbf{r}} = \mathbf{r} - \mathbf{X}(t)$ is the position coordinate in the new frame, and the second term in (237) is the centrifugal or fictitious potential due to motion of the frame. If $\mathbf{X}(t)$ satisfies the classical equation of motion

$$m\ddot{\mathbf{X}}(t) = -\mathbf{K}\cdot\mathbf{X}(t) + \mathbf{F}(t) \tag{238}$$

then one obtains the potential in the moving frame as

$$\bar{v}(\bar{\mathbf{r}}, t) = \tfrac{1}{2}\bar{\mathbf{r}}\cdot\mathbf{K}\cdot\bar{\mathbf{r}} + c(t). \tag{239}$$

This transformed external potential (239) has the same form as the potential for the *undriven* ($\mathbf{F} = \mathbf{0}$) harmonic well problem in the rest frame, except for the term c which depends on time but not on $\bar{\mathbf{r}}$. Furthermore, because the (Coulomb or other) particle-particle interaction is a function only of differences $\mathbf{r}_i - \mathbf{r}_j = \bar{\mathbf{r}}_i - \bar{\mathbf{r}}_j$, the interaction potential is also invariant under the transformation to the accelerated frame. Thus, both classically and quantum mechanically, any state or motion in the rest frame has a counterpart motion with superimposed translation $\mathbf{X}(t)$, provided that (238) is satisfied. In particular

for harmonically-confined interacting systems there exist quantum states in which the ground-state many-body wavefunction is translated rigidly (*up to a phase factor*) *as in classical motion, and hence the ground-state number density* $n_0(\mathbf{r})$ *is replaced by the rigidly moving density* $n(\mathbf{r}, t) = n_0(\mathbf{r} - \mathbf{X}(t))$. (240)

This conclusion is the Harmonic Potential Theorem (HPT). It is an extension of the generalized Kohn Theorem [114]: the latter only refers to the frequency dependence of linear response and does not address the spatial profile of the moving density. The HPT can be also proved more formally [112] by explicit construction of the moving many-body wavefunction as seen in the rest frame. It is important to note that systems confined by a scalar harmonic potential (e.g. quantum dots) are spatially finite in at least one dimension, and have strong spatial inhomogeneity at their edges. Thus the HPT constitutes an *exact* result *beyond the level of linear response* for the *time-dependent* behavior of an *inhomogeneous, interacting many-body system.* As such, it poses an interesting constraint on approximate general theories of time-dependent many-body physics, such as local-density versions of TDDFT. Of course, since the HPT is valid generally it is also valid for linear response. Vignale [107] has shown that the HPT result holds even with the inclusion of a homogeneous magnetic field.

Another closely related constraint is that of Galileian invariance. Suppose that a many-body wavefunction $\Psi(\mathbf{r}_1, \mathbf{r}_2, \dots, \mathbf{r}_N)$ satisfies the time-independent interacting N-particle Schrödinger equation with an external one-particle potential $v(\mathbf{r})$. Then, provided that the inter-particle interaction depends on coordinate differences only, it is readily verified that a boosted wavefunction of the form

$$\exp\left(-iS(t) + i\mathbf{U}\cdot\sum_{j=1}^{N}\mathbf{r}_j\right)\Psi(\mathbf{r}_1 - \mathbf{U}t, \mathbf{r}_2 - \mathbf{U}t, \dots, \mathbf{r}_N - \mathbf{U}t), \tag{241}$$

where $S(t)$ is the corresponding *classical* action [112], satisfies the time-dependent interacting N-body Schrödinger equation with boosted external potential $v(\mathbf{r} - \mathbf{U}t)$. Because the phase factor disappears in forming $|\Psi|^2$, this result implies that all many-body probability densities are rigidly boosted when the external one-body potential is boosted. In particular, the one-particle density is rigidly boosted, and this particular aspect of Galileian invariance should apply to TDDFT which deals directly with densities. In applying this criterion, it will clearly be necessary to relax the condition used in Sect. 5 that the initial many-body wavefunction be the ground-state wavefunction. In fact, to represent a system boosted to constant velocity $\mathbf{U}$, the initial wavefunction must contain the additional phase factor shown in Eq. (241).

The following question now arises: which approximations in TDDFT satisfy the HPT and Galileian invariance? By noting that the ALDA xc potential rigidly follows the density when the latter is rigidly moved, and by examining the TDKS equations for harmonic confinement with and without a driving field Dobson [112] showed that the ALDA satisfies the HPT for motion of arbitrary amplitude. The same proof in fact shows that any approximation to TDDFT satisfies the HPT provided that the xc potential rigidly follows a rigidly translated density. This rigid-following condition will be termed *Generalized Translational Invariance* and can be expressed as:

$$v_{xc}[n'](\mathbf{r}, t) = v_{xc}[n](\mathbf{r} - \mathbf{X}(t)) \tag{242}$$

Here $n(\mathbf{r})$ is an arbitrary time-independent density and $n'(\mathbf{r}, t) = n(\mathbf{r} - \mathbf{X}(t))$ is the same density rigidly boosted. (The otherwise-arbitrary displacement function $\mathbf{X}(t)$ will need to be zero at the initial time t_0, and the initial many body state will need to be the ground-state, in order for v_{xc} in (242) to be defined in the same manner used earlier in this chapter.) Equation (242) simply says that a rigid (possibly accelerated) motion of the density implies a similar rigid motion of the xc potential. Equation (242) was first demonstrated by Vignale [107] from the covariance of the time-dependent Schrödinger equation under transformation to an accelerated reference frame. Vignale also generalized the treatment to include a magnetic field. The same condition (242) with $\mathbf{X}(t) = \mathbf{U}t$ will also ensure that an approximation to v_{xc} satisfies conventional Galileian invariance.

Perhaps surprisingly, the Gross-Kohn approximation (191) unlike the ALDA, does *not* satisfy the HPT constraint. This was proved in Ref. [112] by exhibiting a specific counterexample.

The question now arises how one might correct this situation. One attempt [112] is based on the heuristic picture that, in the rigid HPT motion, all the *relative* particle motions [115], and therefore the exchange and correlation phenomena, are exactly as in the ground-state. In particular, the *static* exchange-correlation kernel $f_{xc}(n, \omega = 0)$ is appropriate for this very special motion, even though the frequency of the HPT motion has the high value $\omega = \omega_P$. This is why the ALDA succeeds with the HPT motion: it uses $f_{xc}(n, \omega = 0)$ in all circumstances and therefore is fortuitously exact for HPT motion. The original GK formalism requires the use of $f_{xc}(n, \omega = \omega_P)$, and this is the core of the difficulty. (A similar difficulty was also demonstrated [112] for hydrodynamic theory of plasmons where, once again, the frequency dependence of a coefficient, $\beta^2(\omega \to \infty) \neq \beta^2(\omega \to 0)$, is to blame, β being the pressure or diffusion coefficient.)

While most motions are not simple rigid displacements, there will be an element of this type of motion, as well as an element of compression, in more general motions provided that the original density is spatially inhomogeneous. In the linear response regime, a well-defined separation between these two components of the motion can be made [112] by first introducing the fluid velocity $\mathbf{u}(\mathbf{r}, t) = \mathbf{J}(\mathbf{r}, t)/n(\mathbf{r}, t)$ where the *exact* current density $\mathbf{J}$ can be obtained from the TDKS orbitals as demonstrated in Sect. 2.3. A fluid element displacement $\mathbf{X}(\mathbf{r}, t)$ is then defined for a general motion by

$$\mathbf{X}(\mathbf{r}, t) = \mathbf{r} + \int_{t_0}^{t} \mathbf{u}(\mathbf{r}, t')\, dt' , \qquad \mathbf{u} = \frac{\partial \mathbf{X}}{\partial t} \tag{243}$$

and by integrating the linearized continuity equation with respect to t at fixed $\mathbf{r}$ we obtain an expression for the perturbation to the equilibrium density $n_0(\mathbf{r})$ in a small motion:

$$n_1(\mathbf{r}, t) = -\nabla \cdot [n_0(\mathbf{r})\, \mathbf{X}(\mathbf{r}, t)] = -n_0(\mathbf{r})\, \nabla \cdot \mathbf{X}(\mathbf{r}, t) - \mathbf{X}(\mathbf{r}, t) \cdot \nabla n_0(\mathbf{r}) . \tag{244}$$

For an arbitrary motion, we denote the first term in (244) as

$$n_{1a}(\mathbf{r}, t) = -n_0(\mathbf{r}) \nabla \cdot \mathbf{X}(\mathbf{r}, t) \tag{245}$$

and interpret n_{1a} as the compressive part of the density perturbation, to be associated with $f_{xc}(\omega)$ where ω is the actual frequency of the linear motion. The other density component from (244) is

$$n_{1b}(\mathbf{r}, t) = -\mathbf{X}(\mathbf{r}, t) \cdot \nabla n_0(\mathbf{r}) \tag{246}$$

and this is the part one would have obtained if the equilibrium density had been rigidly translated, suggesting that it should be associated with a zero-frequency kernel $f_{xc}(\omega = 0)$. These two components make up the total density perturbation,

$$n_1 = n_{1a} + n_{1b}, \tag{247}$$

and the above arguments suggest that the xc potential for small-amplitude motion at frequency ω should be

$$v_{1xc}(\mathbf{r}, \omega) = f_{xc}(n_0(\mathbf{r}), \omega)\, n_{1a}(\mathbf{r}, \omega) + f_{xc}(n_0(\mathbf{r}), \omega = 0)\, n_{1b}(\mathbf{r}, \omega)\,. \tag{248}$$

It is immediately apparent that (248) will give the correct zero-frequency xc potential value for Harmonic Potential Theorem motion. For this motion, the gas moves rigidly implying $\mathbf{X}$ is independent of $\mathbf{r}$ so that the compressive part, n_{1a}, of the density perturbation from (245) is zero. Equally, for perturbations to a uniform electron gas, ∇n_0 and hence n_{1b} is zero, so that (248) gives the uniform-gas xc kernel $f_{xc}(\omega)$ at the actual frequency ω, as required.

A modification similar to (248) was also proposed in [112] for the pressure or diffusion term in hydrodynamics, and the resulting formalism has had some success with a unified description of boundary conditions and plasmon modes on parabolic wells [116].

Since the fluid displacement during linear response at a definite frequency ω is given by $\mathbf{X} = \mathbf{J}/(i\omega n)$, the postulated Eq. (248) suggests that v_{xc} is not a local function of the density but rather of the *current* density $\mathbf{J}$. These are, however, preliminary indications that, for nonlinear phenomena such that a definite frequency cannot be assigned to the motion the fluid displacement $\mathbf{X}$ may yield a more direct formulation of xc phenomena than does the current density (see later in this chapter).

Numerical applications of the new formalism implied by (248) are under development [117]. Preliminary indications are that the ALDA, the Gross-Kohn approximation (191) and (248) will all give substantially different results for at least one of the plasmon modes of a low-density parabolic quantum well, say for $r_s = 6$. (The modes in question are the HPT ("Kohn" or "sloshing") mode, the standing plasmon modes [118], and also the 2D plasmon mode at substantial surface-directed wavenumber $q_{\parallel}$ for which case the frequency is not constrained by model-independent theorems [119].)

Vignale [107] has given an alternative method to ensure that any xc formalism with finite memory (i.e. with frequency-dependent xc kernel) will satisfy the HPT. Starting from a simple *Ansatz* for the action integral, he derived an xc potential

$$v_{1\mathrm{xc}}(\mathbf{r}, t) = \int_{t_0}^{t} f_{\mathrm{xc}}(n_0(\mathbf{r}), t - t')\, \delta n_{\mathrm{rel}}(\mathbf{r}, t')\, dt' \tag{249}$$

where

$$\delta n_{\mathrm{rel}}(\mathbf{r}, t) = n(\mathbf{r} + \mathbf{R}_{\mathrm{cm}}(t)) - n_0(\mathbf{r}) \tag{250}$$

is the density perturbation seen by an observer moving with the global center of mass

$$\mathbf{R}_{\mathrm{cm}}(t) = \frac{1}{N} \int \mathbf{r}\, n(\mathbf{r}, t)\, d^3 r\,. \tag{251}$$

This approach does ensure satisfaction of the HPT. It differs from the method described above in that it is very much less local, requiring a determination of the global center of mass from (251) at each instant t. One can imagine situations where the two formalisms will give very different results. For example, consider two well-separated layers of electron gas confined in parallel parabolic wells. At the Hartree and Hartree-Fock levels there is no interaction between these wells in the absence of significant wavefunction overlap (and in the absence of any perturbation which might break the symmetry in the plane of the electron gas layers). For sufficient separations any residual van der Waals interaction can be made as small as desired, so the $q_{\parallel} = 0$ modes of oscillation of the two wells will be uncoupled. First consider a motion in which the two electron gases execute HPT motion (sloshing sideways) *in phase*. Then the global electronic center of mass also executes HPT motion and the Vignale method will give the correct HPT motion of the combined system. Secondly, however, consider the mode in which the two sloshing motions are 180 degrees out of phase. Then the global center of mass is stationary and the Vignale correction makes no difference, leaving the GK formalism unmodified. But this method is known not give the HPT motion correctly, as discussed above. The method described by Eq. (248), on the other hand, is more local in its effect and it corrects the motion of each well separately, giving the correct HPT motion of each gas even for the 180 degrees phase mode.

In general, for systems far from equilibrium it is not at all clear how one should approximate the full xc potential $v_{\mathrm{xc}}[n](\mathbf{r}, t)$. The most general possible nonlinear dependence of $v_{\mathrm{xc}}[n](\mathbf{r}, t)$ on n must involve at least terms with n evaluated at one space-time point $\xi' \equiv (\mathbf{r}', t')$, terms with n evaluated at two spacetime points ξ' and ξ'', terms with n evaluated at three points ξ', ξ'', ξ''', and so on. (Even this might not cover all possibilities, but the only counterexamples so far noted have involved essentially singular functions [120].) Thus in general we expect to require nonlinear functions $W^{(i)}$ such that

$$v_{\mathrm{xc}}[n](\xi) = \int d\xi'\, W^{(1)}(n(\xi'), \xi, \xi') + \int d\xi'\, d\xi''\, W^{(2)}(n(\xi'), n(\xi''), \xi, \xi', \xi'') + \ldots \tag{252}$$

The functional derivative of (252) is

$$\frac{\delta v_{\mathrm{xc}}(\xi)}{\delta n(\xi')} = \frac{\partial W^{(1)}}{\partial n}(n(\xi'), \xi, \xi') + \int d\xi''\, [W_1^{(2)}(n(\xi'), n(\xi''), \xi, \xi', \xi'') + W_2^{(2)}(n(\xi''), n(\xi'), \xi, \xi'', \xi')] + \ldots \tag{253}$$

where $W_1^{(2)} \equiv \partial W(n', n'', \xi, \xi', \xi'')/\partial n'$ and $W_2^{(2)} \equiv \partial W^{(2)}(n', n'', \xi, \xi', \xi'')/\partial n''$. Considerable simplification is achieved by postulating the following local-density *Ansatz* (254) [121] for the functional derivative

$$\frac{\delta v_{\mathrm{xc}}(\mathbf{r}, t)}{\delta n(\mathbf{r}', t')} \approx f_{\mathrm{xc}}^{\mathrm{hom}}(n(\mathbf{r}', t'), |\mathbf{r} - \mathbf{r}'|, t - t') \tag{254}$$

where $f_{\mathrm{xc}}^{\mathrm{hom}}(n, |\mathbf{r} - \mathbf{r}'|, t - t')$ is the nonlocal, delayed xc kernel of the uniform electron gas of density n. Clearly, in the limit of weakly inhomogeneous systems, i.e., for systems with densities $n(\mathbf{r}', t') \to const$ this *Ansatz* becomes exact. We now seek a functional $v_{\mathrm{xc}}[n](\mathbf{r}, t)$ whose functional derivative $\delta v_{\mathrm{xc}}(\mathbf{r}, t)/\delta n(\mathbf{r}', t')$ is given by (254). The task of finding such a v_{xc} is possible because (254) is a function of density at one point only. Hence the integral terms in (252) and (253) involving $W^{(2)}$ and higher must be discarded and it follows that

$$\frac{\partial W^{(1)}}{\partial n}(n(\mathbf{r}', t'), \mathbf{r}, t, \mathbf{r}', t') = f_{\mathrm{xc}}^{\mathrm{hom}}(n(\mathbf{r}', t'), |\mathbf{r} - \mathbf{r}'|, t - t'). \tag{255}$$

Thus $W^{(1)}$ is a density integral of $f_{\mathrm{xc}}^{\mathrm{hom}}(n, r, \tau)$. Assuming that v_{xc} is zero in a zero-density system, and defining

$$W_{\mathrm{xc}}(n, r, \tau) = \int_0^n f_{\mathrm{xc}}^{\mathrm{hom}}(\rho, r, \tau)\, d\rho \tag{256}$$

we then obtain

$$v_{\mathrm{xc}}[n](\mathbf{r}, t) = \int dt'\, d^3r'\, W_{\mathrm{xc}}(n(\mathbf{r}', t'), |\mathbf{r} - \mathbf{r}'|, t - t'). \tag{257}$$

The *Ansatz* (257) makes $v_{\mathrm{xc}}(\mathbf{r}, t)$ depend principally on the density near the point $\mathbf{r}$, at a range of times t' which are near to, but earlier than, t. In the following we propose to improve this by noting that, if there is streaming in the many-body fluid, the memory of past densities is likely to be greatest when one remains with the same fluid element rather than remaining with the same spatial point $\mathbf{r}$. Thus we propose [122] instead of (257)

$$v_{xc}[n](\mathbf{r}, t) = \int dt'\, d^3r'\, W_{xc}(n(\mathbf{r}', t'), |\mathbf{R}(t'|\mathbf{r}, t) - \mathbf{r}'|, t - t'). \tag{258}$$

where W_{xc} is still given by (256).

In (258) the density from a past time t' which most strongly influences $v_{xc}(\mathbf{r}, t)$ is the density at position $\mathbf{r}' = \mathbf{R}(t'|\mathbf{r}, t)$, where the trajectory function $\mathbf{R}(t'|\mathbf{r}, t)$ of a fluid element is its position at time t', given that its position at time t is $\mathbf{r}$. $\mathbf{R}$ can be defined unambiguously by demanding that its time derivative is the fluid velocity $\mathbf{u} = \mathbf{J}/n$ formed from the current density $\mathbf{J}(\mathbf{r}, t)$:

$$\frac{\partial}{\partial t'}\mathbf{R}(t'|\mathbf{r}, t) = \mathbf{u}(\mathbf{R}, t') \equiv \mathbf{J}(\mathbf{R}, t')/n(\mathbf{R}, t') \tag{259}$$

where all occurrences of $\mathbf{R}$ have the same arguments as on the left-hand side of Eq. (259). The boundary condition on (259) is

$$\mathbf{R}(t|\mathbf{r}, t) = \mathbf{r}. \tag{260}$$

In (258) one acknowledges that the physics of delayed correlation will have its maximum degree of spatial locality if the observer is riding on a fluid element rather than observing from a fixed location $\mathbf{r}$. Eqs. (258)–(260) represent our general expression for the dynamic xc potential. The use of (258) in place of (257) will turn out to provide a nonlinear theory which, regardless of its validity in other respects, at least satisfies both the nonlinear Harmonic Potential Theorem [112] and the requirements of Galileian invariance [107]. To demonstrate that (258) satisfies the HPT, we show that it satisfies the generalized Galileian invariance condition (242). The only difficulty is that (258) has an implicit and highly nonlocal dependence on $n(\mathbf{r}', t)$ via the current density dependence of $\mathbf{R}$. From [112], however, it follows that for HPT motion the TDKS equations involve not only a rigidly boosted density $n'(\mathbf{r}, t) = n_0(\mathbf{r} - \mathbf{X}(t))$, but also a boosted current $\mathbf{J}'(\mathbf{r}, t) = n'(\mathbf{r}, t)\,\dot{\mathbf{X}}(t)$ because of the phase factor introduced by the motion into the KS wavefunctions. Thus the fluid velocity is just $\mathbf{u}(\mathbf{r}, t) = \dot{\mathbf{X}}(t)$. From (259) and (260) it then follows that

$$\mathbf{R}[n'](t'|\mathbf{r}, t) = \mathbf{r} + \mathbf{X}(t') - \mathbf{X}(t). \tag{261}$$

Putting this into (258) we find

$$\begin{aligned} v_{xc}[n'](\mathbf{r}, t) &= \int dt'\, d^3r'\, W_{xc}(n(\mathbf{r}' - \mathbf{X}(t'), t'), |\mathbf{r} + \mathbf{X}(t') - \mathbf{X}(t) - \mathbf{r}'|, t - t') \\ &= \int dt'\, d^3r'\, W_{xc}(n(\mathbf{r}' - \mathbf{X}(t'), t'), |(\mathbf{r} - \mathbf{X}(t)) - (\mathbf{r}' - \mathbf{X}(t'))|, t - t') \\ &= v_{xc}[n](\mathbf{r} - \mathbf{X}(t), t). \end{aligned} \tag{262}$$

Thus (258) satisfies (242). Hence the HPT and Galileian invariance requirements are satisfied.

Apart from its presumed nonlinear capabilities, (258) is also more general than the linear response work in preceding sections because it invokes the *spatial* nonlocality of the uniform gas xc kernel f_{xc}, making it more comparable to the work of Dabrowski [102]. To compare it with the discussions above, we now make the local approximation

$$f_{xc}^{hom}(\mathbf{r}, t; \mathbf{r}', t') \approx \delta(\mathbf{r} - \mathbf{r}')\, f_{xc}^{hom}(n, q = 0, t - t'). \tag{263}$$

If this local approximation is employed in Eqs. (256) and (257), the resulting $v_{xc}[n](\mathbf{r}, t)$ does not satisfy the HPT and the requirements of Galileian invariance because, when linearized, it reduces to the Gross-Kohn form (191) and this is known [112] not to satisfy the HPT. However, combined with Eq. (258), the local approximation (263) leads to

$$v_{xc}[n](\mathbf{r}, t) = \int dt'\, w_{xc}(n(\mathbf{R}(t' \mid \mathbf{r}, t), t'), t - t') \tag{264}$$

where

$$w_{xc}(n, \tau) = \int_0^n f_{xc}^{hom}(\rho, q = 0, \tau)\, d\rho. \tag{265}$$

We now show that the functional (264), when linearized, gives precisely the modified linear-response xc kernel f_{xc} of Eq. (248). To this end we consider small motions around a static equilibrium, in the sense that the displacement

$$\mathbf{x}(\mathbf{r}, t) \equiv \mathbf{R}(t \mid \mathbf{r}, t_0) - \mathbf{r} \tag{266}$$

of each fluid element from its initial ($t = t_0$) position $\mathbf{r}$ is small. Under these circumstances both the fluid displacement $\mathbf{X}$ and the current $\mathbf{J}$ in (259) are small (first-order) quantities. Thus we may write

$$\mathbf{R}(t' \mid \mathbf{r}, t) = \mathbf{r} + O(\mathbf{x}) = \mathbf{R}(t' \mid \mathbf{r}, t_0) + O(\mathbf{x}) \tag{267}$$

and using this in (259) we find

$$\begin{aligned}\frac{\partial}{\partial t'} \mathbf{R}(t' \mid \mathbf{r}, t) &= \mathbf{J}(\mathbf{R}(t' \mid \mathbf{r}, t_0))/n_0(\mathbf{r}) + O(\mathbf{x}^2) \\ &= \frac{\partial}{\partial t'} \mathbf{R}(t' \mid \mathbf{r}, t_0) + O(\mathbf{x}^2).\end{aligned} \tag{268}$$

Integrating both sides of (268) with respect to t', starting from $t' = t$, we find

$$\mathbf{R}(t' \mid \mathbf{r}, t) - \mathbf{R}(t \mid \mathbf{r}, t) = \mathbf{R}(t' \mid \mathbf{r}, t_0) - \mathbf{R}(t \mid \mathbf{r}, t_0) + O(\mathbf{x}^2), \tag{269}$$

so that, by (269) and (266)

$$\mathbf{R}(t' \mid \mathbf{r}, t) = \mathbf{r} + \mathbf{x}(t') - \mathbf{x}(t) + O(\mathbf{x}^2). \tag{270}$$

In the linear limit we can also integrate the linearized continuity equation

$$\frac{\partial n_1}{\partial t} + \nabla \cdot [n_0(\mathbf{r})\, \mathbf{u}(\mathbf{r}, t)] = 0 \tag{271}$$

to give the density perturbation n_1 in terms of the fluid displacement $\mathbf{x}$:

$$n_1(\mathbf{r}, t) = -\nabla \cdot [n_0(\mathbf{r})\, \mathbf{x}(\mathbf{r}, t)] + O(\mathbf{x}^2). \tag{272}$$

We can now use (270) and (272) to expand the density argument of w_{xc} in the nonlinear functional (264):

$$\begin{aligned} n(\mathbf{R}(t'|\mathbf{r}, t), t') &= n(\mathbf{r} + \mathbf{x}(\mathbf{r}, t') - \mathbf{x}(\mathbf{r}, t), t') + O(\mathbf{x}^2) \\ &= n_0(\mathbf{r}) + \nabla n_0(\mathbf{r}) \cdot (\mathbf{x}(\mathbf{r}, t') - \mathbf{x}(\mathbf{r}, t)) + n_1(\mathbf{r}, t') + O(\mathbf{x}^2) \\ &= n_0(\mathbf{r}) - \nabla n_0(\mathbf{r}) \cdot \mathbf{x}(\mathbf{r}, t) - n_0(\mathbf{r})\, \nabla \cdot \mathbf{x}(\mathbf{r}, t') + O(\mathbf{x}^2). \end{aligned} \tag{273}$$

In deriving (273), we used (270) in the first step, standard linearization in the second step and (272) in the third step. Putting (273) into the proposed nonlinear xc potential (264) we find

$$\begin{aligned} v_{xc}(\mathbf{r}, t) = & \int_{-\infty}^{\infty} w_{xc}(n_0(\mathbf{r}), t - t')\, dt' \\ & - \nabla n_0(\mathbf{r}) \cdot \mathbf{x}(\mathbf{r}, t) \int_{-\infty}^{\infty} \frac{\partial w_{xc}}{\partial n}(n_0(\mathbf{r}), t - t')\, dt' \\ & - n_0(\mathbf{r}) \int_{-\infty}^{\infty} \frac{\partial w_{xc}}{\partial n}(n_0(\mathbf{r}), t - t')\, \nabla \cdot \mathbf{x}(\mathbf{r}, t')\, dt' + O(\mathbf{x}^2). \end{aligned} \tag{274}$$

Integrating (256) with respect to time and using $f_{xc}^{hom}(n) = dv_{xc}^{hom}(n)/dn$, we find that the first term in (274) is just the xc potential $v_{xc}^{hom}(n_0(\mathbf{r}))$ of the static, unperturbed problem. The linear correction to this equilibrium value of the xc potential is then, by (274) and (256),

$$\begin{aligned} v_{1xc}(\mathbf{r}, t) = & \left(\int_{-\infty}^{\infty} f_{xc}(n_0(\mathbf{r}), t - t')\, dt' \right) n_{1B}(\mathbf{r}, t) \\ & + \int_{-\infty}^{\infty} f_{xc}(n_0(\mathbf{r}), t - t')\, n_{1A}(\mathbf{r}, t')\, dt' \end{aligned} \tag{275}$$

where

$$n_{1A}(\mathbf{r}, t) = -n_0(\mathbf{r})\, \nabla \cdot \mathbf{x}(\mathbf{r}, t), \quad n_{1B}(\mathbf{r}, t) = -\mathbf{x}(\mathbf{r}, t) \cdot \nabla n_0(\mathbf{r}) \tag{276}$$

Fourier-transforming (275) and writing the terms in reverse order we find

$$v_{1\mathrm{xc}}(\mathbf{r}, \omega) = f_{\mathrm{xc}}(n_0(\mathbf{r}), \omega)\, n_{1A}(\mathbf{r}, \omega) + f_{\mathrm{xc}}(n_0(\mathbf{r}), \omega = 0)\, n_{1B}(\mathbf{r}, \omega). \tag{277}$$

This is identical to the form (248). To summarize, we have proposed a rather bold *Ansatz*, Eq. (258), for the time-dependent xc potential $v_{\mathrm{xc}}(\mathbf{r}, t)$ of an arbitrary system which could be far from equilibrium. This *Ansatz* carries a nonlocal space and time dependence based on uniform-gas data, but accesses the actual system density $n(\mathbf{r}, t)$ in a simple local fashion. The assumption of local space dependence in the uniform gas yields a simpler form again, Eq. (264). It remains to be seen whether our relatively simple forms can cope with the gamut of nonlinear phenomena in systems far from equilibrium. As a first step it would be interesting to investigate second-and higher-order nonlinear susceptibilities described in Sect. 5.2. Computer codes for investigating the fully nonlinear case may be adaptable from the work of Galdrikian et al. [123], who have investigated strongly driven quantum wells. Regardless of the applicability of our method to general nonlinear phenomena, the use of the trajectory function $\mathbf{R}(t'|\mathbf{r}, t)$ in (258) and (264) guarantees two things: firstly, satisfaction of the generalized Galileian invariance condition (242) and hence of the Harmonic Potential Theorem (240), for motion of arbitrarily large amplitude; and secondly, for systems close to equilibrium the more local version (264) reduces to the linear time-delayed or frequency-dependent formalism previously proposed by Dobson [112].

6.2 Time-Dependent Optimized Effective Potential

The approximate xc potentials described so far were derived from the homogeneous electron gas in one or another way. All of them have one deficiency in common: They contain spurious self-interaction contributions. It is known from static DFT that the removal of self-interaction is an important ingredient in the construction of good xc potentials. Various approaches to the construction of self-interaction-free functionals are known in the static case [37, 124–135]. One of these is the so-called optimized potential method (OPM) [133–135]. This method takes as starting point a given expression for the total energy $E[\phi_1 \ldots \phi_N]$ of an N-electron system as a functional of a set of single-particle orbitals $\{\phi_j(\mathbf{r})\}$ (e. g. the Hartree-Fock total energy functional in the exchange-only case). Then, the variationally best *local* effective potential is determined such that, when inserted in a stationary single-particle Schrödinger equation, it yields the set of N eigenfunctions (corresponding to the N lowest eigenvalues) that minimize $E[\phi_1 \ldots \phi_N]$. In practice, the full OPM scheme is computationally quite involved since it requires the numerical solution of an integral equation for $v_{\mathrm{xc}}(\mathbf{r})$. As a consequence, complete OPM calculations have been performed mainly for problems where the potential is a function of a single variable, e.g. for spherically symmetric atoms [134–140]. There exists, however, an approximate OPM scheme, recently proposed by Krieger, Li, and Iafrate (KLI) [141–149], which is numerically as easy to handle as the ordinary KS scheme. This simplified OPM has been applied very successfully to the calculation of atomic properties [6]. In many respects this method is currently the most accurate density-functional scheme.

In the present section we shall describe the construction of a self-interaction-free xc potential which can be viewed as a time-dependent version of the optimized potential method (TDOPM). The approach leads to v_{xc} as a function of $(\mathbf{r}, t)$ rather than to v_{xc} as an explicit functional of the density. In order to derive such a time-dependent generalization of the OPM we consider an N-electron system at some finite time t_0 which, for all times up until t_0, has been in the ground state associated with an external potential $v_0(\mathbf{r})$ (e.g., a nuclear Coulomb potential). We assume that the corresponding stationary OPM problem has been solved for that system, i.e. a local effective potential and a set of N single-particle orbitals $\{\phi_j\}$ (with energy eigenvalues ε_j) minimizing a given energy functional $E[\phi_1 \dots \phi_N]$ are assumed to be known. Again, at $t = t_0$ an additional time-dependent potential $v_1(\mathbf{r}, t)$ is switched on. Our goal is to determine the time evolution of the system under the influence of the total external potential $v(\mathbf{r}, t) = v_0(\mathbf{r}) + v_1(\mathbf{r}, t)$ from t_0 up until an arbitrary later time t_1. To construct an optimized local effective potential we start with the quantum mechanical action [150]

$$\begin{aligned}\mathscr{A}[\varphi_1 \dots \varphi_N] = &\sum_j^N \int_{-\infty}^{t_1} dt \int d^3r\, \varphi_j^*(\mathbf{r}, t) \left(i\frac{\partial}{\partial t} + \frac{\nabla^2}{2} \right) \varphi_j(\mathbf{r}, t) \\ &- \int_{-\infty}^{t_1} dt \int d^3r\, n(\mathbf{r}, t)\, v(\mathbf{r}, t) \\ &- \frac{1}{2} \int_{-\infty}^{t_1} dt \int d^3r \int d^3r' \frac{n(\mathbf{r}, t)\, n(\mathbf{r}', t)}{|\mathbf{r} - \mathbf{r}'|} \\ &- \mathscr{A}_{xc}[\varphi_1 \dots \varphi_N] \end{aligned} \tag{278}$$

written as a functional of N time-dependent single-particle orbitals $\{\varphi_j(\mathbf{r}, t)\}$ where $n(\mathbf{r}, t) = \sum_j^N |\varphi_j(\mathbf{r}, t)|^2$. In a time-dependent exchange-only theory $\mathscr{A}_{xc}[\varphi_1 \dots \varphi_N]$ – the xc part of the quantum mechanical action – would be replaced by the time-dependent Fock expression

$$\mathscr{A}_{xc} \approx \mathscr{A}_x = -\frac{1}{2} \sum_{i,j}^N \delta_{\sigma_i \sigma_j} \int_{-\infty}^{t_1} dt \int d^3r \int d^3r' \frac{\varphi_i^*(\mathbf{r}', t)\, \varphi_j(\mathbf{r}', t)\, \varphi_i(\mathbf{r}, t)\, \varphi_j^*(\mathbf{r}, t)}{|\mathbf{r} - \mathbf{r}'|} \tag{279}$$

(σ_j denotes the spin orientation of the jth orbital). We note that the integrand of (279) is a *local* expression with respect to the time-coordinate, i.e., all orbitals depend on the same time argument t. With approximate functionals of this type, the causality problem described in Sect. 5.2 does not occur. The orbitals are solutions of the time-dependent Schrödinger equation

$$i \frac{\partial}{\partial t} \varphi_j(\mathbf{r}, t) = \left(-\frac{\nabla^2}{2} + v_s(\mathbf{r}, t) \right) \varphi_j(\mathbf{r}, t), \qquad j = 1, \ldots, N, \tag{280}$$

with $\varphi_j(\mathbf{r}, t) = \phi_j(\mathbf{r}) \exp[-i\varepsilon_j(t - t_0)]$ for $-\infty < t \leq t_0$. The local effective potential is given by

$$v_s(\mathbf{r}, t) = v(\mathbf{r}, t) + v_{\mathrm{H}}(\mathbf{r}, t) + v_{\mathrm{xc}}^{\mathrm{TDOPM}}(\mathbf{r}, t), \tag{281}$$

where $v_{\mathrm{H}}(\mathbf{r}, t) = \int d^3 r' \, n(\mathbf{r}', t)/|\mathbf{r} - \mathbf{r}'|$ denotes the time-dependent Hartree potential. The total potential $v_s(\mathbf{r}, t)$ has to be determined in such a way that the $\{\varphi_j(\mathbf{r}, t)\}$, resulting from Eq. (280), render the total action functional $\mathscr{A}[\varphi_1 \ldots \varphi_N]$ stationary. Therefore, we have to solve the following variational problem:

$$\frac{\delta \mathscr{A}[\varphi_1 \ldots \varphi_N]}{\delta v_s(\mathbf{r}, t)} = \sum_j^N \int_{-\infty}^{+\infty} dt' \int d^3 r' \left(\frac{\delta \mathscr{A}[\varphi_1 \ldots \varphi_N]}{\delta \varphi_j(\mathbf{r}', t')} \frac{\delta \varphi_j(\mathbf{r}', t')}{\delta v_s(\mathbf{r}, t)} + \frac{\delta \mathscr{A}[\varphi_1 \ldots \varphi_N]}{\delta \varphi_j^*(\mathbf{r}', t')} \frac{\delta \varphi_j^*(\mathbf{r}', t')}{\delta v_s(\mathbf{r}, t)} \right) = 0 \,. \tag{282}$$

We first compute the functional derivatives $\delta \mathscr{A}/\delta \varphi_j$ and $\delta \mathscr{A}/\delta \varphi_j^*$: defining

$$u_{\mathrm{xc}j}(\mathbf{r}, t) = \frac{1}{\varphi_j^*(\mathbf{r}, t)} \frac{\delta \mathscr{A}_{\mathrm{xc}}[\varphi_1 \ldots \varphi_N]}{\delta \varphi_j(\mathbf{r}, t)} \,, \tag{283}$$

we obtain

$$\frac{\delta \mathscr{A}[\varphi_1 \ldots \varphi_N]}{\delta \varphi_j(\mathbf{r}', t')} = \left[-i \frac{\partial}{\partial t'} - \left(-\frac{\nabla'^2}{2} + v(\mathbf{r}', t') + v_{\mathrm{H}}(\mathbf{r}', t') + u_{\mathrm{xc}j}(\mathbf{r}', t') \right) \right] \varphi_j^*(\mathbf{r}', t') \, \theta(t_1 - t') \tag{284}$$

and an analogous expression for $\delta \mathscr{A}/\delta \varphi_j^*$ which, for all reasonable (i.e. real) functionals $\mathscr{A}[\varphi_1 \ldots \varphi_N]$, is the complex conjugate of (284). $\theta(x)$ denotes the usual step function (1 for $x > 0$, 0 for $x < 0$). To arrive at Eq. (284) the first term of Eq. (278) has to be integrated by parts with respect to the time coordinate. We impose the usual boundary condition on $\varphi_j(\mathbf{r}, t)$ at $t = t_1$, i.e. $\delta \varphi_j(\mathbf{r}, t_1) = 0$, thus obtaining a zero boundary contribution. The other boundary contribution at $t = -\infty$ vanishes, too, because the action functional (278), in order to be well-defined, is to be calculated by introducing the usual factor $e^{\eta t}$ in the integrand and taking $\lim_{\eta \to 0+}$ after the integration. Substituting Eq. (281) into (284) and making use of the fact that φ_j^* solves the complex conjugate of the

Schrödinger equation (280), we find

$$\frac{\mathscr{A}[\varphi_1 \dots \varphi_N]}{\delta\varphi_j(\mathbf{r}', t')} = [v_{xc}^{\text{TDOPM}}(\mathbf{r}', t') - u_{xcj}(\mathbf{r}', t')]\,\varphi_j^*(\mathbf{r}', t')\,\theta(t_1 - t')\,. \tag{285}$$

In order to evaluate $\delta\mathscr{A}/\delta v_s$ from Eq. (282), we further need the functional derivatives $\delta\varphi_j/\delta v_s$ and $\delta\varphi_j^*/\delta v_s$. The stationary OPM eigenfunctions $\{\phi_j(\mathbf{r}), j = 1, \dots, \infty\}$ form a complete orthonormal set, and so do the time-evolved states $\{\varphi_j(\mathbf{r}, t), j = 1, \dots, \infty\}$ for any time $t \in [-\infty, t_1]$, and we denote this set by $\mathbf{\Phi}_t$. Now consider $\mathbf{\Phi}_t$ as unperturbed states, remembering that at $t = t_1$ the orbitals are held *fixed* with respect to variations in the total potential. We therefore start from $t = t_1$, subject the system to an *additional* small perturbation $\delta v_s(\mathbf{r}, t)$ and let it evolve *backward* in time. The corresponding perturbed wave functions $\varphi_j'(\mathbf{r}, t)$ are determined by the backward Schrödinger equation

$$i\frac{\partial}{\partial t}\varphi_j'(\mathbf{r}, t) = \left(-\frac{\nabla^2}{2} + v_s(\mathbf{r}, t) + \delta v_s(\mathbf{r}, t)\right)\varphi_j'(\mathbf{r}, t)\,, \quad j = 1, \dots, N \tag{286}$$

with the initial condition $\varphi_j'(\mathbf{r}, t_1) = \varphi_j(\mathbf{r}, t_1)$. This problem cannot be treated directly with time-dependent perturbation theory as described in standard text books because the unperturbed Hamiltonian is already time-dependent. Nevertheless, Dirac's method of variation of constants can be applied in a straightforward manner. We expand, at each given t, the perturbed wave function $\varphi_j'(\mathbf{r}, t)$ in terms of the set $\mathbf{\Phi}_t$,

$$\varphi_j'(\mathbf{r}, t) = \sum_{k=1}^{\infty} c_{jk}(t)\,\varphi_k(\mathbf{r}, t)\,, \tag{287}$$

and insert this expansion in (286), utilizing Eq. (280). The resulting equation

$$i\sum_{k=1}^{\infty} \dot{c}_{jk}(t)\,\varphi_k(\mathbf{r}, t) = \sum_{k=1}^{\infty} c_{jk}(t)\,\delta v_s(\mathbf{r}, t)\,\varphi_k(\mathbf{r}, t) \tag{288}$$

is then multiplied by $\varphi_l^*(\mathbf{r}, t)$ and integrated over all space; the orthonormality of $\mathbf{\Phi}_t$ yields

$$\dot{c}_{jl}(t) = \frac{1}{i}\sum_{k=1}^{\infty} c_{jk}(t)\int d^3r\,\varphi_l^*(\mathbf{r}, t)\,\delta v_s(\mathbf{r}, t)\,\varphi_k(\mathbf{r}, t)\,. \tag{289}$$

We now make the usual ansatz for a perturbation expansion,

$$c_{jk}(t) = c_{jk}^{(0)}(t) + c_{jk}^{(1)}(t) + \dots \tag{290}$$

and collect corresponding orders on each side of Eq. (289). This yields

$$\dot{c}_{jl}^{(0)}(t) = 0$$

$$\dot{c}_{jl}^{(1)}(t) = \frac{1}{i}\sum_{k=1}^{\infty} c_{jk}^{(0)}(t)\int d^3r\,\varphi_l^*(\mathbf{r}, t)\,\delta v_s(\mathbf{r}, t)\,\varphi_k(\mathbf{r}, t) \tag{291}$$

Since, in our case, the wave function evolves backward from the fixed state $\varphi_j(\mathbf{r}, t_1)$ we find $c_{jk}^{(0)}(t) = \delta_{jk}$ and $c_{jk}^{(1)}(t_1) = 0$, leading to

$$c_{jl}^{(1)}(t) = \frac{1}{i} \int_{t_1}^{t} dt' \int d^3 r \; \varphi_l^*(\mathbf{r}, t') \, \delta v_s(\mathbf{r}, t') \, \varphi_j(\mathbf{r}, t') . \tag{292}$$

It follows that the first-order correction to the wave function $\varphi_j(\mathbf{r}, t)$ under the influence of $\delta v_s(\mathbf{r}, t)$ is given by

$$\begin{aligned} \delta\varphi_j(\mathbf{r}, t) &= \sum_{k=1}^{\infty} c_{jk}^{(1)}(t) \; \varphi_k(\mathbf{r}, t) \\ &= i \sum_{k=1}^{\infty} \int_{t}^{t_1} dt' \int d^3 r' \; \varphi_k^*(\mathbf{r}', t') \, \delta v_s(\mathbf{r}', t') \, \varphi_j(\mathbf{r}', t') \, \varphi_k(\mathbf{r}, t) . \end{aligned} \tag{293}$$

Therefore, the desired functional derivative is

$$\frac{\delta\varphi_j(\mathbf{r}', t')}{\delta v_s(\mathbf{r}, t)} = i \sum_{k=1}^{\infty} \varphi_k^*(\mathbf{r}, t) \, \varphi_j(\mathbf{r}, t) \, \varphi_k(\mathbf{r}', t') \, \theta(t_1 - t) \, \theta(t - t') . \tag{294}$$

Once again, $\delta\varphi_j^* / \delta v_s$ leads to the complex conjugate expression. We can now insert (285) and (294) in the variational equation (282), and the result is the TDOPM integral equation for the local exchange-correlation potential $v_{xc}(\mathbf{r}, t)$:

$$\begin{aligned} & i \sum_{j}^{N} \int_{-\infty}^{t_1} dt' \int d^3 r' \; [v_{xc}^{\mathrm{TDOPM}}(\mathbf{r}', t') - u_{xcj}(\mathbf{r}', t')] \\ & \quad \times \varphi_j(\mathbf{r}, t) \, \varphi_j^*(\mathbf{r}', t') \, K(\mathbf{r}, t, \mathbf{r}', t') + c.c. = 0 \quad . \end{aligned} \tag{295}$$

The kernel

$$K(\mathbf{r}, t, \mathbf{r}', t') = \sum_{k=1}^{\infty} \varphi_k^*(\mathbf{r}, t) \, \varphi_k(\mathbf{r}', t') \, \theta(t - t') \tag{296}$$

can be identified with the Green's function of the system, which satisfies the differential equation

$$\left[i \frac{\partial}{\partial t'} - \left(-\frac{\nabla'^2}{2} + v_s(\mathbf{r}', t') \right) \right] K(\mathbf{r}, t, \mathbf{r}', t') = -i\delta(\mathbf{r} - \mathbf{r}') \, \delta(t - t') \tag{297}$$

with the initial condition $K(\mathbf{r}, t, \mathbf{r}', t') = 0$ for $t' > t$. The TDOPM scheme is now complete: the integral equation (295) has to be solved for $v_{xc}(\mathbf{r}, t)$ in combination with the Schrödinger equation (280) and the differential equation (297) for $K(\mathbf{r}, t, \mathbf{r}', t')$, both with the appropriate initial conditions. It is easy to show that in the time interval $[-\infty, t_1]$ the exchange-correlation potential $v_{xc}(\mathbf{r}, t)$ is only determined up to within a purely time-dependent function $c(t)$ (as expected in view of the time-dependent HK theorem discussed in Sect. 2).

We now demonstrate that for $t < t_0$ or for a time-independent external potential ($v_1(\mathbf{r}, t) \equiv 0$) the TDOPM reduces to the stationary OPM. For this purpose we rewrite Eq. (295) in the following way (using the fact that v_{xc} is real):

$$i \sum_{j}^{N} \int_{-\infty}^{t_1} dt' \int d^3 r' \left[v_{xc}^{\text{TDOPM}}(\mathbf{r}', t') - u_{xcj}(\mathbf{r}', t')\right] \varphi_j(\mathbf{r}, t) \varphi_j^*(\mathbf{r}', t')$$

$$\times \sum_{\substack{k=1 \\ k \neq j}}^{\infty} \varphi_k^*(\mathbf{r}, t) \varphi_k(\mathbf{r}', t') \theta(t - t') + c.c.$$

$$= i \sum_{j}^{N} \varphi_j(\mathbf{r}, t) \varphi_j^*(\mathbf{r}, t) \int_{-\infty}^{t} dt' \int d^3 r' \, (u_{xcj}(\mathbf{r}', t')$$

$$- u_{xcj}^*(\mathbf{r}', t')) \varphi_j(\mathbf{r}', t') \varphi_j^*(\mathbf{r}', t') \,. \tag{298}$$

In the static case, the orbitals $\{\varphi_j(\mathbf{r}, t)\}$ are replaced by $\{\phi_j(\mathbf{r}) \exp[-i\varepsilon_j(t - t_0)]\}$. It is reasonable to assume that the exchange-correlation functional $\mathscr{A}_{xc}$ then becomes

$$\mathscr{A}_{xc}[\varphi_1 \ldots \varphi_N] \rightarrow \int_{-\infty}^{t_1} dt' \; E_{xc}[\varphi_1(t') \ldots \varphi_N(t')], \tag{299}$$

where $E_{xc}[\phi_1 \ldots \phi_N]$ is the corresponding ground state exchange-correlation energy functional. Definition (283) then yields

$$u_{xcj}^{\text{static}}(\mathbf{r}, t) = \left[\frac{1}{\tilde{\phi}_j^*(\mathbf{r})} \frac{\delta E_{xc}[\tilde{\phi}_1 \ldots \tilde{\phi}_N]}{\delta \tilde{\phi}_j(\mathbf{r})}\right]_{\tilde{\phi}_j(\mathbf{r}) = \phi_j(\mathbf{r}) e^{-i\varepsilon_j(t - t_0)}} . \tag{300}$$

We assume that the value of $E_{xc}[\phi_1 \ldots \phi_N]$ remains unchanged if the arguments $\{\phi_j(\mathbf{r})\}$ are multiplied by phase factors $e^{i\alpha_j}$. If this is the case, we can use the identity

$$\varphi_j(\mathbf{r}, t) = \sqrt{\varphi_j(\mathbf{r}, t) \varphi_j^*(\mathbf{r}, t)} \, \frac{\varphi_j(\mathbf{r}, t) \varphi_j^*(\mathbf{r}_0, t)}{|\varphi_j(\mathbf{r}, t) \varphi_j^*(\mathbf{r}_0, t)|} \, e^{i \arg(\varphi_j(\mathbf{r}_0, t))} \tag{301}$$

(where $\mathbf{r}_0$ is an arbitrary reference point) and write E_{xc} in Eq. (299) as a functional of the combinations $\varphi_j(\mathbf{r}, t) \varphi_j^*(\mathbf{r}', t)$. Then it is not difficult to show that u_{xcj}^{static} is independent of time and that the right-hand side of (298) is zero. We therefore obtain

$$i \sum_{j}^{N} \int_{-\infty}^{t_1} dt' \int d^3 r' \left[v_{xc}^{\text{OPM}}(\mathbf{r}') - u_{xcj}^{\text{static}}(\mathbf{r}')\right] \phi_j(\mathbf{r}) \phi_j^*(\mathbf{r}')$$

$$\times \sum_{\substack{k=1 \\ k \neq j}}^{\infty} \phi_k^*(\mathbf{r}) \phi_k(\mathbf{r}') \, e^{-i(\varepsilon_j - \varepsilon_k)(t - t')} \, \theta(t - t') + c.c. = 0 \quad . \tag{302}$$

Performing the integration over t' we find the stationary OPM integral equation [134]

$$\lim_{\eta \to 0^+} \sum_{j}^{N} \int d^3r' \, [v_{xc}^{\text{OPM}}(\mathbf{r}') - u_{xcj}^{\text{static}}(\mathbf{r}')] \, \phi_j(\mathbf{r}) \, \phi_j^*(\mathbf{r}')$$

$$\times \sum_{\substack{k=1 \\ k \neq j}}^{\infty} \frac{\phi_k^*(\mathbf{r}) \phi_k(\mathbf{r}')}{\varepsilon_j - \varepsilon_k - i\eta} + c.c. = 0 \, . \tag{303}$$

The derivation of Eq. (303) shows that in order to recover the static limit from the time-dependent formalism one had to extend the time integral in Eq. (278) to $-\infty$; a *finite* lower time boundary does not correctly account for memory effects in v_{xc} and therefore results in an unphysical time dependence even in the static case.

The numerical implementation of the full TDOPM is an extremely demanding task. It is therefore most desirable to obtain a simplified scheme. To this end we shall perform a transformation of Eq. (295) similar to the one proposed by KLI in the stationary case [146, 149]. This will lead to an alternative but still exact form of the TDOPM scheme which allows one to construct approximations of $v_{xc}(\mathbf{r}, t)$ which are *explicit* functionals of the orbitals $\{\varphi_j\}$, thereby avoiding the need to solve the integral equation. Following Refs. [146, 149], we define

$$p_j(\mathbf{r}, t) = \frac{-i}{\varphi_j^*(\mathbf{r}, t)} \int_{-\infty}^{t_1} dt' \int d^3r' \, [v_{xc}^{\text{TDOPM}}(\mathbf{r}', t') - u_{xcj}(\mathbf{r}', t')] \, \varphi_j^*(\mathbf{r}', t')$$

$$\times \sum_{\substack{k=1 \\ k \neq j}}^{\infty} \varphi_k^*(\mathbf{r}, t) \, \varphi_k(\mathbf{r}', t') \, \theta(t - t') \tag{304}$$

and

$$\bar{u}_{xcj}(t) = \int d^3r \, n_j(\mathbf{r}, t) \, u_{xcj}(\mathbf{r}, t) \tag{305}$$

where $n_j(\mathbf{r}, t) = |\varphi_j(\mathbf{r}, t)|^2$. Eq. (298) can then be written as

$$\sum_{j}^{N} n_j(\mathbf{r}, t) \, p_j(\mathbf{r}, t) + c.c. = -i \sum_{j}^{N} n_j(\mathbf{r}, t) \int_{-\infty}^{t} dt' \, (\bar{u}_{xcj}(t') - \bar{u}_{xcj}^*(t')) \, , \tag{306}$$

and it is easy to show that

$$\int d^3r \, n_j(\mathbf{r}, t) \, p_j(\mathbf{r}, t) = 0 \, . \tag{307}$$

Evaluating $\varphi_j(\mathbf{r}, t) [-i \, \partial/\partial t + \nabla^2/2 - v_s(\mathbf{r}, t)] \, \varphi_j^*(\mathbf{r}, t) \, p_j(\mathbf{r}, t)$ one finds that $p_j(\mathbf{r}, t)$ satisfies the following differential equation:

$$\frac{1}{2} \nabla \cdot (n_j(\mathbf{r}, t) \nabla p_j(\mathbf{r}, t)) - i \, n_j(\mathbf{r}, t) \frac{\partial}{\partial t} p_j(\mathbf{r}, t) - i \, \mathbf{J}_j(\mathbf{r}, t) \cdot \nabla p_j(\mathbf{r}, t)$$

$$= - n_j(\mathbf{r}, t) \, [v_{xc}^{\text{TDOPM}}(\mathbf{r}, t) - u_{xcj}(\mathbf{r}, t) - (\bar{v}_{xcj}(t) - \bar{u}_{xcj}(t))] \tag{308}$$

with the current density $\mathbf{J}_j(\mathbf{r}, t) = (2i)^{-1} (\varphi_j^*(\mathbf{r}, t)\nabla\varphi_j(\mathbf{r}, t) - \varphi_j(\mathbf{r}, t)\nabla\varphi_j^*(\mathbf{r}, t))$ and $\bar{v}_{xcj}(t) = \int d^3r\, n_j(\mathbf{r}, t)\, v_{xc}^{TDOPM}(\mathbf{r}, t)$. Finally, operating with ∇^2 on Eq. (306) and using Eq. (308) we find

$$v_{xc}^{TDOPM}(\mathbf{r}, t) = \frac{1}{n(\mathbf{r}, t)} \sum_j^N n_j(\mathbf{r}, t) \frac{1}{2}(u'_{xcj}(\mathbf{r}, t) + u'^*_{xcj}(\mathbf{r}, t))$$
$$+ \frac{1}{n(\mathbf{r}, t)} \sum_j^N n_j(\mathbf{r}, t)\, [\bar{v}_{xcj}(t) - \tfrac{1}{2}(\bar{u}_{xcj}(t) + \bar{u}^*_{xcj}(t))]$$
$$+ \frac{i}{4n(\mathbf{r}, t)} \sum_j^N \nabla^2 n_j(\mathbf{r}, t) \int_{-\infty}^{t} dt'\, (\bar{u}_{xcj}(t') - \bar{u}^*_{xcj}(t')) \tag{309}$$

where

$$u'_{xcj}(\mathbf{r}, t) = u_{xcj}(\mathbf{r}, t) + \frac{1}{n_j(\mathbf{r}, t)} \left[\frac{1}{2}\nabla \cdot (p_j(\mathbf{r}, t)\nabla n_j(\mathbf{r}, t)) + i\, n_j(\mathbf{r}, t)\frac{\partial}{\partial t} p_j(\mathbf{r}, t) + i\,\mathbf{J}_j(\mathbf{r}, t)\cdot\nabla p_j(\mathbf{r}, t)\right]. \tag{310}$$

Equations (309) and (310) together with the differential equation (308) for $p_j(\mathbf{r}, t)$ and the condition (307) (which can be used to fix the constant left undetermined by Eq. (308)) represent an exact alternative formulation of the TDOPM scheme. The advantage of Eq. (309) lies in the fact that it is a very convenient starting point for constructing approximations of $v_{xc}(\mathbf{r}, t)$ as explicit functionals of the $\{\varphi_j(\mathbf{r}, t)\}$: it is only necessary to approximate $p_j(\mathbf{r}, t)$ in Eq. (310) by a suitably chosen functional of the orbitals. We can then readily solve Eq. (309) *analytically* for $v_{xc}^{TDOPM}(\mathbf{r}, t)$, as we shall show below.

We expect an *approximate* potential $\tilde{v}_{xc}(\mathbf{r}, t)$ defined in this way to be close to the exact $v_{xc}(\mathbf{r}, t)$. This conjecture is based on the observation that the difference between $\tilde{v}_{xc}$ and v_{xc} is entirely accounted for by the differences $u'_{xcj} - u_{xcj}$ which are zero if averaged over the jth orbital, as will be demonstrated in the following. From Eq. (310) we obtain

$$\bar{u}'_{xcj}(t) - \bar{u}_{xcj}(t) = \tfrac{1}{2}\int d^3r\, \nabla\cdot(p_j(\mathbf{r}, t)\nabla n_j(\mathbf{r}, t)) + i\int d^3r \left[n_j(\mathbf{r}, t)\frac{\partial}{\partial t}p_j(\mathbf{r}, t) + \mathbf{J}_j(\mathbf{r}, t)\cdot\nabla p_j(\mathbf{r}, t)\right]. \tag{311}$$

Using the divergence theorem, the first term on the right-hand side can be transformed into a surface integral which vanishes if the time-dependent orbitals decrease exponentially for $r \to \infty$. The contribution to the second integral containing $\mathbf{J}_j \cdot \nabla p_j$ is then integrated by parts. The surface term vanishes due to the same argument as before, and the remaining term is transformed using the continuity equation for the jth orbital to replace $-\nabla\cdot\mathbf{J}_j(\mathbf{r}, t)$ by $\partial n_j(\mathbf{r}, t)/\partial t$.

Hence we find

$$\bar{u}'_{xcj}(t) - \bar{u}_{xcj}(t) = i\frac{\partial}{\partial t}\int d^3r\, n_j(\mathbf{r},t)\, p_j(\mathbf{r},t) = 0\,, \tag{312}$$

where the last equality follows from Eq. (307).

The simplest approximation is obtained by replacing p_j by its average value, i.e. by setting $p_j(\mathbf{r},t) \equiv 0$. The resulting approximate potential will be termed the time-dependent KLI (TDKLI) potential. It is given by the equation

$$\begin{aligned} v_{xc}^{\mathrm{TDKLI}}(\mathbf{r},t) = {} & \frac{1}{n(\mathbf{r},t)}\sum_j^N n_j(\mathbf{r},t)\,\frac{1}{2}\left(u_{xcj}(\mathbf{r},t) + u^*_{xcj}(\mathbf{r},t)\right) \\ & + \frac{1}{n(\mathbf{r},t)}\sum_j^N n_j(\mathbf{r},t)\left[\bar{v}_{xcj}^{\mathrm{TDKLI}}(t) - \tfrac{1}{2}\left(\bar{u}_{xcj}(t) + \bar{u}^*_{xcj}(t)\right)\right] \\ & + \frac{i}{4n(\mathbf{r},t)}\sum_j^N \nabla^2 n_j(\mathbf{r},t)\int_{-\infty}^{t} dt'\left(\bar{u}_{xcj}(t') - \bar{u}^*_{xcj}(t')\right). \end{aligned} \tag{313}$$

This equation is still an integral equation for v_{xc}^{TDKLI}. It can, however, be solved semi-analytically [145]: Multiplying Eq. (313) by $n_k(\mathbf{r},t)$ and integrating over all space yields

$$\bar{v}_{xck}^{\mathrm{TDKLI}}(t) = \bar{w}_{xck}(t) + \sum_j^N M_{kj}(t)\,\bar{v}_{xc\,j}^{\mathrm{TDKLI}}(t)\,, \tag{314}$$

where we have defined

$$\begin{aligned} w_{xc}(\mathbf{r},t) = {} & \frac{1}{n(\mathbf{r},t)}\sum_j^N n_j(\mathbf{r},t)\,\frac{1}{2}\left(u_{xcj}(\mathbf{r},t) + u^*_{xcj}(\mathbf{r},t)\right) \\ & - \frac{1}{n(\mathbf{r},t)}\sum_j^N n_j(\mathbf{r},t)\,\frac{1}{2}\left(\bar{u}_{xcj}(t) + \bar{u}^*_{xcj}(t)\right) \\ & + \frac{i}{4n(\mathbf{r},t)}\sum_j^N \nabla^2 n_j(\mathbf{r},t)\int_{-\infty}^{t} dt'\left(\bar{u}_{xcj}(t') - \bar{u}^*_{xcj}(t')\right) \end{aligned} \tag{315}$$

and

$$M_{kj}(t) = \int d^3r\,\frac{n_k(\mathbf{r},t)\,n_j(\mathbf{r},t)}{n(\mathbf{r},t)}\,. \tag{316}$$

Solving Eq. (314) for $\bar{v}_{xcj}^{\mathrm{TDKLI}}(t)$ requires inversion of the $N \times N$ matrix

$$A_{kj}(t) = \delta_{kj} - M_{kj}(t) \tag{317}$$

and leads to

$$\bar{v}_{xcj}^{\mathrm{TDKLI}}(t) = \sum_k^N \left(A^{-1}(t)\right)_{jk}\,\bar{w}_{xck}(t)\,. \tag{318}$$

When Eq. (318) is substituted into Eq. (313), one obtains $v_{xc}^{TDKLI}(\mathbf{r}, t)$ as an explicit functional of the orbitals $\{\varphi_j(\mathbf{r}, t)\}$. As the exact $v_{xc}(\mathbf{r}, t)$ which follows from Eq. (295), $v_{xc}^{TDKLI}(\mathbf{r}, t)$ is determined by Eq. (313) only up to within a purely time-dependent function $c(t)$.

The last term of Eqs. (313) and (315) vanishes identically for a large class of exchange-correlation functionals $\mathscr{A}_{xc}$. This class includes all functionals depending on $\{\varphi_j\}$ only through the combinations $\varphi_j(\mathbf{r}, t)\, \varphi_j^*(\mathbf{r}', t)$ (such as the time-dependent Fock functional, Eq. (279).

One readily verifies that both the full TDOPM potential and the TDKLI approximation of it satisfy the the generalized translational invariance condition (242) (and hence the harmonic potential theorem) provided that

$$\mathscr{A}_{xc}[\varphi_1' \dots \varphi_N'] = \mathscr{A}_{xc}[\varphi_1 \dots \varphi_N] \tag{319}$$

is satisfied with φ_j' being the orbitals in the accelerated frame:

$$\varphi_j'(\mathbf{r}, t) = \exp(-iS(t) + i\dot{\mathbf{X}} \cdot \mathbf{r})\, \phi_j(\mathbf{r} - \mathbf{X}(t))\,. \tag{320}$$

The TDHF functional (279) is easily seen to satisfy the constraint (319). Equation (319) will be a strong guideline in the proper construction of approximate correlation functionals $\mathscr{A}_c[\varphi_1 \dots \varphi_N]$. Equation (313) combined with the Schrödinger equation (280) represents a time-dependent scheme which is numerically much less involved than, e.g., the time-dependent Hartree-Fock method. Numerical results obtained with this scheme for atoms in strong laser pulses will be described in Sect. 8.

To conclude this section we construct in the following an approximation of the xc kernel f_{xc} on the basis of the TDOPM. A calculation analogous to Eqs. (138)–(152) shows [151, 152] that within TDOPM the central Eq. (152) holds for the quantity f_{xc}^{TDOPM} defined by the integral equation

$$\int_t^\infty dt' \int d^3r' \sum_{jk} [\phi_j(\mathbf{r})\, \phi_j^*(\mathbf{r}')\, \phi_k(\mathbf{r}')\, \phi_k^*(\mathbf{r})\, e^{-i(\varepsilon_j - \varepsilon_k)(t - t')}$$
$$\times (f_j\, f_{xc}^{TDOPM}(\mathbf{y}, \tau, \mathbf{r}', t') - g_{xc}^{(j)}(\mathbf{y}, \tau, \mathbf{r}', t')) - c.c.] = 0 \tag{321}$$

where

$$g_{xc}^{(j)}(\mathbf{y}, \tau, \mathbf{r}', t') = \left[\frac{1}{2\varphi_j^*(\mathbf{r}', t')} \frac{\delta v_{xc}^{TDOPM}(\mathbf{y}, \tau)}{\delta \varphi_j(\mathbf{r}', t')}\right]_{\varphi_j(t) = \phi_j e^{-i\varepsilon_j t}}. \tag{322}$$

Equation (321) has the same algebraic structure as the TDOPM integral equation (295) with the time-dependent orbitals $\varphi_j(\mathbf{r}, t)$ replaced by $e^{-i\varepsilon_j t}\phi_j$ and with $v_{xc}^{TDOPM}(\mathbf{r}', t')$ and $u_{xcj}(\mathbf{r}', t')$ replaced by $f_{xc}^{TDOPM}(\mathbf{y}, \tau, \mathbf{r}', t')$ and $g_{xc}^{(j)}(\mathbf{y}, \tau, \mathbf{r}', t')$, respectively. A simple analytical approximation to $v_{xc}^{TDOPM}(\mathbf{r}, t)$ is given by

$$v_{xc}^{approx}(\mathbf{r}, t) = \sum_j \frac{|\varphi_j(\mathbf{r}, t)|^2}{2n(\mathbf{r}, t)} (u_{xcj}(\mathbf{r}, t) + u_{xcj}^*(\mathbf{r}, t)). \tag{323}$$

Applying this approximation to (321), i.e., setting

$$f_{xc}^{approx}(\mathbf{y}, \tau, \mathbf{r}', t') = \sum_j \frac{|\phi_j(\mathbf{r})|^2}{2n(\mathbf{r})} (g_{xc}^{(j)}(\mathbf{y}, \tau, \mathbf{r}', t') + c.c.) \tag{324}$$

and using the explicit analytical form (323) to evaluate (322) one arrives in the time-dependent x-only limit (279) at the compact expression

$$f_{xc}^{approx}[n_0](\mathbf{r}, \mathbf{r}'; \omega) = -\frac{2|\sum_k f_k \phi_k(\mathbf{r}) \phi_k^*(\mathbf{r}')|^2}{|\mathbf{r}-\mathbf{r}'| n_0(\mathbf{r}) n_0(\mathbf{r}')} . \tag{325}$$

In general, the Fourier transform of the xc kernel defined by Eq. (321) is frequency dependent (even in the TD x-only case), a feature which is not accounted for by the present approximation (325). However, for the special case of a two-electron system treated within TD x-only theory, Eqs. (323) and (325) are the *exact* solutions of the respective integral equations.

7 Applications Within the Perturbative Regime

7.1 Photoresponse of Finite and Infinite Systems

To date, most applications of TDDFT have been in the linear response regime. Calculations of the photoresponse from Eqs. (154) and (155) are, by now, a mature subject. The literature on such calculations is enormous and a whole volume [153] has been devoted to the subject. In this section we shall restrict ourselves to the basic ideas rather than describing the applicational details.

The TDKS formalism has been employed to calculate the photoresponse of atoms [10, 12–14, 154, 155], molecules [156, 157] and clusters [158–168] metallic surfaces [169–175] and semiconductor heterostructures [72, 176–179] bulk semiconductors [180] and bulk metals [181–184].

For simplicity, we consider sufficiently low radiation frequencies, such that the electric field can be assumed to be constant across the atom or molecule. For atoms this is the case for photon energies below 3 keV. The external potential associated with a monochromatic electric field is then given by

$$v_1(\mathbf{r}, t) = Ez\cos\omega t . \tag{326}$$

The induced density change $n(\mathbf{r}, t) - n_0(\mathbf{r})$ (143) can be characterized by a series of multipole moments.

$$p(t) = -\int d^3r\, z(n(\mathbf{r}, t) - n_0(\mathbf{r})) , \tag{327}$$

can be expanded as [185]

$$p(t) = \alpha(\omega) \cdot \mathbf{E} \cos \omega t + \tfrac{1}{2} \beta(0) : \mathbf{EE} + \tfrac{1}{2} \beta(2\omega) : \mathbf{EE} \cos 2\omega t \tag{328}$$

$$+ \tfrac{1}{4} \gamma(\omega) \vdots \mathbf{EEE} \cos \omega t + \tfrac{1}{4} \gamma(3\omega) \vdots \mathbf{EEE} \cos 3\omega t\,, \tag{329}$$

where the notation is meant to indicate the tensorial character of the quantities α, β and γ. The first coefficient, α, is termed the *dipole polarizability*; β and γ denote the second-and third-order *dipole hyperpolarizabilities*. For spherically symmetric ground states β is zero.

The dipole polarizability is related to the frequency-dependent linear density response $n_1(\mathbf{r}, \omega)$ via

$$\alpha(\omega) = -\frac{2}{E} \int d^3 r\, z\, n_1(\mathbf{r}, \omega) \tag{330}$$

and the photoabsorption cross section is given by

$$\sigma(\omega) = \frac{4\pi\omega}{c} \Im \alpha(\omega)\,. \tag{331}$$

Zangwill and Soven [10] have calculated the photoabsorption spectrum of rare-gas atoms from the frequency-dependent KS equations (156)–(157) within the ALDA. As an example for the quality of the results we show, in Fig. 3, the photoabsorption cross section of Xenon just above the 4d threshold. The agreement with experiment is remarkably good. Results of similar quality have been achieved for the photoresponse of small molecules [156, 157].

It should be mentioned that the thresholds characterizing the onset of continuous absorption from the various occupied atomic shells are not well reproduced in the calculations of Zangwill and Soven. The calculated absorption edges are typically several eV below the observed thresholds. While, in principle, TDDFT should yield the correct thresholds, it appears that simple approximations such as the ALDA are not sufficient in this respect.

As a point of practical interest we mention that the KS response function is usually not calculated directly from the KS orbitals as in Eq. (157). Instead, one

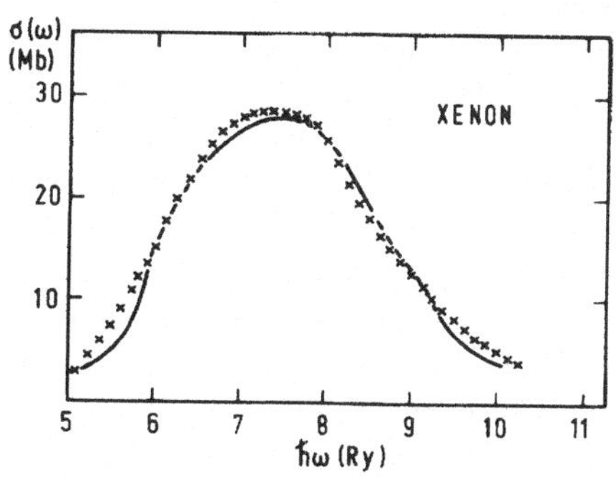

Fig. 3. Total photoabsorption cross section of the Xe atom versus photon energy in the vicinity of the 4d threshold. *Solid line*: self-consistent time-dependent KS calculation from [10]; *crosses*: experimental data from [186].

rewrites the response function in terms of the KS Green's function. The latter is then calculated numerically from the corresponding equation of motion, usually by multipole expansion [10, 187].

The linear photoresponse of metal clusters was successfully calculated for spherical [158–160, 163] as well as for spheroidal clusters [164] within the jellium model [188] using the LDA. The results are improved considerably by the use of self-interaction corrected functionals. In the context of response calculations, self-interaction effects occur at three different levels: First of all, the static KS orbitals, which enter the response function, have a self-interaction error if calculated within LDA. This is because the LDA xc potential of finite systems shows an exponential rather than the correct $-1/r$ behaviour in the asymptotic region. As a consequence, the valence electrons of finite systems are too weakly bound and the effective (ground-state) potential does not support high-lying unoccupied states. Apart from the response function χ_s, the xc kernel $f_{xc}[n_0]$ no matter which approximation is used for it, also has a self-interaction error. This is because $f_{xc}[n_0]$ is evaluated at the unperturbed ground-state density $n_0(\mathbf{r})$, and this density exhibits self-interaction errors if the KS orbitals were calculated in LDA. Finally the ALDA form of f_{xc} itself carries another self-interaction error.

To improve upon these defects, one has to go beyond the LDA: The *(modified) weighted density approximation* [189] retains the correct asymptotic behaviour of v_{xc} and improves the response properties of metal clusters [162, 165]. A different route to improvement provides the *self-interaction correction* (SIC) of Perdew Zunger [37], where the spurious self-interaction of the LDA is compensated by additional terms in the ground-state potential [166] and in the effective perturbing potential as well [167] (Full-SIC).

In most theoretical work on the response of metallic surfaces the ionic potential is replaced by the potential due to a uniform positive charge background in a half space, say $z > 0$. This is the so-called jellium model for metallic surfaces. In this model are two intrinsic microscopic length scales, the inverse Fermi wave-number, k_F^{-1}, and the Thomas-Fermi screening length ($\approx$ surface thickness), k_{TF}^{-1}. Both lengths are typically of the order $a \approx 10^{-8}$ cm. In most applications the perturbing electric potential v_1 and the perturbing vector potential $\mathbf{A}_1$ vary on a length scale ℓ which satisfies $\ell \gg a$. Examples are the scalar potential $v_1(\mathbf{r})$ due to an external charge at a distance $z \gg a$, or the vector potential $\mathbf{A}_1(\mathbf{r}, t)$, associated with a light wave of wavelength $\lambda \gg a$. The corresponding linear responses n_1 and $\mathbf{j}_1$ vary on the scale of ℓ in the $x - y$ plane but, because of the abrupt drop of the unperturbed density at the surface (on the scale of a), they vary on the short scale a in the z-direction. Formal arguments due to Feibelman [190] have shown that, to leading order in a/ℓ, the effect of the surface on the electromagnetic fields far from the surface ($|z| \gg a$) is entirely characterized by two complex frequency-dependent lengths, $d_\parallel(\omega)$ and $d_\perp(\omega)$.

Actually, for the jellium model $d_\parallel(\omega) \equiv 0$. This result has been obtained in the random phase approximation (RPA) in Ref. [190]. It is easily established as a rigorous many-body result for the jellium model [191]. To define $d_\perp(\omega)$ we

Fourier analyze all physical quantities parallel to the surface, in the x–y plane. For example, a Fourier component of the induced charge density becomes

$$n_1(\mathbf{r}, \omega) \equiv n_1(z, \omega)\, e^{i\mathbf{q}_\parallel \cdot \mathbf{r}}\,, \tag{332}$$

where $\mathbf{q}_\parallel = (q_x, q_y, 0)$ (and $|\mathbf{q}_\parallel| a \ll 1$). Then $d_\perp(\omega)$ is given by

$$d_\perp(\omega) \equiv \frac{\int dz\, z\, n_1(z, \omega)}{\int dz\, n_1(z, \omega)}\,, \tag{333}$$

i.e., it is the (complex) center of mass of the induced surface charge. $d_\perp(\omega)$ is the generalization of the static image plane introduced by Lang and Kohn [192].

This $d_\perp(\omega)$ is then the subject of quantitative calculations. They require the density response $n_1(z, \omega)$, to a uniform external electric field perpendicular to the surface. The calculation was first carried out in the RPA equivalent to time-dependent Hartree theory, in which the xc kernel f_{xc} is neglected. These calculations led to very interesting results not present in classical Maxwell theory, such as the surface photo effect and surface plasmons. Plasmons are high-frequency charge-density oscillations of the electron gas. In a bulk material the long-wavelength plasma frequency is $\omega_P = (4\pi n e^2/m)^{1/2}$ in gaussian cgs units. Plasmons occur in the ultraviolet frequency region for metals, but the artificial electron gas in semiconductor quantum wells often has a plasma frequency in the infrared. The confinement of the electron gas at metal edges introduces a range of new plasmon modes at frequencies other than ω_P, and these could potentially yield information about inhomogeneous xc effects. Information about plasmons on films, surfaces and semiconductor wells is most easily available experimentally for small values of the surface-directed wavenumber $q_\parallel$, and unfortunately in this region there are theorems prescribing the plasmon frequencies, *regardless* of the effects of exchange and correlation. A summary of these "no-go" theorems is given in [113] and further review is given in [119] for the case of semiconductor quantum wells.

Early theoretical studies of plasmas used hydrodynamics [193]. These treatments were able to predict the main new feature of the plasmon spectrum at a metal surface due to the strong surface inhomogeneity of the electron gas, namely the surface plasmon. Its frequency approaches $\omega_P/\sqrt{2}$ as the surface directed wavenumber $q_\parallel$ approaches zero, and this is correctly predicted in hydrodynamic and microscopic theories. This result is independent both of the precise electron density edge profile and of the type of xc kernel used, if any [194, 195]. Thus, although the surface plasmon is often the strongest feature in electron energy loss measurements on thin metal films, [196] it is hard to obtain any information from it about dynamic exchange and correlation. To see such effects one needs to measure with great accuracy the *dispersion* of surface plasmons. Only in the last few years has it been possible even to confirm experimentally a result first predicted by Feibelman [190] on the basis of selfconsistent RPA calculations, namely that the dispersion of the surface plasmon on a charge-neutral metal surface is initially *negative*. This result follows basically from the very "soft" or weakly bound nature of the electron gas at a neutral jellium surface, allowing electrons to spill out substantially into the vacuum. For a review of some experimental and theoretical aspects see [197]. While the value of this negative dispersion coefficient does depend to a degree on the xc kernel f_{xc} introduced earlier, it remains to be seen whether experiments on metal films and surfaces can measure this quantity to a useful accuracy. On the theory side, an important observation by Liebsch [198] is that the KS orbitals used to construct the dynamic response must come from a static calculation using a model of exchange and correlation that is consistent with the dynamic xc kernel used in the plasmon calculation. For example, LDA calculation followed by RPA screening (with $f_{xc} = 0$) is not consistent and causes false shifts in predicted surface plasmon frequencies.

The weak binding and wide inhomogeneous density layer at the edge of a neutral metal surface leads to a "multipole" surface plasmon mode in addition to the usual "monopole" surface plasmon [172, 173, 197]. This mode is in principle sensitive to f_{xc} even at $q_{\parallel} = 0$. Gies and Gerhardts [173] and Dobson and Harris [199] investigated this mode both in the ALDA and the frequency dependent parametrization (206)–(210). It was found that, for an aluminium surface, the inclusion of the frequency dependence of f_{xc}. has only a 3% effect on the multipole plasmon frequency, but a 20% effect on the damping of the mode. It seems likely that the frequency dependence of f_{xc} will have a much larger effect on this mode for a low-electron density metal such as Rb, and this may be worth pursuing.

In general, low-dimensional, low-density systems offer the best prospects for strong effects of xc phenomena on plasmon frequencies. A case in point is a pair of parallel quasi-two-dimensional electron layers in a semiconductor double-quantum well experiment. Interesting effects are predicted for this case [200].

Another way of probing dynamic xc effects experimentally is by inelastic X-ray scattering from bulk metals [201–203]. In this way, the so-called dynamical structure factor $S(q, \omega)$ can be measured which is proportional to the imaginary part of the full response function in reciprocal space. With this information at hand and with a first-principles calculation of the non-interacting response function, the connection (159) between f_{xc} and the response functions can be used to deduce information about f_{xc} [204].

All applications quoted so far were for the *linear* response. Very few investigations have dealt with the higher-order response described in Sect. 5.2. The frequency-dependent third-order hyperpolarizabilities of rare-gas atoms were calculated by Senatore and Subbaswamy [86] within the ALDA; the calculated values turned out to bee too large by a factor of two, further indicating the need for self-interaction corrected functionals in the calculation of response properties. The effect of adsorbates on second-harmonic generation at simple metal surfaces was invested by Kuchler and Rebentrost [205, 206]. Most recently, the second-order harmonic generation in bulk insulators was calculated within the ALDA [207].

7.2 Calculation of Excitation Energies

The traditional density-functional formalism of Hohenberg, Kohn and Sham [1, 2] is a powerful tool in predicting ground-state properties of many-electron systems [3–5]. The description of excited-state properties within density-functional theory, however, is notoriously difficult. One might be tempted to interpret the Kohn-Sham single-particle energy differences $\omega_{jk} := \varepsilon_j - \varepsilon_k$ as excitation energies. This interpretation, however, has no rigorous basis and in practice the Kohn-Sham orbital energy differences ω_{jk} deviate by 10–50% from the true excitation energies $\Omega_m := E_m - E_0$. Several extensions of ground-state DFT have been devised to tackle excited states. They are based either on the Rayleigh-Ritz principle for the lowest eigenstate of each symmetry class [208–210] or on a variational principle for ensembles [211–222]. A fundamental difficulty is that the xc energy functionals appearing in these approaches depend on the symmetry labels of the state considered or on the particular ensemble, respectively. Until today very little is known on how these excited-state xc functionals differ from the ordinary ground-state xc energy.

In this section we are going to develop a different approach to the calculation of excitation energies which is based on TDDFT [69, 84, 152]. Similar ideas were recently proposed by Casida [223] on the basis of the one-particle density matrix. To extract excitation energies from TDDFT we exploit the fact that the frequency-dependent linear density response of a finite system has discrete poles at the excitation energies of the unperturbed system. The idea is to use the formally exact representation (156) of the linear density response $n_1(\mathbf{r}, \omega)$, to calculate the shift of the Kohn-Sham orbital energy differences ω_{jk} (which are the poles of the Kohn-Sham response function) towards the true excitation energies Ω_m in a systematic fashion.

The spin-dependent generalization [59] of TDDFT described in Sect. 4.1 leads to the following analogue of Eq. (156) for the linear density response of electrons with spin σ:

$$n_{1\sigma}(\mathbf{r}, \omega) = \sum_{\nu} \int d^3y\, \chi_{s\sigma\nu}(\mathbf{r}, \mathbf{y}; \omega)\, v_{1\nu}(\mathbf{y}, \omega)$$

$$+ \sum_{\nu, \nu'} \int d^3y \int d^3y'\, \chi_{s\sigma\nu}(\mathbf{r}, \mathbf{y}; \omega)$$

$$\times \left(\frac{1}{|\mathbf{y} - \mathbf{y}'|} + f_{xc\,\nu\nu'}(\mathbf{y}, \mathbf{y}'; \omega) \right) n_{1\nu'}(\mathbf{y}', \omega). \tag{334}$$

Here the spin-dependent exchange-correlation kernel is given by the Fourier transform of

$$f_{xc\,\sigma\sigma'}(\mathbf{r}, t, \mathbf{r}', t') := \left. \frac{\delta v_{xc\sigma}[n_\uparrow, n_\downarrow](\mathbf{r}, t)}{\delta n_{\sigma'}(\mathbf{r}', t')} \right|_{n_{0\uparrow},\, n_{0\downarrow}} \tag{335}$$

with respect to $(t - t')$. Note that the spin-dependent response-function of noninteracting particles

$$\chi_{s\,\sigma\sigma'}(\mathbf{r}, \mathbf{r}'; \omega) = \delta_{\sigma\sigma'} \sum_{j,k} (f_{k\sigma} - f_{j\sigma}) \frac{\phi_{j\sigma}(\mathbf{r})\,\phi^*_{k\sigma}(\mathbf{r})\,\phi^*_{j\sigma}(\mathbf{r}')\,\phi_{k\sigma}(\mathbf{r}')}{\omega - (\varepsilon_{j\sigma} - \varepsilon_{k\sigma}) + i\eta} \tag{336}$$

is diagonal in the spin variable and exhibits poles at frequencies $\omega_{jk\sigma} \equiv \varepsilon_{j\sigma} - \varepsilon_{k\sigma}$ corresponding to single-particle excitations within the same spin space. In order to calculate the shifts towards the true excitation energies Ω of the interacting system, we rewrite Eq. (334) as

$$\begin{aligned}\sum_{\nu'} \int d^3y' \Bigg(\delta_{\sigma\nu'}\delta(\mathbf{r} - \mathbf{y}') - \sum_{\nu} \int d^3y\, \chi_{s\,\sigma\nu}(\mathbf{r}, \mathbf{y}; \omega) \\ \times \left(\frac{1}{|\mathbf{y} - \mathbf{y}'|} + f_{xc\,\nu\nu'}(\mathbf{y}, \mathbf{y}'; \omega)\right)\Bigg)\, n_{1\nu'}(\mathbf{y}', \omega) \\ = \sum_{\nu} \int d^3y\, \chi_{s\,\sigma\nu}(\mathbf{r}, \mathbf{y}; \omega)\, v_{1\nu}(\mathbf{y}, \omega).\end{aligned} \tag{337}$$

Since, in general, the true excitation energies Ω are not identical with the Kohn-Sham excitation energies $\omega_{jk\sigma}$, the right-hand side of Eq. (337) remains finite for $\omega \to \Omega$. In contrast, the *exact* spin-density response $n_{1\sigma}$, has poles at the true excitation energies $\omega = \Omega$. Hence the integral operator acting on $n_{1\sigma}$ on the left-hand side of Eq. (337) cannot be invertible for $\omega \to \Omega$. If it were invertible one could act with the inverse operator on both sides of Eq. (337) leading to a *finite* result for $\omega \to \Omega$ on the right-hand side in contradiction to the fact that $n_{1\sigma}$, on the left-hand side, has a pole at $\omega = \Omega$.

The true excitation energies Ω can therefore be characterized as those frequencies where the eigenvalues of the integral operator acting on the spin-density vector in Eq. (337) vanish or, if the integration over the delta-function is performed, where the eigenvalues $\lambda(\omega)$ of

$$\begin{aligned}\sum_{\nu'} \int d^3y' \sum_{\nu} \int d^3y\, \chi_{s\sigma\nu}(\mathbf{r}, \mathbf{y}; \omega) \left(\frac{1}{|\mathbf{y} - \mathbf{y}'|} + f_{xc\,\nu\nu'}(\mathbf{y}, \mathbf{y}'; \omega)\right) \zeta_{\nu'}(\mathbf{y}', \omega) \\ = \lambda(\omega)\, \zeta_{\sigma}(\mathbf{r}, \omega)\end{aligned} \tag{338}$$

satisfy

$$\lambda(\Omega) = 1\,. \tag{339}$$

This condition rigorously determines the true excitation spectrum of the interacting system considered.

To simplify the notation, we now introduce double indices $q \equiv (j, k)$ so that $\omega_{q\sigma} \equiv \varepsilon_{j\sigma} - \varepsilon_{k\sigma}$ denotes the excitation energy of the single-particle transition $(j\sigma \rightarrow k\sigma)$. Moreover, we define

$$\Phi_{q\sigma}(\mathbf{r}) := \phi_{k\sigma}(\mathbf{r})^* \, \phi_{j\sigma}(\mathbf{r}), \tag{340}$$

$$\alpha_{q\sigma} := f_{k\sigma} - f_{j\sigma} \tag{341}$$

and set

$$\xi_{q\sigma}(\omega) := \sum_{\nu'} \int d^3y' \sum_{\nu} \int d^3y \, \delta_{\sigma\nu} \Phi_{q\nu}(\mathbf{y})^* \times \left(\frac{1}{|\mathbf{y} - \mathbf{y}'|} + f_{\mathrm{xc}\,\nu\nu'}(\mathbf{y}, \mathbf{y}'; \omega) \right) \zeta_{\nu'}(\mathbf{y}', \omega) \,. \tag{342}$$

With these definitions, Eq. (338) takes the form

$$\sum_{q} \frac{\alpha_{q\sigma} \Phi_{q\sigma}(\mathbf{r})}{\omega - \omega_{q\sigma} + i\eta} \, \xi_{q\sigma}(\omega) = \lambda(\omega) \, \zeta_{\sigma}(\mathbf{r}, \omega) \,. \tag{343}$$

Solving this equation for $\zeta_\sigma(\mathbf{r}, \omega)$ and reinserting the result on the right-hand side of Eq. (342) leads to

$$\sum_{\sigma'} \sum_{q'} \frac{M_{q\sigma\, q'\sigma'}(\omega)}{\omega - \omega_{q'\sigma'} + i\eta} \, \xi_{q'\sigma'}(\omega) = \lambda(\omega) \, \xi_{q\sigma}(\omega) \tag{344}$$

with the matrix elements

$$M_{q\sigma\, q'\sigma'}(\omega) = \alpha_{q'\sigma'} \int d^3r \int d^3r' \, \Phi^*_{q\sigma}(\mathbf{r}) \times \left(\frac{1}{|\mathbf{r} - \mathbf{r}'|} + f_{\mathrm{xc}\sigma\sigma'}(\mathbf{r}, \mathbf{r}'; \omega) \right) \Phi_{q'\sigma'}(\mathbf{r}') \,. \tag{345}$$

Note that the summation in Eq. (344) extends over all single-particle transitions $q'\sigma'$ between occupied and unoccupied Kohn-Sham orbitals, including the continuum states. Up to this point, no approximations have been made. In order to actually calculate $\lambda(\omega)$, the eigenvalue problem (344) has to be truncated in one way or another. One possibility is to expand all quantities in Eq. (344) about one particular KS-orbital energy difference $\omega_{p\tau}$:

$$\xi_{q\sigma}(\omega) = \xi_{q\sigma}(\omega_{p\tau}) + \left. \frac{d\xi_{q\sigma}(\omega)}{d\omega} \right|_{\omega_{p\tau}} (\omega - \omega_{p\tau}) + \ldots \tag{346}$$

$$\lambda(\omega) = \frac{A(\omega_{p\tau})}{\omega - \omega_{p\tau}} + B(\omega_{p\tau}) + \ldots \tag{347}$$

The matrix elements with $(\omega_{p\tau} \neq \omega_{q'\sigma'})$ can be written as

$$\frac{M_{q\sigma\, q'\sigma'}(\omega)}{\omega - \omega_{q'\sigma'} + i\eta} = \frac{M_{q\sigma\, q'\sigma'}(\omega_{p\tau})}{\omega_{p\sigma} - \omega_{q'\sigma'} + i\eta} + \frac{d}{d\omega}\left[\frac{M_{q\sigma\, q'\sigma'}(\omega)}{\omega - \omega_{q'\sigma'} + i\eta}\right]_{\omega_{p\tau}} (\omega - \omega_{p\tau}) + \dots \tag{348}$$

whereas if $(\omega_{p\tau} = \omega_{q'\sigma'})$,

$$\frac{M_{q\sigma\, q'\sigma'}(\omega)}{\omega - \omega_{q'\sigma'} + i\eta} = \frac{M_{q\sigma\, q'\sigma'}(\omega_{p\tau})}{\omega - \omega_{p\tau} + i\eta} + i\eta + \left.\frac{dM_{q\sigma\, q'\sigma'}(\omega)}{d\omega}\right|_{\omega_{p\tau}} + \dots \tag{349}$$

Inserting Eqs. (346)–(349) in Eq. (344) the coefficients A and B are readily identified. If the pole $\omega_{p\tau}$ is non-degenerate, one finds:

$$A(\omega_{p\tau}) = M_{p\tau\, p\tau}(\omega_{p\tau}) \tag{350}$$

and

$$B(\omega_{p\tau}) = \left.\frac{dM_{p\tau p\tau}}{d\omega}\right|_{\omega_{p\tau}} + \frac{1}{M_{p\tau p\tau}(\omega_{p\tau})} \sum_{q'\sigma' \neq p\tau} \frac{M_{p\tau\, q'\sigma'}(\omega_{p\tau})\, M_{q'\sigma'\, p\tau}(\omega_{p\tau})}{\omega_{p\tau} - \omega_{q'\sigma'} + i\eta}. \tag{351}$$

The corresponding eigenvector (in lowest order) is given by

$$\xi_{q\sigma} = \frac{1}{A(\omega_{p\tau})} M_{q\sigma\, p\tau}(\omega_{p\tau})\, \xi_{p\tau} \tag{352}$$

with $(p\tau)$ fixed. The number $\xi_{p\tau}$ is free and can be chosen to properly normalize the vector ξ.

If the pole $\omega_{p\tau}$ is $\wp$-fold degenerate,

$$\omega_{p_1\tau_1} = \omega_{p_2\tau_2} = \dots = \omega_{p_\wp \tau_\wp} \equiv \omega_0\,, \tag{353}$$

the lowest-order coefficient A in Eq. (347) is determined by the following matrix equation

$$\sum_{k=1}^{\wp} M_{p_i\tau_i p_k\tau_k}(\omega_0)\, \xi^{(n)}_{p_k\tau_k} = A_n(\omega_0)\, \xi^{(n)}_{p_i\tau_i}\,, \quad i = 1 \dots \wp\,. \tag{354}$$

In general, one obtains $\wp$ different eigenvalues $A_1 \cdots A_\wp$. Then the remaining components of the corresponding eigenvectors $\xi^{(n)}$ can be calculated from

$$\xi^{(n)}_{q\sigma} = \frac{1}{A_n(\omega_0)} \sum_{k=1}^{\wp} M_{q\sigma\, p_k\tau_k}(\omega_0)\, \xi^{(n)}_{p_k\tau_k}\,, \tag{355}$$

once the eigenvalue problem (354) has been solved. Assuming that the true excitation energy Ω is not too far away from ω_0 it will be sufficient to consider only the lowest-order terms of the above Laurent expansions. In particular, we set

$$\lambda_n(\omega) \approx \frac{A_n(\omega_0)}{\omega - \omega_0} . \tag{356}$$

The condition (339) and its complex conjugate, $\lambda^*(\Omega) = 1$, then lead to

$$\Omega_n = \omega_0 + \Re A_n(\omega_0) \tag{357}$$

This is the central result of our analysis. Eq. (357) shows that a single KS pole can lead to several many-body excitation energies. The corresponding oscillator strengths can be obtained [152] from the eigenvectors $\xi^{(n)}$ and the KS oscillator strengths.

In the following, we exclusively consider closed-shell systems. For these systems, the Kohn-Sham orbital eigenvalues are degenerate with respect to the spin variable, which implies a lack of spin-multiplet structure. In what follows, we demonstrate how this is restored by the lowest-order corrections (357). Assuming that there are no further degeneracies besides the spin degeneracy, Eq. (354) reduces to the following (2×2) eigenvalue problem:

$$\sum_{\sigma' = \uparrow, \downarrow} M_{p\sigma p\sigma'}(\omega_0)\, \xi_{p\sigma'}(\omega_0) = A\, \xi_{p\sigma}(\omega_0) . \tag{358}$$

For spin-saturated systems, $M_{p\uparrow p\uparrow} = M_{p\downarrow p\downarrow}$ and $M_{p\uparrow p\downarrow} = M_{p\downarrow p\uparrow}$, so that the eigenvalues of Eq. (358) are given by

$$A_{1,2} = M_{p\uparrow p\uparrow} \pm M_{p\uparrow p\downarrow} . \tag{359}$$

By Eq. (357), the resulting excitation energies are:

$$\Omega_1 = \omega_0 + \Re\{M_{p\uparrow p\uparrow} + M_{p\uparrow p\downarrow}\} \tag{360}$$

$$\Omega_2 = \omega_0 + \Re\{M_{p\uparrow p\uparrow} - M_{p\uparrow p\downarrow}\} . \tag{361}$$

Inserting the explicit form of the matrix elements (345) one finds

$$\Omega_1 = \omega_0 + 2\Re \int d^3r \int d^3r'\, \Phi_p^*(\mathbf{r}) \left(\frac{1}{|\mathbf{r} - \mathbf{r}'|} + f_{xc}(\mathbf{r}, \mathbf{r}'; \omega_0) \right) \Phi_p(\mathbf{r}') \tag{362}$$

$$\Omega_2 = \omega_0 + 2\Re \int d^3r \int d^3r'\, \Phi_p^*(\mathbf{r})\, \mu_0^2\, G_{xc}(\mathbf{r}, \mathbf{r}'; \omega_0)\, \Phi_p(\mathbf{r}') \tag{363}$$

where, for simplicity, we have dropped the spin-index of $\Phi_{p\sigma}$.[4] Obviously, the xc

[4] This is possible only if the unperturbed KS ground-state determinant is spin-saturated since, in this case, $\phi_{j\uparrow}(\mathbf{r}) = \phi_{j\downarrow}(\mathbf{r})$ for all j.

kernel appearing in Eq. (362),

$$f_{xc}(\mathbf{r}, \mathbf{r}'; \omega) = \frac{1}{4} \sum_{\sigma, \sigma' = \pm 1} f_{xc\sigma\sigma'}(\mathbf{r}, \mathbf{r}'; \omega) \tag{364}$$

is identical with the one already defined in Sect. 5.1. On the other hand, Eq. (363) exhibits the kernel

$$G_{xc}(\mathbf{r}, \mathbf{r}'; \omega) = \frac{1}{4\mu_0^2} \sum_{\sigma, \sigma' = \pm 1} (\sigma \cdot \sigma') f_{xc\sigma\sigma'}(\mathbf{r}, \mathbf{r}'; \omega). \tag{365}$$

This quantity gives rise to exchange and correlation effects in the Kohn-Sham equation for the linear response of the frequency-dependent magnetization density $m(\mathbf{r}, \omega)$ [59]. The fact that the magnetization density response naturally involves spin-flip processes, suggests that Ω_2 represents the spin triplet excitation energies of many-electron systems with spin-saturated ground states. The corresponding spin singlet excitation energies, on the other hand, are given by Ω_1. This assignment will be given further evidence by the numerical results presented at the end of this section.

Apart from the truncation of the Laurent series, two further approximations are necessary:

(i) The frequency-dependent xc kernels f_{xc} and G_{xc} have to be approximated.
(ii) The static Kohn-Sham orbitals entering Eqs. (363) and (363) (cf. Eq. (340) have to be calculated with an *approximate* (static) potential v_{xc}^{stat}.

As an application of the method, we consider the lowest excitation energies of the alkaline earth elements and the zinc series. Here, in addition to the degeneracy with respect to the spin index, the $s \rightarrow p$ transitions under consideration are threefold degenerate in the magnetic quantum number m of the "final" state. Hence, we have six degenerate poles and Eq. (354) is a (6×6) eigenvalue problem. In our case, however, the matrix $M_{p_i \tau_i p_k \tau_k}$ in Eq. (354) consists of (three) identical (2×2) blocks, leading only to two distinct corrections, independent of m, as it should be.

Tables 1–3 show the results of calculations based on Eqs. (362) and (363). The calculation of Table 1 employs the ordinary local density approximation (LDA) for v_{xc}^{stat} and the adiabatic LDA (188) for f_{xc} (both using the parametrization of Vosko, Wilk and Nusair [90]). In this limit, the kernel G_{xc} is approximated by [103]

$$G_{xc}^{ALDA}[n](\mathbf{r}, \mathbf{r}'; \omega) = \delta(\mathbf{r} - \mathbf{r}') \frac{1}{\mu_0^2 n(\mathbf{r})} \alpha_{xc}(n(\mathbf{r})) \tag{366}$$

The xc contribution to the so called "spin-stiffness coefficient" α_{xc} is also approximated within the LDA of [90].

The calculation of Table 2 uses the x-only optimized effective potential (OPM) for v_{xc}^{stat} in the approximation of Krieger, Li and Iafrate (KLI) [224] and

Table 1. The lowest $S \rightarrow P$ excitation energies of various atoms. The experimental values (first column) [226] are compared with results calculated from Eq. (362) for the singlet and from Eq. (363) for the triplet (second column) and with ordinary Δ_{SCF} values (third column). The LDA was employed for v_{xc} and the ALDA for the xc kernels. The corresponding Kohn-Sham orbital-energy differences ω_0 are shown in the last column (All values in rydbergs)

Atom	State	Ω_{exp}	Ω^{LDA}	$\Omega(\Delta_{SCF})$	ω_0^{LDA}
Be	$^1P_1$	0.388	0.399	0.331	0.257
	$^3P_0$	0.200			
	$^3P_1$	0.200	0.192	0.181	0.257
	$^3P_2$	0.200			
Mg	$^1P_1$	0.319	0.351	0.299	0.249
	$^3P_0$	0.199			
	$^3P_1$	0.199	0.209	0.206	0.249
	$^3P_2$	0.200			
Ca	$^1P_1$	0.216	0.263	0.211	0.176
	$^3P_0$	0.138			
	$^3P_1$	0.139	0.145	0.144	0.176
	$^3P_2$	0.140			
Zn	$^1P_1$	0.426	0.477	0.403	0.352
	$^3P_0$	0.294			
	$^3P_1$	0.296	0.314	0.316	0.352
	$^3P_2$	0.300			
Sr	$^1P_1$	0.198	0.241	0.193	0.163
	$^3P_0$	0.130			
	$^3P_1$	0.132	0.136	0.135	0.163
	$^3P_2$	0.136			
Cd	$^1P_1$	0.398	0.427	0.346	0.303
	$^3P_0$	0.274			
	$^3P_1$	0.279	0.269	0.272	0.303
	$^3P_2$	0.290			

for f_{xc} the TDOPM kernel (325) derived in Sect. 6.2. Concerning the singlet spectrum, the OPM values are clearly superior to the LDA results and are also better than the usual Δ_{SCF} values. The unoccupied orbitals and their energy eigenvalues are very sensitive to the behavior of the potential far from the nucleus. Thus one major reason for the superiority of the optimized effective potential is the fact that it is self-interaction free and therefore has the correct $-1/r$ tail (while the LDA potential falls off exponentially). An important point to note is that the optimized effective potential decreases *correctly* for *all* orbitals. For this reason, the x-only optimized effective potential is also superior to the Hartree-Fock (HF) potential which is self-interaction free only for the occupied orbitals but not for the unoccupied ones. As a consequence, HF orbital-energy differences are typically too large. However, in spite of the fact that the OPM provides self-interaction free orbitals, it reproduces the triplet spectrum less accurately. This hinges on the approximation the xc *kernel* is based on. Substituting the TD-Fock expression (279) for the xc action functional

Table 2. The lowest $S \to P$ excitation energies of various atoms. The experimental values (first column) [226] are compared with results calculated from Eq. (362) for the singlet and from Eq. (363) for the triplet (second column) and with ordinary Δ_{SCF} values (third column). The optimized effective potential was used for v_{xc} and the approximate OPM kernel (325) for f_{xc} and G_{xc}. The corresponding Kohn-Sham orbital-energy differences ω_0 are shown in the last column (All values in rydbergs)

Atom	State	Ω_{exp}	Ω^{OPM}	$\Omega(\Delta_{SCF})$	ω_0^{OPM}
Be	$^1P_1$	0.388	0.392	0.331	0.259
	$^3P_0$	0.200			
	$^3P_1$	0.200	0.138	0.181	0.259
	$^3P_2$	0.200			
Mg	$^1P_1$	0.319	0.327	0.299	0.234
	$^3P_0$	0.199			
	$^3P_1$	0.199	0.151	0.206	0.234
	$^3P_2$	0.200			
Ca	$^1P_1$	0.216	0.234	0.211	0.157
	$^3P_0$	0.138			
	$^3P_1$	0.139	0.090	0.144	0.157
	$^3P_2$	0.140			
Zn	$^1P_1$	0.426	0.422	0.403	0.314
	$^3P_0$	0.294			
	$^3P_1$	0.296	0.250	0.316	0.314
	$^3P_2$	0.300			
Sr	$^1P_1$	0.198	0.210	0.193	0.141
	$^3P_0$	0.130			
	$^3P_1$	0.132	0.081	0.135	0.141
	$^3P_2$	0.136			
Cd	$^1P_1$	0.398	0.376	0.346	0.269
	$^3P_0$	0.274			
	$^3P_1$	0.279	0.211	0.272	0.269
	$^3P_2$	0.290			

defined in (278) leads to a xc kernel *diagonal* in spin space, because the correlation between antiparallel spins is neglected. Accordingly, from Eqs. (364) and (365) we have, within the x-only TDOPM

$$G_{xc}^{TDOPM}[n](\mathbf{r}, \mathbf{r}'; \omega) = \frac{1}{\mu_0^2} f_{xc}^{TDOPM}[n](\mathbf{r}, \mathbf{r}'; \omega). \tag{367}$$

This should be cured by adding appropriate correlation terms to the xc part of the action functional, which is further backed by the observation that when combining the advantage of approximating v_{xc}^{stat} by the optimized effective potential, together with a local density prescription of exchange *and* correlation in the xc kernels f_{xc} and G_{xc}, both singlet and the triplet spectrum are reproduced well by Eqs. (362) and (363), as can be seen from Table 3.

In spite of the fact that we focused our attention to the situation of closed shells and spin-multiplets, the method is also capable of dealing with open-shell systems and spatial multiplets. More details can be found in [152].

Table 3. The lowest $S \to P$ excitation energies of various atoms. The experimental values (first column) [226] are compared with results calculated from Eq. (362) for the singlet and from Eq. (363) for the triplet (second column) and with ordinary Δ_{SCF} values (third column). The optimized effective potential was used for v_{xc} and the ALDA for the xc kernels. The corresponding Kohn-Sham orbital-energy differences ω_0 are shown in the last column (All values in rydbergs)

Atom	State	Ω_{exp}	$\Omega^{OPM+ALDA}$	$\Omega(\Delta_{SCF})$	ω_0^{OPM}
Be	$^1P_1$	0.388	0.398	0.331	0.259
	$^3P_0$	0.200			
	$^3P_1$	0.200	0.196	0.181	0.259
	$^3P_2$	0.200			
Mg	$^1P_1$	0.319	0.329	0.299	0.234
	$^3P_0$	0.199			
	$^3P_1$	0.199	0.196	0.206	0.234
	$^3P_2$	0.200			
Ca	$^1P_1$	0.216	0.236	0.211	0.157
	$^3P_0$	0.138			
	$^3P_1$	0.139	0.129	0.144	0.157
	$^3P_2$	0.140			
Zn	$^1P_1$	0.426	0.417	0.403	0.314
	$^3P_0$	0.294			
	$^3P_1$	0.296	0.280	0.316	0.314
	$^3P_2$	0.300			
Sr	$^1P_1$	0.198	0.211	0.193	0.141
	$^3P_0$	0.130			
	$^3P_1$	0.132	0.117	0.135	0.141
	$^3P_2$	0.136			
Cd	$^1P_1$	0.398	0.370	0.346	0.269
	$^3P_0$	0.274			
	$^3P_1$	0.279	0.239	0.272	0.269
	$^3P_2$	0.290			

We emphasize that the calculation of excitation energies from Eqs. (362) and (363) involves only *known ground-state* quantities, i.e., the ordinary static Kohn-Sham orbitals and the corresponding Kohn-Sham eigenvalues. Thus the scheme described here requires only one selfconsistent Kohn-Sham calculation, whereas the so-called Δ_{SCF} procedure involves linear combinations of two or more selfconsistent total energies [209]. So far, the best results are obtained with the optimized effective potential for v_{xc}^{stat} in the KLI x-only approximation. Further improvement is expected from the inclusion of correlation terms [6, 225] in the OPM.

7.3 Van der Waals Interactions

While TDDFT has its main applications in time-dependent phenomena, and in the calculation of excitation or promotion energies, certain aspects of *ground-*

state energy calculations are also assisted by TDDFT. This development principally concerns the van der Waals (vdW) or dispersion-force component of the groundstate energy. The usual groundstate LDA and its various gradient extensions [227] do not give an adequate description of vdW forces [228], presumably because these forces arise (in one picture at least: see below) from the correlations between dynamic electron density fluctuations in widely separated positions. This makes the usual local or near-local approximations invalid. The approach to be introduced here facilitates the derivation of van der Waals functionals via a frequency integration over dynamic susceptibilities.

(i) *vdW interactions for widely-separated fragments*: Perhaps the most familiar example of a dispersion interaction is the attractive mutual energy of a pair of neutral spherical atoms separated by a large distance R, an interaction which forms the tail of the well-known Lennard-Jones potential. To lowest order this interaction energy falls off [229] as R^{-6}. This form of dispersion interaction is readily derived for a general pair of non-overlapping electronic systems by regarding the electrons on the first system as distinguishable from those on the second system. One then obtains the R^{-6} dispersion energy (in addition to some "polarization " terms relating to any static electric moments [230]) by performing second-order Rayleigh-Schrödinger perturbation theory, treating the Coulomb interaction between the two groups of electrons as the perturbation Hamiltonian . (For very large separations R the retardation of the electromagnetic interactions between the systems cannot be ignored. In this regime the R^{-6} law just quoted is replaced [229] by R^{-7}. This retarded form takes over whenever $R \gg c/\omega$, where ω is a characteristic response or fluctuation frequency of the electronic systems. We will consider only the non-retarded case here).

From the work of Casimir, Lifshitz, London and many others [229] we know that the perturbation expression for the dispersion interaction between separated systems can be related to the electric polarizabilities of the interacting species, and also to the correlation of fluctuating electric multipoles on the two systems. In the Present TDDFT context, a useful polarizability form for the second-order dispersion interaction was given by Zaremba and Kohn [231] who derived it directly from second-order perturbation theory:

$$E^{(2)} = -\frac{1}{2\pi}\int d^3r_1 \int d^3r_2 \int d^3r'_1 \int d^3r'_2 \,\frac{1}{|\mathbf{r}_1-\mathbf{r}_2|}\,\frac{1}{|\mathbf{r}'_1-\mathbf{r}'_2|}$$

$$\times \int_0^\infty du\, \chi_a(\mathbf{r}_1,\mathbf{r}'_1,iu)\,\chi_b(\mathbf{r}_2,\mathbf{r}'_2,iu). \tag{368}$$

Here $\chi_a(\mathbf{r},\mathbf{r}',\omega)$ and $\chi_b(\mathbf{r},\mathbf{r}',\omega)$ are the exact density-density response functions (157) of each separate system in the absence of the other. χ_a is defined by the linear density response $n_{1a}(\mathbf{r})\exp(ut)$ of the electrons in system a to an externally applied electron potential energy perturbation $V_1^{ext}(\mathbf{r})\,e^{ut}$:

$$n_{1a}(\mathbf{r}) = \int d^3r'\,\chi_a(\mathbf{r},\mathbf{r}',iu)\,V_1^{ext}(\mathbf{r}') \tag{369}$$

and similarly for χ_b. It is important to note that χ_a includes the electron-electron interaction amongst the electrons of system a to all orders, and similarly for χ_b. (Note also that, unlike Ref. [231], we have referred the space arguments of χ_a and χ_b in (368) to a common origin.)

The expression (368) is more general than the familiar asymptotic R^{-6} form. It applies to neutral quantal systems of any shape (not necessarily spherical) provided that R is still large enough that the electron densities do not overlap and that the inter-system Coulomb interaction can be treated in second order. We can recover the R^{-6} form by assuming that $R \gg A, B$ where A and B are the spatial dimensions of the two systems. Then one can expand the Coulomb interactions in (368) in multipoles. The lowest nonvanishing term gives, with the "3" axis chosen along $\mathbf{R}$,

$$E^{(2)}(R) \sim \frac{-1}{2\pi} \sum_{i,j=1}^{3} \frac{(1-3\delta_{3i})\,(1-3\delta_{3j})}{R^6} \int_0^\infty du\, \alpha_{ij}^{(a)}(iu)\, \alpha_{ij}^{(b)}(iu), \qquad R \gg A, B \tag{370}$$

where, for each system

$$\alpha_{ij}(\omega) = \int d^3 r \int d^3 r' \,(r_i - X_i)\,(r'_j - X_j)\, \chi(\mathbf{r}, \mathbf{r}', \omega) \tag{371}$$

is the dipole polarizability tensor and $\mathbf{X}$ is the centre of electronic charge of the system. When the polarizabilities are isotropic so that $\alpha_{ij} = \alpha\delta_{ij}$, (370) reduces to the more familiar London form [229]

$$E^{(2)}(R) \sim \frac{-3}{\pi R^6} \int_0^\infty du\, \alpha^{(a)}(iu)\, \alpha^{(b)}(iu). \tag{372}$$

Van Gisbergen, Snijders and Baerends [232] have evaluated a formula equivalent to (370) for diatomic and polyatomic molecules, using the ALDA to obtain the $\{\alpha_{ij}\}$. They find that, for the *isotropic* part of the vdW interaction, ALDA gives errors of similar size (but mostly opposite sign) to time-dependent Hartree Fock theory (except for the smallest atoms). This was achieved with much less computational effort than in the time-dependent Hartree Fock approach. The isotropic vdW coefficients, like the static and dynamic polarizabilities, were found to be somewhat too large. For the *anisotropic* part of the interaction, they found that ALDA compares favourably with both Hartree-Fock and Many-Body Perturbation Theory. Scince the ALDA contains unphysical orbital self-interaction, one can speculate that the use of self-interaction corrected (SIC) functionals might further improve the accuracy of the method employed in Ref. [232]. These unphysical self-interactions cause orbitals to "see" an incorrectly large charge from the other electrons, causing orbitals to be too spatially extended and hence too polarizable. This presumably has effects mainly at the isotropic level. Furthermore, SIC phenomena are known to be strongest for small systems with highly localized orbitals. Both of these considerations can be expected to cause difficulties in the very cases where van Gisbergen, Snijders and Baerends observed the least favourable ALDA results in comparison with other methods. This SIC explanation gains further support from the work of Pacheco and Ekardt [166] on alkali metal microclusters. Their static and dynamic SIC terms [233] were found to have significant effects on the polarizability and vdW interaction for small clusters and even, to a lesser degree, for quite large ones.

Van Gisbergen et al. [232] commented that their numerical method could accommodate more sophisticated forms of TDDFT than simply the ALDA, and in particular, considering temporal and spatial nonlocality in the xc kernel, they felt that the latter might be the more important.

Before leaving the discussion of vdW interactions in non-overlapping systems, we mention that the exact second-order dispersion formula (368) can be used [234] to derive a class of approximate vdW expressions for the groundstate energy as an explicit but highly nonlocal functional of the groundstate density. The idea is to make a direct local density approximation for the interacting susceptibilities χ_a and χ_b in (368). Extreme care is needed, however, to ensure one does not violate the charge conservation condition

$$\int d^3 r\, \chi(\mathbf{r}, \mathbf{r}', \omega) = 0 \tag{373}$$

or the reciprocity condition

$$\chi(\mathbf{r}, \mathbf{r}', iu) = \chi(\mathbf{r}', \mathbf{r}, -iu) \text{ for real } u. \tag{374}$$

An *Ansatz* satisfying these conditions and based on the simplest, pressure-free hydrodynamic analysis of the uniform electron gas was given in [234]:

$$\chi_{\text{local}}^{\text{inhom}}(\mathbf{r}, \mathbf{r}', \omega) = \nabla_{r} \cdot \nabla_{r'} \left[\frac{n(\mathbf{r})\ \delta(\mathbf{r} - \mathbf{r}')}{\omega^2 - \omega_P^2(n(\mathbf{r}))} \right]. \tag{375}$$

When this is substituted into (368) for each of χ_a and χ_b one obtains

$$E^{(2)} = -\frac{3}{32\pi^2} \int d^3 r_1 \int d^3 r_2 \frac{1}{r_{12}^6} \frac{\omega_1 \omega_2}{(\omega_1 + \omega_2)} \tag{376}$$

where $\omega_1 = \omega_{P1} = (4\pi n_a(\mathbf{r}_1)/m)^{1/2}$ is the local plasma frequency at an arbitrary point $\mathbf{r}_1$ inside system a, and similarly for ω_2. Equation (376) constitutes a very nonlocal groundstate density functional, and it clearly provides a systematic basis for the much-used [229] simple notion of pairwise addition of R^{-6} vdW contributions. It is interesting that the integrand in (376) is proportional to the *harmonic mean*, $\omega_1\omega_2/(\omega_1 + \omega_2)$, of the two local plasma frequencies. The same formula (376) was very recently postulated [235] by Andersson, Langreth and Lundqvist on different grounds. They obtained (376) by examining limiting cases and so modifying a somewhat similar formula previously postulated by Rapcewicz and Ashcroft [236] on the basis of diagrammatic arguments. The

Rapcewicz-Ashcroft formula differs from (376) only in the replacement of $\omega_1 + \omega_2$ by $2\sqrt{\omega_1\omega_2}$ on the denominator of (376). It was shown in [236] and [235] that these simple formulae give quite good answers for the isotropic R^{-6} dispersion coefficient for various atomic pairs, provided that one uses an appropriate cutoff in the low-density tails of the electron distributions.

The derivation of (376) given in [234] promises to be extendable to more sophisticated local approximations for χ_a and χ_b in (368), based perhaps on hydrodynamics with the inclusion of pressure (Thomas-Fermi hydrodynamics [112]) or of pressure plus density gradient (Thomas-Fermi-Weizsäcker hydrodynamics [237]). With suitable care to satisfy the constraints (373) and (374), one may thereby hope to obtain a more accurate extension of (376) involving gradients of the groundstate density and, possibly, having less dependence on spatial cutoffs.

(ii) *vdW interactions in closely juxtaposed or overlapping systems*: The work of van Gisbergen et al. [232] and Pacheco and Ekardt [166], discussed in the previous section, shows that TDDFT, at least in the form of the ALDA, can represent the state of the art in evaluating van der Waals interactions in *well-separated* systems that are too large for methods such as the Configuration Interaction approach. What of more general cases where the electron clouds *overlap* or where no large separation exists? To study this for large systems, we seek a density functional approach, but first we need to appreciate the origin of the vdW force in terms of correlation physics.

In essence, dispersion forces arise from the correlation between dynamic charge density fluctuations in two different systems or in distant parts of one system. The difficulty [228] in describing vdW forces in the static LDA or gradient approaches is therefore not surprising since in a highly inhomogeneous system (exemplified by, but not limited to, a pair of separated subsystems) these correlations may be quite different from those in the uniform or near-uniform electron gas upon which the LDA and the various gradient approximations are based.

The previous section applied only to well-separated subsystems. The necessary correlations between distant fluctuations were generated by the application of second-order perturbation theory, and the TDLDA aspect of the calculation was not called upon to produce the vdW correlations directly. For overlapping systems (and for some closely juxtaposed systems), low-order perturbation theory in the Coulomb potential is not appropriate. The present section will outline an approach, currently under development, which does generate such long-ranged correlations in a natural fashion by the solution of a highly nonlocal real-space screening integral equation. Nevertheless, local density approximations are made wherever possible for the independent-electron susceptibility χ_s and the exchange-correlation kernel f_{xc}, neither of which needs to be long-ranged in order to generate the basic long-ranged vdW correlations.

The starting point for the proposed new approach is an exact formula [238], [239], based on the adiabatic connection formula and the zero-temperature fluctuation-dissipation theorem, relating the groundstate xc energy to the interacting susceptibility χ:

$$E_{xc} = -\frac{1}{2}\int_0^1 d\lambda \int d^3r \int d^3r' \frac{1}{|\mathbf{r}-\mathbf{r}'|}\left[\left(\frac{1}{\pi}\int_0^\infty du\, \chi(\lambda, \mathbf{r}, \mathbf{r}', iu)\right) + n(\mathbf{r})\,\delta(\mathbf{r}-\mathbf{r}')\right]. \tag{377}$$

Here $\chi(\lambda, \mathbf{r}, \mathbf{r}', \omega)$ is the interacting susceptibility defined as before but with a reduced Coulomb interaction λ/r acting between electrons. It was shown in Ref. [234] that *the charge conservation condition* (373) *for* χ *implies xc hole normalization.* Use of the independent-electron Kohn-Sham susceptibility $\chi_s = \chi(\lambda = 0, \mathbf{r}, \mathbf{r}', iu)$ rather than $\chi(\lambda, \mathbf{r}, \mathbf{r}', iu)$ in (377) yields the exact exchange energy. Subtraction of this exchange energy expression from the above xc energy yields the correlation energy

$$E_c = -\frac{1}{2\pi}\int_0^1 d\lambda \int d^3r \int d^3r' \frac{1}{|\mathbf{r}-\mathbf{r}'|}\int_0^\infty du\,(\chi(\lambda, \mathbf{r}, \mathbf{r}', iu) - \chi_s(\mathbf{r}, \mathbf{r}', iu))\,. \tag{378}$$

Equation (378) is required to produce the dispersion interactions under study. Petersilka, Gossmann and Gross [69] have shown that χ and χ_s are related exactly by a Dyson-type equation involving the dynamic nonlocal xc kernel f_{xc} as well as the Coulomb kernel (cf. Equation (152)):

$$\chi(\mathbf{r}, \mathbf{r}', \omega) = \chi_s(\mathbf{r}, \mathbf{r}', \omega) + \int d^3x \int d^3x'\, \chi_s(\mathbf{r}, \mathbf{x}, \omega)\left(\frac{1}{|\mathbf{x}-\mathbf{x}'|} + f_{xc}(\mathbf{x}, \mathbf{x}', \omega)\right)\chi(\mathbf{x}', \mathbf{r}', \omega). \tag{379}$$

Equations (377) and (379) are of course exact provided that f_{xc} is exact, and so they contain *inter alia* the exact vdW interaction. Consider first a homogeneous electron gas. If f_{xc} is arbitrarily set to zero, and (379) is Fourier-transformed with use of the convolution theorem, (379) is then seen to be the equation for the RPA response function χ in terms of the bare (dynamic Kohn-Sham-Lindhard) response χ_s. Again with the assumption $f_{xc} = 0$, but with the homogeneous assumption removed, (377) and (379) merely represent the inhomogeneous generalization of the well-known RPA groundstate correlation energy of the homogeneous electron gas. This case of zero f_{xc} already has some rather useful properties with respect to the vdW interaction. It has been shown in detail [239] that, when the correlation energy recipe (377), (379) with $f_{xc} = 0$ is applied to an arbitrary pair of widely-separated systems, the Zaremba-Kohn second-order vdW energy expression (368) is reproduced, with the following exception: the susceptibilities χ_a and χ_b are the approximate RPA-interacting susceptibilities of each system, rather than including the exact interactions within each subsystem. Thus the full inhomogeneous RPA correlation energy already contains the essence of the vdW interaction, and will produce an R^{-6} dependence in the appropriate limit. An examination of the detailed proof in [239] further shows that the long-ranged vdW interaction achieved in the RPA does NOT arise because of any long-ranged behaviour of the independent-electron susceptibility χ_s (indeed χ_s is not normally long-ranged). Rather, the long range of the vdW interaction comes from the long range of the Coulomb interaction in the screening Eq. (379). Thus a local density approximation for χ_s will not spoil the vdW properties, but may slightly alter the interacting susceptibilities χ_a and χ_b in the asymptotic form (368). Furthermore the reintroduction of f_{xc} within a local approximation can also be seen, from the working of Ref. [239], to maintain the form (368) in the separated limit, but the individual susceptibilities χ_a and χ_b will now involve f_{xc} and hence will be closer to the required interacting susceptibilities.

To summarize the previous paragraph: If we make short-ranged local-density or gradient approximations for χ_s and f_{xc} in the exact groundstate energy scheme (377), (379), we obtain an approximate and highly nonlocal prescription for the groundstate correlation energy, with the groundstate density $n(\mathbf{r})$ as the only input. This scheme is expected to produce a rather good approximation to the long-ranged vdW dispersion interaction between widely separated subsystems, a result due principally to the retention, in full, of the nonlocal coulomb kernel in the real-space screening integral equation (379).

What is now required is a sufficient set of constraints on the kernel f_{xc} so that the *short-ranged* aspects of the groundstate correlation energy are also reproduced by (377), (379) at a level of approximation comparable, say, to the groundstate LDA or GGA. If this can be achieved, we will have a "seamless" scheme, equally reasonable for chemically bonded systems, metals etc., and also for fully or partly subdivided systems at all separations. This should allow investigation of the intermediate region of interaction where both short-ranged and long-ranged correlations are significant, even for systems too large for traditionally accurate methods such as CI or Møller-Plesset perturbation theory. (Recall that wavefunctions are not needed for the present scheme, only groundstate densities, so that one may perform "real-space quantum chemistry" without basis-set problems).

The details of this scheme are currently being worked out. Although it aims for a groundstate energy functional, it depends heavily on time-dependent density functional theory in the sense that the properties of the *dynamic* TDDFT xc kernel $f_{xc}(\mathbf{r}, \mathbf{r}', iu)$ for inhomogeneous systems are of the essence. Further details of some constraints to be obeyed by f_{xc} are discussed in Ref. [239]. Some supporting evidence for the utility of a local approximation for χ_s in a highly inhomogeneous system are given in [113]. Current indications are that, for jellium slab situations where the inhomogeneity is only one-dimensional, the complete scheme (377), (379) (even with the exact Kohn-Sham χ_s but with a local approximation for f_{xc}) can be computed on a single-processor 1- MFlop workstation in 10^2 hours or less. With Monte Carlo methods for the integrations in (377), and/or faster or parallel machines, more involved geometries should be tractable.

8 Applications Beyond the Perturbative Regime: Atoms in Strong Femto-Second Laser Pulses

Owing to rapid experimental progress in the field of laser physics, ultra-short laser pulses of very high intensity have become available in recent years. The electric field produced in such pulses can reach or even exceed the strength of the static nuclear Coulomb field. If an atomic system is placed in the focus of such a laser pulse one observes a wealth of new phenomena [240] which cannot be explained by perturbation theory. In this case a non-perturbative treatment, i.e., the solution of the full TDKS equations (39)–(41) is mandatory. The total *external* potential seen by the electrons is given by

$$v(\mathbf{r}, t) = -\frac{Z}{r} + E_0 f(t)\, z \sin(\omega_0 t) \tag{380}$$

where Z is the nuclear charge. The second term on the right-hand side of Eq. (380) is the potential due to the laser field in dipole approximation, written in the length form. Since the wavelength of currently used lasers is almost always very large compared to any characteristic length associated with an atomic system, the dipole approximation turns out to be very good in practice [241]. E_0 denotes the peak electric field strength and $f(t)$ characterizes the envelope function of the pulse which, in the calculations described below, is linearly ramped to its peak value over the first 10 cycles and then held constant. The field is assumed to be polarized along the z-direction.

In the following, we compare the results of a TDKLI calculation using the approximate potential (313) with an ALDA calculation using the potential (186), both for the exchange-only case [242, 243]. The numerical procedure [244] to solve the TDKS equations is similar to the one developed by Kulander [245, 246], who solved the time-dependent Schrödinger equation for hydrogen and the time-dependent Hartree equation for helium in a laser pulse. The spin orbitals are expressed in cylindrical coordinates and, due to the linear polarization of the field, the spin as well as the angular part of the orbitals are preserved. Consequently, a fully three-dimensional treatment only requires a two-dimensional grid for the numerical integration. In the following, the time-dependent orbitals will always be characterized by the indices indicating the *initial* state of the respective orbital; e.g., $\varphi_{2s}(\mathbf{r}, t)$ describes an electron which initially was in a 2*s* spin orbital: $\varphi_{2s}(\mathbf{r}, t = 0) = \phi_{2s}(\mathbf{r})$. The integration of the single-particle equations is performed using a finite-difference representation of the kinetic energy operator. A Crank-Nicholson technique is employed for the (unitary) time propagation of the orbitals.

Once a numerical solution of the TDKS equations has been obtained, the resulting time-dependent density is sufficient to calculate any desired observable of the system. Some quantities are easily calculated while others (such as ATI spectra) are harder to extract from the density. But, as demonstrated in Sect. 2, *all* physical observables can be calculated from the density, in principle. In the following we shall describe the calculation of two different quantities, namely the *harmonic spectrum* and the *ionization yields*.

To obtain the harmonic spectrum, we calculate the induced dipole moment

$$d(t) = \int d^3 r\, z\, n(\mathbf{r}, t) \tag{381}$$

which is then Fourier transformed over the last 5 cycles of the constant-intensity interval. The square of the resulting Fourier transform, $|d(\omega)|^2$, has been shown [247] to be proportional to the experimentally observed harmonic distribution to within a very good approximation. Figure 4 shows the result of a simulation for the helium atom at a laser wavelength of $\lambda = 616$ nm and peak intensity of $I = 3.5 \times 10^{14}$ W/cm^2. The calculation was made with the TDKLI scheme which, for two electrons in the x-only limit, reduces to the ordinary time-dependent Hartree method. One observes peaks in the energy-resolved photon spectrum at odd multiples of the external laser frequency. From perturbation theory one would expect an exponential decrease of the peak intensities. Figure 4, however, shows a plateau of peak intensities up until roughly the 47th harmonic. This plateau is a typical nonlinear phenomenon. The squares in Fig. 4 indicate experimental results [248] obtained with the same laser frequency at an intensity of 1.4×10^{14} W/cm^2. Various calculations were performed with different peak intensities, but the best agreement with the experiment was achieved in the calculation for $I = 3.5 \times 10^{14}$ W/cm^2 shown in Fig. 4. The discrepancy between this intensity and the experimental intensity of

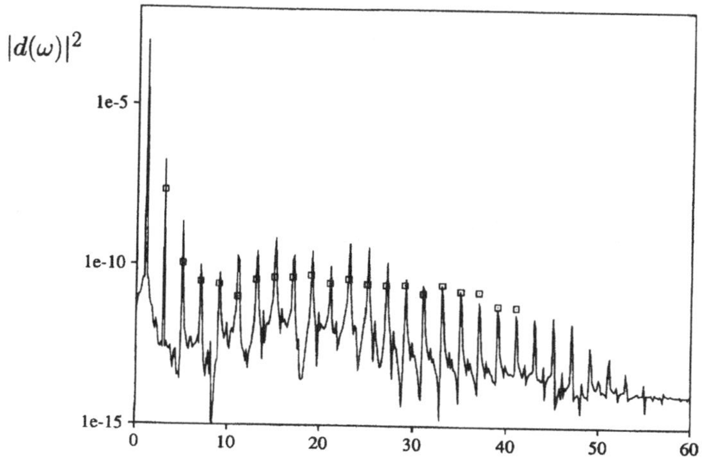

Fig. 4. Harmonic spectrum for He at $\lambda = 616$ nm, $I = 3.5 \times 10^{14}$ W/cm^2. The *squares* represent experimental data taken from Ref. [248] normalized to the value of the 33rd harmonic of the calculated spectrum. The experiment was performed with a peak intensity of 1.4×10^{14} W/cm^2.

1.4×10^{14} W/cm^2 might be due to the uncertainty of the experimentally determined peak intensity which can be as high as a factor of two.

The harmonic generation of helium in a strong two-color laser field has also been studied [243, 108]. The two lasers with frequencies ω_0 and $2\omega_0$, respectively, are operated with the same peak intensity and a constant relative phase difference φ. This results in a total external potential of the form

$$v(\mathbf{r}, t) = -\frac{Z}{r} + E_0 f(t)\, z[\sin(\omega_0 t) + \sin(2\omega_0 t + \varphi)] \tag{382}$$

where both fields are linearly polarized along the z-axis.

Calculated harmonic distributions induced by a two-colour field with different relative phases are shown in Fig. 5. To avoid overcrowding, only the calculated peak intensities are plotted and connected with straight lines. The fundamental wave length is again 616 nm and the intensity is 3.5×10^{14} W/cm^2 for both frequency components. We also show the one-colour spectrum for $\lambda = 616$ nm calculated with the *same* total intensity as the two-colour field, i.e. $I = 7 \times 10^{14}$ W/cm^2. In the two-colour spectrum, harmonics at all higher multiples (including even multiples) of the fundamental frequency ω_0 occur due to nonlinear mixing processes of the two fields [249]. Most of the harmonics produced by the two-colour field in the plateau region are one to two orders of magnitude more intense than those obtained in the one-colour calculation although the total intensity of the external laser field is the same in all cases.

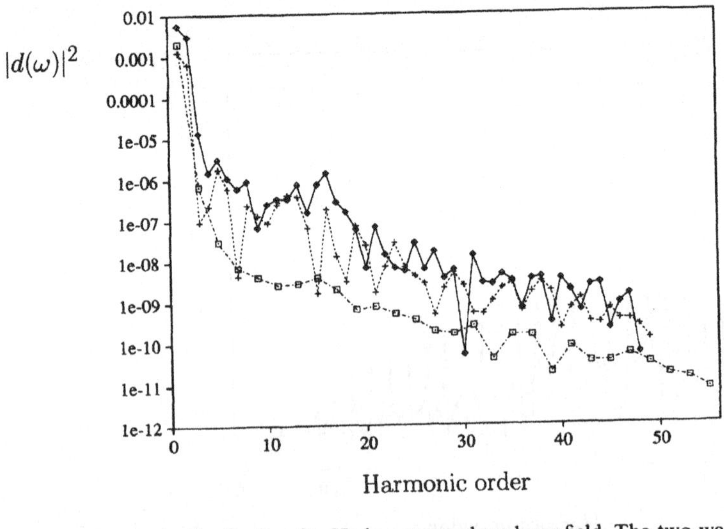

Fig. 5. Harmonic distribution for He in a two-colour laser field. The two wavelengths are 616 nm and 308 nm, and the intensity is 3.5×10^{14} W/cm^2 for both of them. *Crosses* are the results for $\varphi = 0$ and *diamonds* denote the values obtained with phase shift $\varphi = 0.7\pi$. For comparison, the *squares* indicate the harmonic distribution for He in a one-colour field with $\lambda = 616$ nm and $I = 7 \times 10^{14}$ W/cm^2.

Similar results have recently been found for hydrogen in a two-colour field [250]. One possible reason for this remarkable enhancement is that in a two-colour field one specific high-order harmonic can be generated by a large number of different mixing processes [249].

In order to simulate ionization, the grid contains an absorbing boundary to remove the flux of electrons leaving the nucleus. When some portion of the wave function propagates to the outer edges of the grid it is absorbed. We assume this flux corresponds to the ionized part of the wave function. Strictly speaking, such a criterion is meaningful only after long times when the respective contributions have propagated very far away from the nucleus. For the wave lengths considered here, a cylindrical grid of 20 × 60 a.u. was found to be sufficient. As time proceeds, more and more electrons will be removed from the atom and, accordingly, the norm of the TDKS orbitals taken over the finite volume of the grid,

$$N_{j\sigma}(t) = \int_{\substack{\text{finite} \\ \text{volume}}} d^3r \, |\varphi_{j\sigma}(\mathbf{r}, t)|^2 , \tag{383}$$

decreases with time.

Figure 6 compares the results of a TDKLI and an ALDA calculation [243] for Ne exposed to a laser field with wavelength $\lambda = 248$ nm and intensity $I = 3 \times 10^{15}\text{W/cm}^2$. Figure 6 shows the norm (383) of those orbitals which were initially in the Ne 2*s*, $2p_0$ and $2p_1$ states. The 1s electrons have been frozen, i.e., only the 2s and 2p electrons are propagated by solving the TDKS equations, whereas the time evolution of the 1s electrons is given by

$$\varphi_{1s}(\mathbf{r}, t) = \phi_{1s}(\mathbf{r}) \, e^{-i\varepsilon_{1s}(t - t_0)} \quad . \tag{384}$$

As expected, among the Ne 2*s*, $2p_0$ and $2p_1$ orbitals, the 2*s* orbital is the least ionized one because it is initially more strongly bound (by roughly a factor of 2) than the 2*p* orbitals. A little surprising at first sight, the $2p_0$ and $2p_1$ orbitals differ by about an order of magnitude in their degree of ionization (60% for the $2p_0$ orbital compared to only 4.75% for the $2p_1$ orbital within TDKLI, and 56% for the $2p_0$ compared to 7.7% for the $2p_1$ orbital within the ALDA). This difference was observed before by Kulander [251, 252] for the case of xenon (in a single-active-electron calculation). It is due to the fact that the $2p_0$ orbital is oriented along the polarization direction of the laser field, which makes it easier for the electrons to escape the nuclear attraction than for the case of the $2p_1$ orbital, which is oriented perpendicularly to the field polarization.

To explain the difference between the results obtained within the TDKLI and ALDA schemes shown in Fig. 6, we observe that the initial 2*s* and $2p_0$, $2p_1$ orbital energies in LDA differ quite considerably from those obtained with the KLI method: It takes 5 photons to ionize the 2*p* orbitals in KLI compared to only 3 photons in LDA. Similarly, it takes 11 photons to ionize the 2*s* orbital in KLI and only 9 in LDA. The difference between the curves in Fig. 6A and C is thus hardly surprising. On the other hand, it seems quite unexpected that the

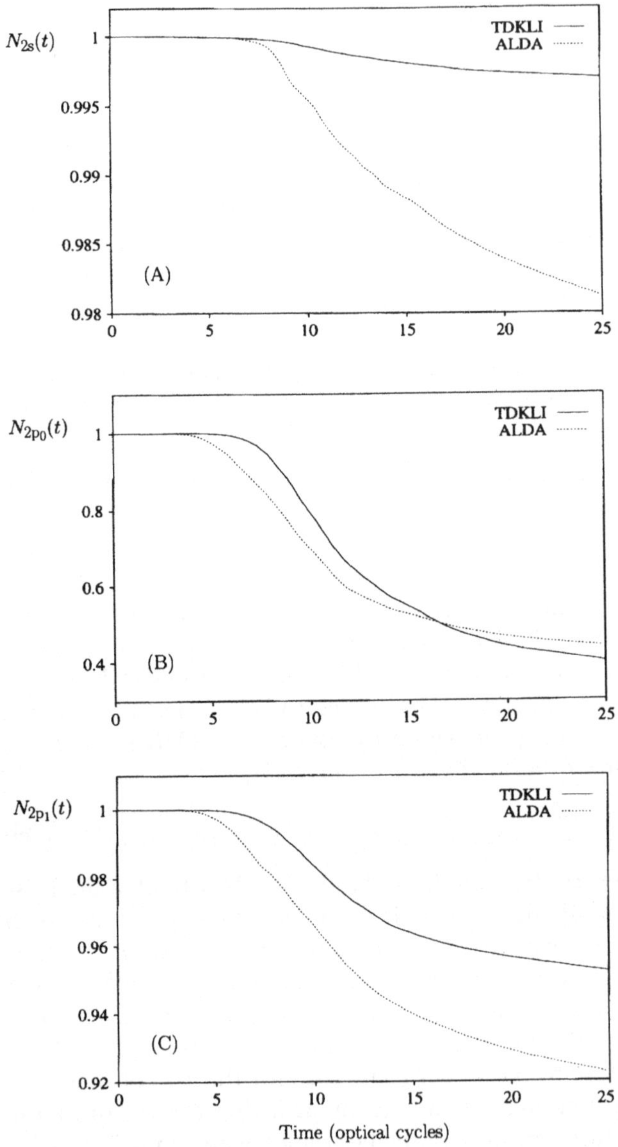

Fig. 6. Time evolution of the norm of the Ne 2*s* orbital **(A)**, the Ne $2p_0$ orbital **(B)** and the *Ne* $2p_1$ orbital **(C)**, calculated in the x-only TDKLI and ALDA schemes. Laser parameters: $\lambda = 248$ nm, $I = 3 \times 10^{15}\mathrm{W/cm^2}$, linear ramp over the first 10 cycles. One optical cycle corresponds to 0.82 femtoseconds

ALDA and TDKLI curves cross in Fig. 6B so that the ALDA curve comes to lie *above* the TDKLI curve. This behavior can be attributed to the fact that the other orbitals are ionized much more strongly in ALDA than in TDKLI, so that their electron density near the nucleus (and therefore their screening of the nuclear charge) is decreased. This makes it more difficult for the $2p_0$ electrons to escape within the ALDA scheme.

Figure 6 clearly shows the superiority of the TDKLI approach over the ALDA. The spurious self-interaction present in the ALDA causes the orbitals to be too weakly bound and hence the ALDA is not reliable in the calculation of ionization.

The probabilities of finding neutral, singly, doubly, etc. ionized atoms at time t are readily expressed in terms of the norms (383). For instance, in the case of helium, one has

$$N_{1s\uparrow}(t) = N_{1s\downarrow}(t) = \int_{\substack{\text{finite}\\ \text{volume}}} d^3r \, \frac{n(\mathbf{r}, t)}{2} \equiv N_{1s}(t) , \tag{385}$$

and the probabilities for neutral, singly and doubly charged helium are

$$P^0(t) = N_{1s}(t)^2 \tag{386}$$

$$P^{+1}(t) = 2N_{1s}(t)\,(1 - N_{1s}(t)) \tag{387}$$

$$P^{+2}(t) = (1 - N_{1s}(t))^2 \quad . \tag{388}$$

For many-electron atoms similar combinatorical considerations [244] are performed to determine the probabilities for the various charged ions. Figure 7 shows the probabilities of finding neutral, singly, doubly and triply charged Ne as calculated from the norms of Fig. 6.

These probabilities as a function of time cannot be compared directly with experiment. This is because the laser focus, in addition to the temporal pulse

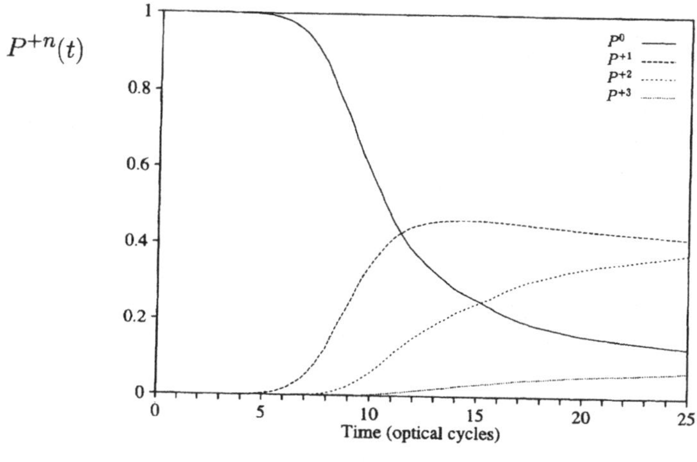

Fig. 7. Population of the differently charged states of Ne. Laser parameters as in Fig. 6 ($\lambda = 248$ nm, $I = 3 \times 10^{15}$ W/cm^2, linear ramp over the first 10 cycles)

shape, has a *spatial* intensity profile due to which not all atoms in the laser focus experience the same intensity. Hence a realistic calculation of ion yields requires many runs at various peak intensities. Work along these lines remains an important field for the future. In this way one might be able to understand the structures in the strong field ionization spectra of He [253] which have been the subject of heated discussions in recent years.

Acknowledgments. We thank the Deutsche Forschungsgemeinschaft and the Australian Research Council for separate financial support to EKUG and JFD respectively. Both JFD and EKUG gratefully acknowledge a DEET/ARC International Fellowship which supported a visit by EKUG to Griffith University where part of this work was done. We thank S. Erhard and C.A. Ullrich for many helpful discussions and S. Erhard for a critical reading of the manuscript.

9 References

1. Hohenberg P, Kohn W (1964) Phys Rev 136: B864
2. Kohn W, Sham LJ (1965) Phys Rev 140: A1133
3. Parr RG, Yang W (1989) Density-functional theory of atoms and molecules. Oxford University Press, New York
4. Dreizler RM, Gross EKU (1990) Density functional theory, An approach to the quantum many-body problem. Springer, Berlin Heidelberg, New York
5. Density functional theory (1995) Vol. 337 of NATO ASI Series B, Gross EKU, Dreizler RM (eds) Plenum Press, New York
6. Tobias Grabo, Gross EKU (1995) Chem Phys Lett 240: 141; 241: 635(E)
7. Runge E, Gross EKU (1984) Phys Rev Lett 52: 997
8. Bloch F (1993) Z Physik 81: 363
9. Peuckert V (1978) J Phys C 11: 4945
10. Zangwill A, Soven P (1980) Phys Rev A 21: 1561
11. Deb BM, Ghosh SK (1982) J Chem Phys 77: 342
12. Ghosh SK, Deb BM (1982) Chem Phys 71: 295
13. Ghosh SK, Deb BM (1983) Theoret Chim Acta 62: 209
14. Ghosh SK, Deb BM (19883) J Mol Struct 103: 163
15. Bartolotti LJ (1981) Phys Rev A 24: 1661
16. Bartolotti LJ (1982) Phys Rev A 26: 2243
17. Bartolotti LJ (1984) J Chem Phys 80: 5687
18. Bartolotti LJ (1987) Phys Rev A 36: 4492
19. Gross EKU, Kohn W (1990) Adv Quant Chem 21: 255
20. Lieb EH (1985) in: Dreizler RM, da Providencia J (eds) Density functional methods in Physics. Plenum, New York
21. Levy M (1982) Phys Rev A 26: 1200
22. Levy M (1979) Proc Natl Acad Sci USA 76: 6062
23. Kohl H, Dreizler RM (1986) Phys Rev Lett 56: 1993
24. Ghosh SK, Dhara AK (1989) Phys Rev A 38: 1149.
25. Tiszauer D, Kulander KC (1984) Phys Rev A 29: 2909
26. Errea LF, Méndez L, Riera A, Yáñez M, Hanssen J, Harel C, Salin A (1985) J. Physique 46: 719
27. Cooper IL, Dickinson AS, Sur SK, Ta CT (1987) J Phys B 20: 2005
28. Henne A, Lüdde H-J, Toepfer A, Dreizler RM (1987) Phys Lett A 124: 508
29. Fritsch W, Lin CD (1987) Phys Lett A 123: 128
30. Reading JF, Ford AL (1987) Phys Rev Lett 58: 543
31. Reading JF, Ford AL (1987) J Phys B 20: 3747
32. Tang X, Rudolph H, Lambropoulos P (1991) Phys Rev A 44: R6994
33. Harriman JE (1981) Phys Rev A 24: 680
34. Zumbach G, Maschke K (1983) Phys Rev A 28: 544; (1984) 29: 1585(E)
35. Li TC, Tong PQ (1986) Phys Rev A 34: 529

36. Ullrich CA, Gross EKU, Comments At. Mol. Phys. in press
37. Perdew JP, Zunger A (1981) Phys Rev B 23: 5048
38. Theilhaber J (1992) Phys Rev B 46: 12990
39. Delos JB, Thorson WR, Knudson SK (1972) Phys Rev A 6: 709
40. Rajagopal AK (1994) Phys Lett A 193: 363
41. Rajagopal AK, Buot FA (1994) Phys Rev E 50: 721
42. Rajagopal AK, Buot FA (1995) Phys Rev A 51: 1883
43. Saalmann U, Schmidt R, to be published
44. Car R, Parrinello M (1985) Phys Rev Lett 55: 2471
45. Remler DK, Madden PA (1990) Mol Phys 70: 921
46. Pastore G, Smargiassi E, Buda F (1991) Phys Rev A 44: 6334
47. Bréchignac C, Cahuzac P, Carlier F, de Frutos M, Barnett RN, Landman U (1994) Phys Rev Lett 72: 1636
48. Barnett RN, Landman U, Nitzan A, Rajagopal G (1991) J Chem Phys 94: 608
49. Barnett RN, Landman U, Rajagopal G (1991) Phys Rev Lett 22: 3058
50. Payne MC, Teter MP, Allan DC, Arias TA, Joannopoulos JD (1992) Rev Mod Phys **64**: 1045
51. Schmidt R, Seifert G, Lutz HO (1991) Phys Lett A 158: 231
52. Seifert G, Schmidt R, Lutz HO (1991) Phys Lett A 158: 237
53. Schmidt R, Schulte J, Knospe O, Seifert G (1994) Phys Lett A 194: 101
54. Schulte J, Knospe O, Seifert G, Schmidt R (1995) Phys Lett A 198: 51
55. Rosker MJ, Rose TS, Zewail AH (1988) Chem Phys Lett 146: 175
56. Engel V, Metiu H (1989) J Chem Phys 90: 6116
57. Engel V, Baumert T, Meier Ch, and Gerber G (1993) Z Phys D 28: 37
58. Femtosecond chemistry, (1995) Manz J, Wöste L (eds) VCH, Weinheim
59. Liu KL, Vosko SH (1989) Can J Phys 67: 1015
60. Wacker OJ, Kümmel R, Gross EKU (1984) Phys Rev Lett 73: 2915
61. Vignale J, Rasolt M (1987) Phys Rev Lett 59: 2360
62. Vignale G. Rasolt M. (1988) Phys Rev B 37: 10685
63 Erhard S, Gross EKU (1996) Phys Rev A 53: R5
64. Rajagopal AK, Buot FA (1995) Phys Rev B 52: 6769
65. Li T, Tong P (1985) Phys Rev A 31: 1950
66. Li T, Li Y (1985) Phys Rev A 31: 3970
67. Rajagopal AK, Buot FA (1994) Phys Lett A 195: 312
68. Rajagopal AK (1994) Phys Rev A 50: 3759
69. Petersilka M, Gossmann UJ, Gross EKU (1996) Phys Rev Lett 76: 1212
70. Kugler AA (1975) J Stat Phys 12: 35
71. Gross EKU, Kohn W (1985) Phys Rev Lett 55: 2850; (1986) 57: 923(E)
72. Ando T (1977) Z Phys B 26: 263
73. Ng TK, Singwi KS (1987) Phys Rev Lett 59: 2627
74. Yang W (1988) Phys Rev A 38: 5512
75. Ng TK (1989) Phys Rev Lett 62: 2417
76. Ichimaru S (1982) Rev Mod Phys 54: 1017
77. Gorobchenko VD, EG Maksimov (1980) Sov Phys Usp 23: 35
78. Singwi KS, Tosi MP (1981) Solid State Physics 36: 177
79. Ashcroft NW (1995) in: Gross EKU, Dreizler RM (eds) Density functional theory. Vol. 337 of NATO ASI Series B, Plenum Press, New York
80. Mearns D, Kohn W (1987) Phys Rev A 35: 4796
81. Gross EKU, Mearns D, Oliveira LN (1988) Phys Rev Lett 61: 1518
82. Ng TK, Singwi KS (1989) Phys Rev Lett 61: 1519
83. Wloka J (1971) Funktionalanalysis und Anwendungen. de Gruyter, Berlin
84. Petersilka M, (1993) Diplomarbeit, Universität Würzburg
85. Wehrum RP, Hermeking H (1974) J Phys C 7: L107
86. Senatore G. Subbaswamy KR, (1987) Phys Rev A 35: 2440
87. Gonze X, Vigneron JP (1989) Phys Rev B 39: 13120
88. Iwamoto N, Gross EKU (1987) Phys Rev B 35: 3003
89. Ceperley DM (1978) Phys Rev B 18: 3126
90. Vosko SH, Wilk L, Nusair M (1980) Can J Phys 58: 1200
91. Holas A (1987) in Rogers FJ, DeWitt HE (eds) Strongly coupled plasma physics. Plenum, New York
92. Moroni S, Ceperley DM, Senatore G (1995) Phys Rev Lett 75: 689

93. Niklasson G (1974) Phys Rev B 10: 3052
94. Sturm K (1995) Phys Rev B 52: 8028
95. Holas A. Singwi KS (1989) Phys Rev B 40: 158
96. Glick AJ, Long WF (1971) Phys Rev B 4: 3455
97. Dubois DF, Kivelson MG (1969) Phys Rev 186: 409
98. Hasegawa M, Watabe M, (1969) J Phys Soc Jap 27: 1393
99. Iwamoto N (1984) Phys Rev A 30: 2597
100. Iwamoto N (1984) Phys Rev A 30: 3289
101. Byrd PF, Friedman MD (1954) Handbook of Elliptic Integrals for Engineers and Physicists. Springer, Berlin
102. Dabrowski B (1986) Phys Rev B 34: 4989
103. Liu KL (1991) Can J Phys 69: 573
104. Jones RO, Gunnarsson O (1989) Rev Mod Phys 61: 689
105. Görling A. Levy M (1992) Phys Rev A 45: 1509
106. Kadanoff LP, Martin PC (1963) Ann Phys New York 24: 419
107. Vignale G (1995) Phys Rev Lett 74: 3233
108. Erhard S (1996) Diplomarbeit, Universität Würzburg
109. Erhard S, Gross EKU, to be published.
110. Levy M, Perdew JP (1985) Phys Rev A 32: 2010
111. Vignale G (1996) Phys Rev B in press.
112. Dobson JF (1994) Phys Rev Lett 73: 2244
113. Dobson JF in Density Functional Theory, Vol. 337 of NATO ASI Series B, edited by E.K.U. Gross and R.M. Dreizler (Plenum Press, New York, 1995), p. 393
114. Brey L, Johnson NF, Halperin BI (1989) Phys Rev B 40: 10647
115. Yip SK (1991) Phys Rev B 43: 1707
116. Dobson JF, Le H, unpublished
117. Dobson JF, to be published
118. Schaich WL DobsonJF (1994) Phys Rev B 49: 14700
119. Dobson JF (1993) Australian J. Phys. 46: 391
120. O'Connor AJ (1995) private communication
121. Bünner M, Diplomarbeit (1992) Universität Würzburg
122. Bünner M, Dobson JF, Gross EKU, to be published
123. Galdrikian B, Sherwin M, Birnir B (1994) Phys Rev B 49: 13744
124. Perdew J (1979) Chem Phys Lett 64: 127
125. Dobson JF, Rose JH (1982) J Phys C 15: L1183
126. Dobson JF (1991) J Chem Phys 94: 4328
127. Dobson JF (1992) J Phys Cond Matter 4: 7877
128. Dobson JF (1993) J Chem Phys 98: 8870
129. Heaton RA, Harrison JG, Lin CC (1983) Phys Rev B 28: 5992
130. Svane A, Gunnarsson O (1990) Phys Rev Lett 65: 1148
131. Szotek Z, Temmerman WM, Winter H (1993) Phys Rev B 47: 4092
132. Szotek Z, Temmerman WM, Winter H (1994) Phys Rev Lett 72: 1244
133. Sharp RT, Horton GK (1953) Phys Rev 90: 317
134. Talman JD, Shadwick WF (1976) Phys Rev A 14: 36
135. Kotani T (1995) Phys Rev Lett 74: 2989
136. Norman MR, Koelling DD (1984) Phys Rev B 30: 5530
137. Talman JD (1989) Comput Phys Commun 54: 85
138. Wang Y. Chevary JA, Macdonald LD, Vosko SH (1990) Phys Rev A 41: 78
139. Engel E, Chevary JA, Macdonald LD, Vosko SH (1992) Z Phys D 23: 7
140. Engel E, Vosko SH (1993) Phys Rev A 47: 2800
141. Krieger JB, Li Y, Iafrate GJ (1990) Phys Lett A 146: 256
142. Krieger JB, Li Y, Iafrate GJ (1990) Phys Lett A 148: 470
143. Li Y, Krieger JB, Norman MR, Iafrate GJ (1991) Phys Rev B 44: 10437
144. Krieger JB, Li Y, Iafrate BJ (1992) Int J Quantum Chem 41: 489
145. Krieger JB, Li Y, Iafrate GJ (1992) Phys Rev A 45: 101
146. Krieger JB, Li Y, Iafrate GJ (1992) Phys Rev A 46: 5453
147. Krieger JB, Li Y, Iafrate GJ (1992) Chem Phys Lett 191: 38
148. Li Y, Krieger JB, Iafrate GJ (1993) Phys Rev A 47: 165
149. Krieger JB, Li Y, Iafrate GJ (1995) in: Gross EKU, Dreizlu RM (eds) Density functional theory. Vol. 337 of NATO ASI Series B, Plenum Press, New York, p 191

150. Ullrich CA, Gossmann UJ, Gross EKU (1995) Phys Rev Lett 74: 872
151. Gossmann UJ (1994) Diplomarbeit, Universität Würzburg
152. Petersilka M, Gossmann UJ, Gross EKU, to be published
153. Mahan GD, Subbaswamy KR (1990) Local Density Theory of Polarizability. Plenum Press, New York
154. Zangwill A, Soven P (1980) Phys Rev Lett 45: 204
155. Zangwill A, Soven P (1981) Phys Rev B 24: 4121
156. Levine ZH, Soven P (1983) Phys Rev Lett 50: 2074
157. Levine ZH, Soven P (1984) Phys Rev A 29: 625
158. Ekardt W (1984) Phys Rev Lett 52: 1925
159. Puska MJ, Nieminen RM, Manninen M (1985) Phys Rev B 31: 3486
160. Ekardt W (1985) Phys Rev B 31: 6360
161. Beck DE (1987) Phys Rev B 35: 7352
162. Rubio A, Balbás LC, Alonso JA (1992) Phys Rev B 46: 4891
163. Ekardt W (1985) Phys Rev B 32: 1961
164. Ekardt W, Penzar Z (1991) Phys Rev B 43: 1322
165. Alonso JA, Rubio A, Balbás LC, (1994) Phil Mag B 69: 1037
166. Pacheco JM, Ekardt W (1992) Z Phys D 24: 65
167. Pacheco JM, Ekardt W (1992) Ann Physik 1: 254
168. Pacheco JM, Ekardt W (1993) Phys Rev B 47: 6667
169. Liebsch A (1987) Phys Rev B 36: 7378
170. Dobson JF, Harris GH (1986) J Phys C 19: 3971
171. Dobson JF, Harris GH (1987) J Phys C 20: 6127
172. Dobson JF, Harris GH (1989) J Phys C 21: L729
173. Gies P, Gerhardts RR (1987) Phys Rev B 36: 4422
174. Gies P, Gerhardts RR (1988) Phys Rev B 37: 10020
175. Kempa K, Schaich WL (1988) Phys Rev B 37: 6711
176. Ando T (1977) Solid State Commun 21: 133
177. Yi KS, Quinn JJ (1983) Phys Rev B 27: 2396
178. Wendler L. Grigoryan VG (1989) Solid State Commun 71: 527
179. Schmitt W (1993) Ph.D. thesis, Universität Würzburg
180. Levine ZH, Allan DC (1989) Phys Rev Lett 63: 1719
181. Bross H, Belhachemi O, Mekki B, Seoud A (1990) J Phys Condens Matter 2: 3919
182. Quong AA, Eguiluz AG (1993) Phys Rev Lett 70: 3955
183. Maddocks NE, Godby RW, Needs RJ (1994) Europhys Lett 27: 681
184. Fleszar A, Quong AA, Eguiluz AG (1995) Phys Rev Lett 74: 590
185. Langhoff PW, Epstein ST, Karplus M (1972) Rev Mod Phys 44: 602
186. Haensel R, Keitel G, Schreiber P, Kunz C (1969) Phys Rev 188: 1375
187. Levine ZH (1984) Phys Rev A 30: 1120
188. Brack M (1993) Rev Mod Phys 65: 677
189. Przybylski H Borstel G (1984) Solid State Commun 49: 317
190. Feibelman PJ (1982) Prog Surf Sci 12: 287
191. Kohn W, unpublished
192. Lang ND, Kohn W (1973) Phys Rev B 7: 3541
193. Barton G (1979) Rep Prog Phys 42: 964
194. Feibelman PJ (1970) Phys Rev B 2: 220
195. Harris J, Griffin A (1971) Phys Lett A 34: 51
196. Pines D, (1963) Elementary Excitations in Solids. W.A. Benjamin, New York, fig. 4.4
197. Tsuei KD et al. (1991) Surf Sci 247: 302
198. Liebsch A (1985) Phys Rev B 32: 6255
199. Dobson JF, Harris GH, O'Connor AJ (1988) J Phys Condens Matter 2: 6461
200. Neilson D, Swierkowski L, Szymanski J, Liu L (1993) Phys Rev Lett 71: 4035
201. Platzman PM, Isaacs ED, Williams H, Zschak P, E. Ice E (1992) Phys Rev B 46: 12943
202. Schülke W, Schulte-Schrepping H, Schmitz JR (1993) Phys Rev B 47: 12426
203. Schülke W, Schmitz JR, Schulte-Schrepping H, Kaprolat A (1995) Phys Rev B 52: 11721
204. Larson BC, Tischler JZ, Isaacs ED, Zschak P, Fleszar A, Eguiluz AG, to be published.
205. Kuchler M, Rebentrost F (1993) Phys Rev Lett 71: 2662
206. Kuchler M, Rebentrost F, (1994) Phys Rev B 50: 5651
207. Dal Corso A, Mauri F, Rubio A, to be published
208. Gunnarsson O, Lundqvist BI, (1976) Phys Rev B 13: 4274

209. Ziegler T, Rauk A, Baerends EJ (1977) Theoret Chim Acta 43: 261
210. von Barth U (1979) Phys Rev A 20: 1693
211. Theophilou A (1979) J Phys C 12: 5419
212. Hadjisavvas N, Theophilou A (1985) Phys Rev A 32: 720
213. Kohn W (1986) Phys Rev A 34: 5419
214. Gross EKU, Oliveira LN, Kohn W (1988) Phys Rev A 37: 2805
215. Gross EKU, Oliveira LN, Kohn W (1988) Phys Rev A 37: 2809
216. Oliveira LN, Gross EKU, Kohn W (1988) Phys Rev A 37: 2821
217. Nagy A (1990) Phys Rev A 42: 4388
218. Nagy A (1991) J Phys B 24: 4691
219. Nagy A (1994) Phys Rev A 49: 3074
220. Nagy A (1995) Int J Quantum Chem Symp 29
221. Levy M (1995) Phys Rev A 52: R4313
222. Chattaraj PK, Ghosh SK, Liu S, Parr RG, Int. J. Quant. Chem, in press.
223. Casida ME, to be published.
224. Krieger JB, Li Y, Iafrate GJ (1992) Phys Rev A 45: 101
225. Gross EKU, Petersilka M, Grabo T (1996) in: Ziegler (ed) Density functional methods in Chemistry. American Chemical Society, Washington, DC
226. Moore CE (1971) Nat Stand Ref Data Ser (US Nat Bur Stand) 35, Vol I II III
227. Perdew JP, Chevary JA, Vosko SH, Jackson KA, Pederson MR, Singh DJ, Fiolhas C (1992) Phys Rev B 46: 6671
228. Perez-Jorda JM, Becke AD (1995) Chem Phys Lett 233: 134
229. Mahanty J, Ninham BW (1976) Dispersion Forces. Academic Press, London
230. Jeziorski B, Moszynski R, Szalewicz K ((1994) Chem Rev 94: 1887
231. Zaremba E, Kohn W (1976) Phys Rev B 13: 2270
232. van Gisbergen SJA, Snijders JG, Baerends EJ (1995) J Chem Phys 103: 9347
233. Pacheco JM, Ekardt W (1993) Mod Phys Lett B 7: 573
234. Dobson JF, Dinte BP, Phys. Rev. Lett., in press
235. Andersson Y, Langreth DC, Lundqvist BI (1996) Phys Rev Lett 76: 102
236. Rapcewicz K, Ashcroft NW (1991) Phys Rev B 44: 4032
237. Zaremba E, Tso HS (1994) Phys Rev B 49; 8147
238. Harris J, Griffin A (1975) Phys Rev B 11: 3669
239. Dobson JF (1994) in Topics in Condensed Matter Physics, edited by M.P. Das (Nova, New York)
240. Atoms in Intense Laser Fields, edited by M. Gavrila (Academic Press, Boston, 1992).
241. Faisal FHM (1987) Theory of Multiphoton Processes Plenum Press, New York
242. Ullrich CA, Gossmann UJ, Gross EKU (1995) Ber Bunsenges Phys Chem 99: 488
243. Ullrich CA, Erhard S, Gross EKU (1996) in Super Intense Laser Atom Physics IV, edited by H.G. Muller (Kluwer, Dordrecht, 1996)
244. Ullrich CA (1995) Ph.D. thesis. Universität Würzburg
245. Kulander KC (1987) Phys Rev A 35: 445
246. Kulander KC (1987) Phys Rev A 36: 2726
247. L'Huiller A, Lompré LA, Mainfray G, Manus C (1992) in Atoms in Intense Laser Fields, edited by M. Gavrila (Academic Press, Boston, 1992), p 139.
248. Miyazaki K, Sakai H (1992) J Phys B 25: L83
249. Eichmann H, Egbert A, Nolte S, Momma C, Wellegehausen B, Becker W, Long S, McIver JK (1995) Phys Rev A 51: R3414
250. Protopapas M, Sanpera A, Knight PL, Burnett K (1995) Phys Rev A 52: R2527
251. Kulander KC (1988) Phys Rev A 38: 778
252. Kulander KC, Shore BW (1990) J Opt Soc Am B 7: 502
253. Walker B, Sheehy B, DiMauro LF, Agostini P, Schafer KJ, Kulander KC (1994) Phys Rev Lett 73: 1227

Generalized Functional Theory of Interacting Coupled Liouvillean Quantum Fields of Condensed Matter

A. K. Rajagopal and F. A. Buot

Naval Research Laboratory, Washington D.C. 20375, USA

Table of Contents

A large class of time-dependent quantum problems involves strongly interacting coupled fields requiring self-consistent non-perturbative and non-adiabatic approaches. We present here a general framework for analyzing these, based on Liouvillean Quantum Field Dynamics. Thus a multi-functional extension of the time-dependent density functional approach to many-body problems is set up. We also generalize the time-dependent non-equilibrium Green function method of Schwinger and Keldysh including all the relevant fields. We thus enlarge the context of the Φ-derivable, effective action functional theory of Baym for transport phenomena involving all relevant fields consistent with the conservation laws. This formalism is essential in understanding very-short time quantum dynamics of coupled fields of electrons, ions, and electromagnetic fields, as in nanoelectronic and optoelectronic devices, quantum dots in intense laser beams, micro-cavity quantum electrodynamics, as well as strongly coupled electron-phonon (high T_C), electron-photon (laser-atom systems), and molecular dynamic systems.

1 Introductory Remarks

Density functional theory (DFT) for stationary states or ensembles is a formulation of many-body theory in terms of particle density. This is a mature subject with many successful applications. This formalism is based on the works of Hohenberg and Kohn (HK) [1] and Kohn and Sham (KS) [2] and many subsequent extensions (see for example extensive reviews by Rajagopal [3], Callaway and March [4], and Dreizler and Gross [5]). The success is due to a strong interplay of formal theory, phenomenology, and computational methods. Thus DFT is an approach to the quantum many-body problem and provides a realistic basis for detailed computation of the properties of condensed matter. This is in contrast to the approaches which may be characterized as "model calculations of idealized systems" e.g., Ising and Heisenberg models; the Hubbard model; tight binding models, etc.; special one- and two-dimensional models, etc.

DFT is an approach which in principle incorporates the many-body techniques into a practical scheme, e.g. band structure and Fermi surface of bulk solids; surface physics problems [6]; problems in chemistry [7], etc. Thus over the last three decades or so, DFT has progressed as a full grown methodology covering almost all aspects of condensed matter physics, e.g. ground state, equilibrium, finite ensembles, and recently time-dependent aspects, ranging from atoms to solids (both bulk and surface-dependent properties, clusters, etc.). These advances were possible because the methods have been generalized (we give here references to review articles where available to keep the list to a reasonable size): (a) to incorporate magnetic (spin) properties developed independently by von Barth and Hedin, and, Rajagopal and Callaway, contained in a review article [3]; (b) to relativistic theory developed independently by Rajagopal and Ramana, MacDonald and Vosko, and Engel and Dreizler and reviewed in [8, 9]; (c) temperature-dependent generalization by Mermin and developed further independently by Gupta and Rajagopal, Dharmawardana and Perrot, and reviewed in [10], and a more recent work in [11]; (d) to finite ensembles due to Theophilou, Gross, Kohn, and Oliveira, and recently put into a unified framework by Rajagopal and Buot [12]; (e) to a generalization to incorporate superconductivity properties by Gross, Kohn, and Oliveira, for equilibrium and by Wacker, Kümmel, and Gross for time-dependent properties of the superconducting state, and, fully generalized recently by Rajagopal and Buot [13]; (f) to time-dependent density functional developed principally by Runge, Gross, and Kohn, and reviewed by Gross and Kohn [14, 15]. In this paper, we will give an account of a general functional theory of coupled fields of electrons, ions, and electromagnetic fields for studying the time-dependent and non-equilibrium situations, recently developed by us [16].

So far the applications of the time-dependent (TD) formalism are not as extensive as those of the stationary theory partly because of limitations of computation. This is expected to change in the near future not only because of advances in computer hardware and software but also due to a demand for understanding new experimental avenues open in nanostructure systems. The new phenomena encountered involves shorter temporal and spatial scales than before, displaying nonlinear behavior, as for example in resonant tunneling devices under the action of TD-bias. Here self-consistency is crucial without which one gets contradictory results [17]. Our new generalization of the TD functional formalism goes beyond the original DFT framework because it incorporates self-consistent ionic motion and electrodynamics and includes both equilibrium and non-equilibrium conditions. Thus the new functional theory involves a coupled self-consistent set of TD-equations for electrons, a dynamical equation of motion for ions, and a field equation for the electrodynamic variables. The physical quantities of interest involved in this theory are TD-potential, TD-force, TD-current generated from a universal functional of TD-density, TD-ion position, and TD-vector potential respectively, which now depends on the initial given conditions.

In recent years schemes for the dynamical simulation of matter from 'first principles' have been based on the Car-Parrinello [18] method which is founded on a fictitious Lagrangian molecular dynamics framework coupled with density-functional theory [19, 20]. Our formalism provides a firm physical and mathematical basis for performing time-dependent simulation procedures such as those based on the Car-Parrinello approach by replacing the fictitious Lagrangian by a precise action principle. Some possible examples of applications of the new TD-functional theory are (a) changes in phonon frequencies due to superconducting transition in high T_C systems; (b) electrodynamics of type II superconductivity in high T_C systems with unusual vortex structures; and (c) nanoelectronic systems subjected to TD-bias.

In this review, we begin with a treatment of the functional theory employing as basis the maximum entropy principle for the determination of the density matrix of equilibrium ensembles of any system. This naturally leads to the time-dependent functional theory which will be based on the TD-density matrix which obeys the von Neumann equation of motion. In this way, we present a unified formulation of the functional theory of a condensed matter system for both equilibrium and non-equilibrium situations, which we hope will give the reader a complete picture of the functional approach to many-body interacting systems of interest to condensed matter physics and chemistry.

2 Functional Theory of Many-Body Systems

We first recapitulate the known procedures employed in the original density functional theory and then present our unified approach. For the ground state of the many-body system, Hohenberg and Kohn used the well known Rayleigh-Ritz variational principle. For excited states and degenerate states, a mixed state description was used by Theophilou (equi-ensembles), and Gross, Kohn, and Oliveira (general ensembles). For equilibrium statistical thermodynamic ensembles Mermin used the Gibbs free energy minimum principle employing a density matrix of large systems. For studying the time-dependence of the state of a system for a given initial pure state Peuckert, Runge and Gross, and Gross and Kohn based the theory on the stationary action principle of Dirac and Frenkel. For the theory of the time-dependent behavior of a system for a given initial thermodynamic equilibrium state, a stationary action principle in Liouville space for the density matrix was recently put forward by Rajagopal and Buot [16]. The Liouville space quantum theory was originally developed by Umezawa [21]. Within this framework there is also an extension of a variational principle for one-particle Green function for both equilibrium and non-equilibrium situations originally due to Baym.

2.1 Basic Methodology – Mapping Theorems

In this section, we first give a brief précis of the basic principles involved in the DFT and then develop our unified theory.

Theorem I: (Mapping theorem) $V(\boldsymbol{r}) \to n(\boldsymbol{r})$. The ground state energy of a many-electron system is a universal functional of the ground state density, $n(\boldsymbol{r})$, only for a non-degenerate ground state (Hohenberg and Kohn, (HK)).

Theorem II: (Minimum energy principle, HK)

$$\frac{\delta E[n]}{\delta n} = 0.$$

Theorem III: [Effective Schrödinger's equation, Kohn and Sham (KS)]
Extensions of these to spin density, relativistic systems, and equilibrium finite temperatures have all been worked out as mentioned in the Introduction. It may be of interest to point out that the above three theorems hold for a finite temperature theory if we make the following changes: (1) change the ground state wave function to the equilibrium density matrix; and (2) replace the minimum energy principle by the minimum free energy principle. The TD density functional theory comparable to HK for arbitrary TD-systems was developed by Runge and Gross (RG) (see review article by Gross and Kohn

[14]). The central result is a set of TD-equations which includes all many-body correlations through a local TD exchange-correlation potential. In this time-dependent theory, one replaces the ground state density by the time-dependent density. We note:

(a) Theorem I is replaced by a weaker one, due to the dependence on the given initial condition.
(b) Theorem II is replaced also by a weaker one — by a stationary action principle.
(c) A different form of Theorem III is obtained.

These results were obtained by using the time-dependent quantum mechanical evolution of a state vector. We have generalized these to non-equilibrium situations [16] with the given initial state in a thermodynamic equilibrium state. This theory employs the density matrix which obeys the von Neumann equation. To incorporate the thermodynamic initial condition along with the von Neumann equation, it is advantageous to go to Liouville (L) space instead of the Hilbert (H) space in which DFT is formulated. This L-space quantum theory was developed by Umezawa over the last 25 years. We have adopted this theory to set up a new action principle which leads to the von Neumann equation. Appropriate variants of the theorems above are deduced in this framework.

2.2 Variational Principles for Ensembles of Quantum States

We begin by giving here a generalized Hohenberg-Kohn theorem by giving the variational principles for equilibrium ensembles of quantum states. We consider a many-electron system with Hamiltonian

$$H = T + U + V, \tag{1}$$

where

$$T = \frac{1}{2m} \int \nabla\psi^\dagger(r) \cdot \nabla\psi(r) d^3r, \tag{2}$$

$$U = \frac{e^2}{2} \iint \frac{\psi^\dagger(r)\psi^\dagger(r')\psi(r')\psi(r)}{|r - r'|} d^3r d^3r', \tag{3}$$

$$V = \int \rho(r) v(r) d^3r, \tag{4}$$

and

$$\rho(r) = \psi^\dagger(r)\psi(r). \tag{5}$$

Here, $v(r)$ denotes the external potential, $\psi(r)$ is the usual fermion field operator, and $\rho(r)$ is the particle density operator.

Theorem: We consider the density matrix written in terms the many-body states $\{|\Phi_m\rangle\}$ which minimizes the "free energy" associated with the Hamiltonian given above:

$$D = \sum_{m=1}^{M} w_m |\Phi_m\rangle\langle\Phi_m|, \quad \langle\Phi_m|\Phi_{m'}\rangle = \delta_{m,m'},$$
$$\mathrm{tr}D = 1 = \sum_{m=1}^{M} w_m. \tag{6}$$

which leads to the ensemble particle density

$$n(r) = \sum_{m=1}^{M} w_m \langle\Phi_m|\rho(r)|\Phi_m\rangle, \tag{7}$$

The choice of weights ensures that

$$\int n(r) d^3r = N, \tag{8}$$

provided the states chosen contain N particles.
If now

$$n'(r) = \sum_{m=1}^{M} w'_m \langle\Phi'_m|\rho(r)|\Phi'_m\rangle \tag{9}$$

is calculated with the states $|\Phi'_m\rangle$ of the Hamiltonian $H' = T + U + V'$, where the corresponding density matrix operator is given by

$$D' = \sum_{m=1}^{M} w'_m |\Phi'_m\rangle\langle\Phi'_m|, \ \langle\Phi'_m|\Phi'_{m'}\rangle = \delta_{m,m'}, \mathrm{tr}D' = 1 = \sum_{m=1}^{M} w'_m \tag{10}$$

which minimizes the "free energy" associated with H', then we prove that $n'(r) \neq n(r)$, provided V and V' differ by more than a constant.

Proof: The proof of this statement differs from the original argument of Mermin for the thermal ensembles in an important way when applied to the density operators D and D' defined above. Unlike the Mermin proof for thermal ensembles, we cannot use the **same** Lagrange parameter ξ to characterize both ensembles; in general ξ and ξ' are different parameters which we use as subscripts in the foregoing to indicate this dependence.

The density operator D' would then give us a larger "free energy" when used as a trial density operator to estimate $F[D]$:

$$F_\xi[D'_{\xi'}] > F_\xi[D_\xi] \Rightarrow$$
$$F_{\xi'}[D'_{\xi'}] + \int [v(r) - v'(r)] n'(r) d^3r - (\xi - \xi') S_{\xi'}[D'_{\xi'}] > F_\xi[D_\xi]. \tag{11}$$

Here we manipulated the left side of the inequality by using the corresponding definition and the Lagrange parameter associated with the Hamiltonian operator H'. For the sake of clarity, we have here exhibited the parameter dependence of the appropriate ensemble density operators along with those associated with the respective minimum "free energies" and entropies corresponding to the two Hamiltonians.

The reverse argument involving interchange of primed and unprimed quantities leads to the inequality

$$F_{\xi}[D_{\xi}] + \int[v'(r) - v(r)]n(r)d^3r - (\xi' - \xi)S_{\xi}[D_{\xi}] > F_{\xi'}[D'_{\xi'}]. \qquad (12)$$

With the choice of the weight parameters the entropies are constructed to be the same for the two ensembles, and we see at once that if two different external potentials give rise to the same ensemble particle density, the above two inequalities lead to a contradiction. Note that in the Mermin proof, thermal ensembles used are such that the temperature is fixed and hence $\xi' = \xi$ and therefore the entropy terms do not contribute. This proves the statement made above concerning the one-to-one mapping of density and potential. This theorem reduces to

(a) the original Hohenberg-Kohn's for a pure state density matrix;
(b) Mermin's for thermal density matrix; and
(c) Gross-Oliviera-Kohn's for a general mixed state density matrix.

For fixed w, i.e., for fixed ξ, the potential V' and hence the eigenstates $\{|\Phi'_m\rangle\}$ are uniquely determined by $n'(r)$, so that the ensemble expectation value of $T + U + k\xi \log D$ is a functional of the particle density,

$$\mathcal{F}[w; n'] \equiv trD\{T + U + k\xi \log D\}. \qquad (13)$$

It follows from the variational principle derived above that the functional

$$F_v[w; n'] \equiv \int n'(r)v(r)d^3r + \mathcal{F}[w; n'] \qquad (14)$$

has the following property:

$$F_v[w; n'] > F_v[w; n], \text{ for } n'(r) \neq n(r), \qquad (15)$$

where $n(r)$ is the average ensemble particle density associated with $v(r)$. This is our generalization of the Hohenberg-Kohn theorem for ensembles of states characterized by a density matrix D.

A further use of the variational freedom in the choice of the states $\{|\Phi_m\rangle\}$ in terms of the Slater determinants of one particle-like states may be used to derive the Kohn-Sham-type equations for the one particle states. We may state here that this can be generalized to include other variables such as magnetization (spin) etc. in a straightforward way.

2.3 Non-Equilibrium Quantum Field Dynamics

In order to develop our general TD theory, we now give a brief outline of Liouvillean quantum dynamics (LQD). For purposes of illustration of the method, we first consider a system of interacting electrons (non-relativistic) expressed in the usual Hilbert (H) space in terms of the creation ($\psi^{\dagger}_{\sigma}(\boldsymbol{r})$) and annihilation ($\psi_{\sigma}(\boldsymbol{r})$) operators in position space obeying the usual anticommutation rule

$$[\psi_\sigma(\mathbf{r}), \psi^\dagger_{\sigma'}(\mathbf{r}')]_+ = \delta_{\sigma\sigma'}\delta(\mathbf{r}-\mathbf{r}'),$$
$$[\psi^\dagger_\sigma(\mathbf{r}), \psi^\dagger_{\sigma'}(\mathbf{r}')]_+ = 0 = [\psi_\sigma(\mathbf{r}), \psi_{\sigma'}(\mathbf{r}')]_+ . \tag{16}$$

The system Hamiltonian is

$$H_t = H + H_t^{\mathrm{ext}} \equiv H^{(0)} + H^{(1)} + H_t^{\mathrm{ext}}, \tag{17}$$

where

$$H^{(0)} = \sum_\sigma \int d^3r \psi^\dagger_\sigma(\mathbf{r}) \left(-\frac{\hbar^2}{2m} \nabla^2 + V(\mathbf{r}) \right) \psi_\sigma(\mathbf{r}), \tag{18}$$

$$H^{(1)} = 1/2 \sum_{\sigma\sigma'} \iint d^3r d^3r' \psi^\dagger_\sigma(\mathbf{r}) \psi^\dagger_{\sigma'}(\mathbf{r}') V_c(\mathbf{r}-\mathbf{r}') \psi_{\sigma'}(\mathbf{r}') \psi_\sigma(\mathbf{r}), \tag{19}$$

and

$$H_t^{\mathrm{ext}} = \sum_\sigma \int d^3r V_e(\mathbf{r}t) \psi^\dagger_\sigma(\mathbf{r}) \psi_\sigma(\mathbf{r}). \tag{20}$$

Here $V(\mathbf{r})$ is the one-particle potential in which the electron moves and $V_c(\mathbf{r}-\mathbf{r}')$ is the bare Coulomb potential between the electrons. $V_e(\mathbf{r}t)$ is the external space- and time-dependent potential which acts as a source coupled to the electron density operator. σ is the spin index, and is such that

$$H_t^{\mathrm{ext}} = 0, \quad \text{for } t < t_0,$$
$$= H_t^{\mathrm{ext}}, \quad \text{for } t \geq t_0. \tag{20'}$$

The electron density operator is

$$\rho(\mathbf{r}) = \sum_\sigma \psi^\dagger_\sigma(\mathbf{r}) \psi_\sigma(\mathbf{r}), \tag{21}$$

with the total number operator given by $N = \int d^3r \rho(\mathbf{r})$. The current operator is

$$\mathbf{j}(\mathbf{r}) = \frac{\hbar}{2mi} \sum_\sigma \{ \psi^\dagger_\sigma(\mathbf{r})(\nabla\psi_\sigma(\mathbf{r})) - (\nabla\psi^\dagger_\sigma(\mathbf{r}))\psi_\sigma(\mathbf{r}) \}. \tag{22}$$

Later, similar considerations will be given to other dynamical fields associated with ion and electrodynamic variables as well as relativistic electrons. Quite often it is convenient to consider a relativistic electron system due to the elegance of presentation and due to formal ease with which to formulate TD-functional theory which includes spin and electrodynamics together. The details given here for electrons should be sufficient for understanding the needed procedures of LQD in these contexts. For completeness and for ease of following the LQD method, we first give a brief account of the traditional theory in H-space.

If $D(t)$ is the TD-density matrix of the system obeying the usual von Neumann equation, given that at the initial time the system is in thermal equilibrium described by the density matrix D_{eq}, then we have

$$i\hbar \frac{\partial D(t)}{\partial t} = [H_t, D(t)]_-, D(t_0) \equiv D_{eq} = \exp(-\beta(H - \mu N))/Z(\beta),$$
$$Z(\beta) = \mathrm{Tr}\exp(-\beta(H - \mu N)), \text{ with } \mathrm{Tr}D(t) = 1 \text{ for all } t, \tag{23}$$

where β is the usual inverse temperature, $(k_B T)^{-1}$ and μ is the chemical potential corresponding to the fixed number of electrons in the system, N being the total number operator. Tr stands for the trace over the entire system. It may be recalled that D_{eq} is determined from a free energy minimum principle. The well-known formal solution of Eq. (23) is

$$D(t) = U(tt_0)D(t_0)U(t_0t), \quad \text{where } i\hbar \frac{\partial}{\partial t} U(tt_0) = H_t U(tt_0),$$
$$\text{with } U(t_0t_0) = 1. \tag{24}$$

Thus $U(tt_0)$ represents the unitary time evolution according to the given system Hamiltonian. The time evolution of the expectation value of an arbitrary Hermitian TD operator $A(t)$ which has an intrinsic time-dependence of its own representing a physical quantity evolves accordingly by the rule

$$\langle A(t)\rangle = \mathrm{Tr}\{D(t)A(t)\} = \mathrm{Tr}\{D(t_0)A(t, t_0)\},$$
$$\text{where } A(t, t_0) = U(t_0t)A(t)U(tt_0). \tag{25}$$

Thus $A(tt_0)$ contains both the Hamiltonian and its own intrinsic evolution built in, and this may be recognized as defining the Heisenberg representation of any arbitrary operator. A general time correlation function associated with two general TD operators $A(t)$ and $B(t')$ representing two different physical entities at different times t and t′, is defined then by the formula

$$C_{t_0}(tt') = \mathrm{Tr}\{D(t_0)T(A(tt_0)B(t't_0))\}. \tag{26}$$

where T is the usual time-ordering operator. If the external potential is taken to be time-independent, then this correlation function obeys the well-known periodicity condition in the imaginary time domain.

2.4 Non-Equilibrium Stationary Action Principle

At the risk of being redundant, we may state here the salient features of the TD-functional formalism. The first requirement is a variational principle, and for a time-dependent quantum description only a stationary action principle is available. With this a mapping theorem is established which turns the action functional into a functional of relevant physical quantities (which are the expectation values), and the condition of stationarity is now in terms of these variables instead of the entire density matrix. Thus the stationary property with respect to the density matrix now becomes one with respect to all the variables which characterize the density matrix solution. Further developments in setting up effective equations determining the physical quantities follow from this. We thus obtain a *multivariable stationary property of the action functional*, leading to a coupled set of equations. The H-space formalism does not lend itself to such a framework because a stationary action principle for deducing the equation of motion for the density matrix given by Eq. (23) does not exist, and hence the usefulness of the functional method we wish to develop. We will now reformulate the above theory in L-space which meets all our needs in the construction of a TD-functional approach to the problem. This will be the non-equilibrium TD-functional theory which runs parallel to the pure-state formalism of Gross and Kohn in ordinary H-space provided we make the necessary translation of the H-space theory to the L-space formalism.

We give here a brief account of the quantum dynamics in L-space explicitly for the system described above. The relation between H- and L-space is summarized in Table 1. The creation operator in the H-space becomes a pair of creation superoperators, and similarly for the annihilation operator, in L-space. They are denoted by

$$\begin{aligned} &\psi_\sigma^\dagger(\boldsymbol{r}) \to (\hat{\psi}_\sigma^\dagger(\boldsymbol{r}), \tilde{\psi}_\sigma^\dagger(\boldsymbol{r})),\\ &\psi_\sigma(\boldsymbol{r}) \to (\hat{\psi}_\sigma(\boldsymbol{r}), \tilde{\psi}_\sigma(\boldsymbol{r})).\\ &(H\text{-space}) \to (L\text{-space}). \end{aligned} \tag{27}$$

These superoperators obey the anticommutation relations given by:

$$[\hat{\psi}_\sigma(\boldsymbol{r}), \hat{\psi}_{\sigma'}^\dagger(\boldsymbol{r}')]_+ = \delta_{\sigma\sigma'}\delta(\boldsymbol{r}-\boldsymbol{r}') = [\tilde{\psi}_\sigma(\boldsymbol{r}), \tilde{\psi}_{\sigma'}^\dagger(\boldsymbol{r}')]_+, \tag{28a}$$

$$\begin{aligned} &[\hat{\psi}_\sigma(\boldsymbol{r}), \tilde{\psi}_{\sigma'}^\dagger(\boldsymbol{r}')]_+ = 0,\\ &[\hat{\psi}_\sigma(\boldsymbol{r}), \hat{\psi}_{\sigma'}(\boldsymbol{r}')]_+ = 0 = [\tilde{\psi}_\sigma(\boldsymbol{r}), \tilde{\psi}_{\sigma'}(\boldsymbol{r}')]_+. \end{aligned} \tag{28b}$$

and their Hermitian conjugates. A general operator A in the H-space which is a function of ψ and $\psi^\dagger$ is represented by the superoperator twin set $(\hat{A}, \tilde{A})$ and the algebra of the superoperators $\hat{A}(\tilde{A})$ is a linear (antilinear) representation of the algebra of the operators A. The operators such as the Hamiltonian, number, and current in the H-space become respectively twin-pairs of superoperators in the L-space:

$$H_t \to (\hat{H}_t, \tilde{H}_t),\ \rho \to (\hat{\rho}, \tilde{\rho}), \boldsymbol{j} \to (\hat{\boldsymbol{j}}, \tilde{\boldsymbol{j}}). \tag{29}$$

Defining the unit superket

$$\begin{aligned} |1\rangle\!\rangle = \sum_\alpha |\alpha, \alpha\rangle\!\rangle, \quad \text{such that} \quad &\hat{H}_t|1\rangle\!\rangle = \tilde{H}_t|1\rangle\!\rangle\\ &= \langle\!\langle 1|\hat{H}_t = \langle\!\langle 1|\tilde{H}_t, \end{aligned} \tag{30}$$

where $|\alpha, \alpha\rangle\!\rangle = \| \alpha\rangle\langle\alpha|\rangle\!\rangle$ with $\{|\alpha\rangle\}$ any complete set of kets in the ordinary H-space, at the same time t as appears in Eq. (30), then an operator A in the

Table 1. Correspondence between H- and L-spaces

H-space theory	L-space theory
(1) Destruction and creation operators: $\psi_\sigma(\mathbf{r})$, $\psi_\sigma^\dagger(\mathbf{r})$ becomes	(1) doublets of destruction and creation operators: $(\hat{\psi}_\sigma(\mathbf{r}), \tilde{\psi}_\sigma(\mathbf{r}))$ $(\hat{\psi}_\sigma^\dagger(\mathbf{r}), \tilde{\psi}_\sigma^\dagger(\mathbf{r}))$
(2) An arbitrary operator A becomes	a superket $\|A\rangle\!\rangle$ defined by (2) $\|A\rangle\!\rangle = \hat{A}\|1\rangle\!\rangle = \tilde{A}^\dagger\|1\rangle\!\rangle$ where $\|1\rangle\!\rangle$ is a unit superket.
(3) More generally, a product of two arbitrary operators A and B here becomes	(3) a superoperator acting on a superket: $\|AB\rangle\!\rangle = \hat{A}\|B\rangle\!\rangle$, $\|BA\rangle\!\rangle = \tilde{A}^\dagger\|B\rangle\!\rangle$
(4) The von Neumann equation for the density matrix $i\hbar\frac{\partial}{\partial t}D(t) = [H_t, D(t)]_-$ becomes	(4) a Schrödinger-like equation for the superket: $i\hbar\partial_t\|D(t)\rangle\!\rangle = (\hat{H}_t - \tilde{H}_t)\|D(t)\rangle\!\rangle \equiv \bar{H}_t\|D(t)\rangle\!\rangle$, where $i\hbar\partial_t = \widehat{\left(i\hbar\frac{\partial}{\partial t}\right)} - \widetilde{\left(i\hbar\frac{\partial}{\partial t}\right)}$
(5) The average of any physical quantity given by the trace $\langle A(\mathbf{r},t)\rangle = \mathrm{Tr}A(\mathbf{r})D(t)$ becomes	(5) the matrix element $\langle A(\mathbf{r},t)\rangle = \langle\!\langle 1\|\hat{A}(\mathbf{r})\|D(t)\rangle\!\rangle$
(6) No action principle for deducing the von Neumann equation given above, whereas	(6) the superket equation above can be deduced from a stationary action principle for a functional W: $W(t_0, t_1) = \frac{1}{2}\int_{t_0}^{t_1} dt \langle\!\langle\Phi(t)\|(i\hbar\partial_t - \bar{H}_t)\|D(t)\rangle\!\rangle$

H-space becomes a superket in the L-space defined by

$$|A\rangle\!\rangle = \hat{A}|1\rangle\!\rangle = \tilde{A}^\dagger|1\rangle\!\rangle. \tag{31a}$$

More generally, a product of any two operators, A and B, in the H-space becomes a superoperator acting on a superket in the following manner:

$$|AB\rangle\!\rangle = \hat{A}|B\rangle\!\rangle,\ |BA\rangle\!\rangle = \tilde{A}^\dagger|B\rangle\!\rangle. \tag{31b}$$

[For fermions, A has to be a number conserving operator for Eq. (31b) to hold.]

The superket in this space, $|D(t)\rangle\!\rangle$, associated with the density matrix, is here a vector corresponding to the density matrix in the usual H-space, and obeys the equation corresponding to Eq. (23) which in the L-space assumes the form

$$\left| i\hbar \frac{\partial D(t)}{\partial t} \right\rangle\!\!\right\rangle \equiv i\hbar\partial_t|\check{\rho}(t)\rangle\!\rangle = |[H_t, D(t)]_-\rangle\!\rangle \equiv \bar{H}_t|D(t)\rangle\!\rangle, \tag{32}$$

where $\bar{H}_t = \hat{H}_t - \tilde{H}_t$ and $i\hbar\partial_t \equiv \widehat{\left(i\hbar\frac{\partial}{\partial t}\right)} - \widetilde{\left(i\hbar\frac{\partial}{\partial t}\right)}$ in the same sense as Schmutz [22], with the condition that the state is initially in thermal equilibrium as described above, namely, $|D(t_0)\rangle\!\rangle = |D_{eq}\rangle\!\rangle$. The solution of Eq. (32) corresponding to Eq. (24) is

$$\begin{aligned} &|D(t)\rangle\!\rangle = \bar{\mathscr{U}}(tt_0)|D(t_0)\rangle\!\rangle, \quad \text{where } i\hbar\frac{\partial}{\partial t}\bar{\mathscr{U}}(tt_0) = \bar{H}_t\bar{\mathscr{U}}(tt_0), \\ &\text{with } \bar{\mathscr{U}}(t_0t_0) = 1. \end{aligned} \tag{33}$$

As stated earlier, this initial state, $|D(t_0)\rangle\!\rangle$, is itself determined from a free energy minimum principle which will be incorporated in this sequel. This procedure is important in maintaining self-consistency of the formalism at all levels and on equal footing.

The time derivative operator in the H-space now becomes a superoperator, ∂_t, in view of Eq. (31b), defined in the sense shown on the left-hand side of Eq. (32). This was missed in earlier work [21] even though its existence was surmised. It is here found to be crucial in the development of the stationary action principle in the L-space. When the system is considered to be in a pure state, this gives a transparent reduction to the pure-state TD-functional theory. We may then state the usual normalization condition on the density matrix in the form of a matrix element in the superspace:

$$\mathrm{Tr}D(t) = \langle\!\langle 1|D(t)\rangle\!\rangle = 1. \tag{34}$$

We refer to Eq. (34) as the asymmetric formulation of the L-space quantum dynamics. Also the expectation value of any TD operator $A(\mathbf{r}t)$ in the ordinary H-space is just a matrix element of either of its twin operators $\hat{A}(\mathbf{r}t)$ or $\tilde{A}(\mathbf{r}t)$ in the L-space:

$$\langle A(\mathbf{r}t)\rangle = \mathrm{Tr}(D(t)A(\mathbf{r}t)) = \langle\!\langle 1|\hat{A}(\mathbf{r}t)|D(t)\rangle\!\rangle = \langle\!\langle 1|\tilde{A}^\dagger(\mathbf{r}t)|D(t)\rangle\!\rangle. \tag{35}$$

Equivalently these may be written in convenient forms to exhibit the averaging in terms of the given initial equilibrium state:

$$\begin{aligned} \langle A(\mathbf{r}t)\rangle &= \mathrm{Tr}(D(t_0)U(t_0t)A(\mathbf{r}t)U(tt_0)) \\ &= \langle\!\langle 1|\bar{\mathscr{U}}(t_0t)\hat{A}(\mathbf{r}t)\bar{\mathscr{U}}(tt_0)|D(t_0)\rangle\!\rangle \\ &\equiv \langle\!\langle 1|\hat{A}(\mathbf{r}tt_0)|D(t_0)\rangle\!\rangle. \end{aligned} \tag{36}$$

Eq. (36) is a version of the Heisenberg representation of an operator in the L-space, which will be useful in later manipulations. Similarly the time correlation functions between two operators given in Eq. (26) in H-space may also be re-expressed as a matrix element of the above form in the L-space. Thus in the

Liouvillean space the expectation values become simple matrix elements, no longer a trace, and may even be viewed as "thermal vacuum" expectation values with respect to the given initial thermodynamic state. We can now restate Eq. (32) as a **stationary action principle** in this superspace:

$$W(t_0, t_1) = 1/2 \int_{t_0}^{t_1} dt \langle\langle \Phi(t)|(i\hbar\bar{\partial}_t - \bar{H}_t)|D(t)\rangle\rangle \tag{37}$$

subject to the initial condition, given in Eq. (32), by varying the left supervector and setting the result equal to zero. Also by varying the right supervector, we obtain the corresponding equation for the superbra. The factor 1/2 in Eq. (37) is chosen, to account for the presence of the twins, so that Eq. (37) reduces to the correct action in the special case when the density matrix represents a pure state and left superbra is the unit vector introduced above. This is the reason for introducing $\bar{\partial}_t$ in place of $\partial/\partial t$. Furthermore, the physical action functional is obtained when we choose the left superbra to be the unit supervector and the right supervector to be that associated with the density matrix, in view of Eq. (34).

We will now prepare the initial density matrix superket to be the equilibrium state or equivalently the equilibrium density matrix, by a minimum free energy principle, with the free energy given by:

$$\Omega = \mathrm{Tr} D_{\mathrm{eq}}(\theta)(H_{t_0} + \beta^{-1} \ln D_{\mathrm{eq}}(\theta)) \equiv \langle\langle 1|(\hat{H}_{t_0} - T\hat{S})|D_{\mathrm{eq}}(\theta)\rangle\rangle, \tag{38}$$

where $|D_{\mathrm{eq}}(\theta)\rangle\rangle$ is the variational thermal state. The entropy superoperator in the L-space is correspondingly given by

$$\hat{S} = -k_B(\hat{\ln} D). \tag{39}$$

In Eq. (38), the first term is the well-known free energy expression in the usual H-space. The variation of Ω with respect to $D_{\mathrm{eq}}(\theta)$ yields the equilibrium state and the value of Ω at the minimum is the free energy of the system given initially. Therefore, a functional theory holds for this equilibrium state also and thus the two principles together lead to a new procedure which maintains self-consistency at all levels and treats in tandem the stationary action principle and its initial state specification on equal footing.

With this at hand, it is now straightforward to formulate the appropriate generalization of the time-dependent functional theory of Gross and coworkers, generalized to include the initial thermal equilibrium condition. We enunciate here the essential theorems of this formalism without giving the details of proof, except in one typical case. We thus exhibit the advantages of the new method outlined here. We choose to work in this asymmetric formulation of the L-space quantum dynamics since it has the distinct advantage that the traditional real-time Green functions of the many-particle systems defined in this space go over to the closed-time path ordered matrix Green functions as was shown by Schmutz [21]. It thus makes it possible to derive the matrix Green function equations in the functional formalism which is useful for formulating the transport equation and in turn is amenable to numerical implementation.

Theorem I′ (mapping theorem): For every one-electron external potential, $a_0^{\text{ext}}(\boldsymbol{r}t)$, force $\boldsymbol{F}^{\text{ext}}(\ell\kappa t)$ on the ℓ-th ion of species κ, external vector potential $\boldsymbol{a}^{\text{ext}}(\boldsymbol{r}t)$, and external four current $j_\mu^{\text{ext}}(\boldsymbol{r}t)$, which can be expanded into respective Taylor series in the time coordinate around the initial time t_0, a map

$$\begin{aligned}
&(a_\mu^{\text{ext}}(\boldsymbol{r}t)) \to (J_\mu(\boldsymbol{r}t))\\
&(\boldsymbol{F}^{\text{ext}}(\ell\kappa t)) \to (\boldsymbol{R}(\ell\kappa t))\\
&(j_\mu^{\text{ext}}(\boldsymbol{r}t)) \to (A_\mu(\boldsymbol{r}t))
\end{aligned} \tag{40}$$

is defined by solving the time-dependent Liouville equation with the given initial thermodynamic state and calculating the corresponding electron density, $n(\boldsymbol{r}t)$, ion-position, $\boldsymbol{R}(\ell\kappa t)$ the electron current density, $\boldsymbol{J}(\boldsymbol{r}t)$, and electromagnetic field $A_\mu(\boldsymbol{r}t)$ given by:

$$\begin{aligned}
J_\mu(\boldsymbol{r}t) &= \mathrm{Tr}D(t)j_\mu(\boldsymbol{r}) = \langle\!\langle 1|j_\mu(\boldsymbol{r})|D(t)\rangle\!\rangle,\\
A_\mu(\boldsymbol{r}t) &= \mathrm{Tr}D(t)A_\mu(\boldsymbol{r}) = \langle\!\langle 1|\hat{A}_\mu(\boldsymbol{r})|D(t)\rangle\!\rangle;\\
\boldsymbol{R}(\ell\kappa t) &= \mathrm{Tr}D(t)\boldsymbol{R}(\ell\kappa) = \langle\!\langle 1|\hat{\boldsymbol{R}}(\ell\kappa)|D(t)\rangle\!\rangle.
\end{aligned} \tag{41}$$

This map is invertible under certain simple conditions.

It is often useful to consider externally specified density, $n_{\text{ext}}(\boldsymbol{r}t)$, and/or current density, $\boldsymbol{j}_{\text{ext}}(\boldsymbol{r}t)$, distributions for which the mapping theorem reads

$$\begin{aligned}
&(n_{\text{ext}}(\boldsymbol{r}t)) \to (A_0(\boldsymbol{r}t))\\
&(\boldsymbol{j}_{\text{ext}}(\boldsymbol{r}t)) \to (\boldsymbol{A}(\boldsymbol{r}t))
\end{aligned} \tag{42}$$

The gauge invariance is assured since one always takes the four-divergence of the four vectors to be zero and so, within this class, the mapping is assured.

Proof: We give here a proof of this theorem in a typical case. We must point out that the minimum free energy principle assures us of the above mapping at the initial time. The Taylor expansion then provides the mapping for subsequent times. We shall not give the first step based on the minimum property of free energy associated with the initial density matrix as this has already been proved earlier in this section. From this follows the important conclusion that the initial density matrix can be considered as a functional of the initial values of the various averages. We choose for the purpose of illustrating our method of proof the ion-coordinate.

If the forces $\boldsymbol{F}^{\text{ext}}(\ell\kappa t)$ can be expanded into a Taylor series in the time coordinate around the initial time t_0, a map

$$(\boldsymbol{F}^{\text{ext}}(\ell\kappa t)) \to (\boldsymbol{R}(\ell\kappa t)) \tag{43}$$

is defined by solving the time-dependent von Neumann equation in L-space with a given fixed initial density matrix $|D(t_0)\rangle\!\rangle$, and calculating the corresponding coordinates $\boldsymbol{R}(\ell\kappa t)$. Assume that there exist two unequal forces, $\boldsymbol{F}^{\text{ext}}(\ell\kappa t)$ and $\boldsymbol{F}^{\text{ext}\prime}(\ell\kappa t)$, both of which are given to be equal at $t = t_0$. Assume that these can be expanded in respective Taylor series around $t = t_0$.

We will now show that $\boldsymbol{R}(l\kappa t)$ and $\boldsymbol{R}'(l\kappa t)$ are different, corresponding to $\boldsymbol{F}^{\text{ext}}(\ell\kappa t)$ and $\boldsymbol{F}^{\text{ext}'}(\ell\kappa t)$ as long as there exists a non-negative integer k such that

$$(\partial/\partial t)^k [\boldsymbol{F}^{\text{ext}}(l\kappa t) - \boldsymbol{F}^{\text{ext}'}(l\kappa t)]_{t=t_0} \neq 0. \tag{44}$$

In other words, the ion positions are tracked by the forces.

We first obtain

$$\frac{d}{dt}(\boldsymbol{R}(l\kappa t) - \boldsymbol{R}'(l\kappa t)) = \frac{1}{M_\kappa}(\boldsymbol{P}(l\kappa t) - \boldsymbol{P}'(l\kappa t)), \tag{45}$$

which follows by a direct calculation. We have, similarly

$$i\hbar \frac{d}{dt}(\boldsymbol{P}(l\kappa t) - \boldsymbol{P}'(l\kappa t))_{t=t_0} = \langle\!\langle 1 | [\hat{\boldsymbol{P}}(l\kappa), \bar{H}(t_0) - \bar{H}'(t_0)]_- | D(t_0) \rangle\!\rangle$$

$$= i\hbar(\boldsymbol{F}(l\kappa t_0) - \boldsymbol{F}'(l\kappa t_0)). \tag{46}$$

Thus if $\boldsymbol{F}^{\text{ext}}(\ell\kappa t)$ and $\boldsymbol{F}^{\text{ext}'}(\ell\kappa t)$ differ at t_0 then for $k = 0$, $\boldsymbol{P} \neq \boldsymbol{P}'$ for t infinitesimally later than t_0. If $k > 0$, then applying the differentiation k times and we get

$$\left(i\hbar \frac{d}{dt}\right)^{k+1} (\boldsymbol{P}(l\kappa t) - \boldsymbol{P}'(l\kappa t))_{t=t_0} = i\hbar \left\{\left(i\hbar \frac{\partial}{\partial t}\right)^k (\boldsymbol{F}(l\kappa t) - \boldsymbol{F}'(l\kappa t))\right\}_{t=t_0} \neq 0, \tag{47}$$

implying that $\boldsymbol{P} \neq \boldsymbol{P}'$ for t infinitesimally later than t_0. These in turn lead to the statement on $\boldsymbol{R}$'s:

$$\left(\frac{d}{dt}\right)^{k+2} (\boldsymbol{R}(l\kappa t) - \boldsymbol{R}'(l\kappa t))_{t=t_0} = \frac{1}{M_\kappa}\left(\frac{\partial}{\partial t}\right)^k (\boldsymbol{F}(l\kappa t) - \boldsymbol{F}'(l\kappa t))_{t=t_0}. \tag{48}$$

Thus the ionic coordinates become different for t infinitesimally later than t_0.

Since we consider, by construction, only density matrices that evolve from the same initial state, the average positions as well as the momenta of the ions are identical at the initial time and evolve subsequently to be different as demonstrated above. Thus we have established the 1–1 correspondence between the time-dependent positions of the ions and the external forces; the external forces on the other hand, uniquely determine the time-dependent density matrix, which can therefore be considered as a functional of the time-dependent positions. As a consequence, the average value of any quantum mechanical operator is a unique functional of the ionic positions.

Before we proceed to state the theorems we give here a brief description of the Hamiltonian for a system consisting of electrons, ions, and electromagnetic fields in the Hilbert space. From this, the procedure to go to the Liouvillean space follows along the lines given earlier. We will choose to express the electrons in terms of the Dirac field to keep the discussion general as well as elegant from which the non-relativistic results can be extracted by known procedures, and treat the ions in the non-relativistic framework as this suffices for most problems of interest in condensed matter. We choose to be in the Coulomb gauge in describing the electromagnetic fields as it is convenient for quantization. Since the whole theory is gauge invariant, the questions concerning the choice of gauge etc. need not concern us here. Also we ignore the coupling of the ions to the electromagnetic fields in this development. The Hamiltonian is thus written in the form

$$H = H_e + H_i + H_f + H_{\mathrm{int}(e-f)} + H_{\mathrm{int}(e-i)} + H_{\mathrm{ext}(i)}(t) + H_{\mathrm{ext}(e,f)}(t) + E_{\mathrm{vac}}. \tag{49}$$

E_{vac} in Eq. (49), is the vacuum expectation value of the total Hamiltonian which we ignore in further considerations. The appearance of the normal ordering of operators is a consequence of this term. Here the electron Hamiltonian H_e is composed of its kinetic, rest, and mutual Coulomb interaction terms:

$$H_e = T_e + W_{ee},\ T_e = \int d^3r : \bar{\psi}(\boldsymbol{r})(-i\gamma\cdot\boldsymbol{\nabla} + m_0c^2(1-\gamma_0))\psi(\boldsymbol{r}):, \tag{50a}$$

$$W_{ee} = \frac{1}{2}\iint d^3r d^3r' \frac{e^2}{|\boldsymbol{r}-\boldsymbol{r}'|} : \bar{\psi}(\boldsymbol{r})\gamma_0\psi(\boldsymbol{r})::\bar{\psi}(\boldsymbol{r}')\gamma_0\psi(\boldsymbol{r}'):.$$

Here $\bar{\psi} = \psi^\dagger\gamma_0$ in the standard notation (see for example, Kaku [23]). The ion-Hamiltonian H_i consists of its kinetic and mutual interaction terms with $\boldsymbol{P}(\ell\kappa)$ representing the ion momentum operator conjugate to the ion position operator $\boldsymbol{R}(l\kappa)$, which in the Heisenberg representation is given by $M_k\dot{\boldsymbol{R}}(\ell\kappa t) = \boldsymbol{P}(\ell\kappa t)$. The overdot here and elsewhere stands for the time derivative of the operator in the usual way. Here ℓ and κ stand for the ion index and species respectively.

$$H_i = T_i + W_{ii},\ T_i = \sum_{\ell\kappa}\frac{\boldsymbol{P}^2(\ell\kappa)}{2M_\kappa}, \tag{50b}$$

$$W_{ii} = \frac{1}{2}\sum_{\ell\kappa}\sum_{\ell'\kappa'} w_{ii}(\boldsymbol{R}(\ell\kappa) - \boldsymbol{R}(\ell'\kappa')).$$

The electromagnetic field energy in the Coulomb gauge is

$$H_f = \frac{1}{2}\int d^3r:(\boldsymbol{\pi}_{tr}^2(\boldsymbol{r}) + \boldsymbol{B}^2(\boldsymbol{r})):, \tag{50c}$$

$$\text{with } \boldsymbol{\pi}_{tr}(\boldsymbol{r}) = -\boldsymbol{E}_{tr}(\boldsymbol{r}) = \frac{1}{c}\boldsymbol{A}(\boldsymbol{r}) \quad \text{and} \quad \boldsymbol{B}(\boldsymbol{r}) = \boldsymbol{\nabla}\times\boldsymbol{A}(\boldsymbol{r}),$$

where $\boldsymbol{\pi}_{tr}$ is the conjugate field operator to the vector field $\boldsymbol{A}(\boldsymbol{r})$ and its familiar form in terms of the time derivative is to be understood in the sense of the Heisenberg representation of the operators. The Hamiltonian representing electron-field interaction is

$$H_{\mathrm{int}(e-f)} = -\frac{e}{c}\int d^3r : \boldsymbol{j}(\boldsymbol{r})\cdot\boldsymbol{A}(\boldsymbol{r}):, \tag{50d}$$

and that representing the electron and ion is

$$H_{\text{int}(e-i)} = -\sum_{\ell\kappa} \int d^3r w_{ei}(\boldsymbol{r} - \boldsymbol{R}(\ell\kappa)) \colon \bar{\psi}(\boldsymbol{r})\gamma_0\psi(\boldsymbol{r})\colon. \tag{50e}$$

The time-dependent external source fields coupled to the respective operators of the system are represented by the Hamiltonian:

$$H_{\text{ext}(e,f)}(t) = -\frac{e}{c}\int d^3r(\colon j_\mu(\boldsymbol{r})\colon a^\mu_{\text{ext}}(\boldsymbol{r}t) + j^{\text{ext}}_\mu(\boldsymbol{r}t)\colon A^\mu(\boldsymbol{r})\colon). \tag{50f}$$

We choose to employ external forces acting on the ion-position operators instead of the conventional ion potential acting on the ion density operator so as to capture the explicit dynamics of the ionic motion:

$$H_{\text{ext}(i)}(t) = \sum_{\ell\kappa} \boldsymbol{F}^{\text{ext}}(\ell\kappa t)\cdot \boldsymbol{R}(\ell\kappa). \tag{50g}$$

The four-current operator is defined in the standard way [23] by:

$$j_\mu(\boldsymbol{r}) = \colon \bar{\psi}(\boldsymbol{r})\gamma_\mu\psi(\boldsymbol{r})\colon. \tag{51}$$

All the operators appearing here are in the Schrödinger representation. The physical quantities of our theory involve the expectation values [Eq. (25)] of the various operators defined above, which also serves to define the Heisenberg representations of the operators defined here [see Eq. (41)]. The various field operators obey the standard equal time commutation rules among themselves while they commute between them. We give here the one for the vector fields in the Heisenberg representation, representing the electromagnetic fields in the Coulomb gauge, to keep the discussion clear [23]:

$$[\dot{A}_i(\boldsymbol{r}t), A_j(\boldsymbol{r}'t)] = -i(\delta_{ij} - \nabla^{-2}\nabla_i\nabla_j)\delta^{(3)}(\boldsymbol{r} - \boldsymbol{r}'). \tag{52a}$$

As is well known, in this gauge, the scalar potential is not an independent variable, as it is determined by the instantaneous charge distribution, $n(\boldsymbol{r}t)$, by the relation $A_0(\boldsymbol{r}t) = e\int d^3r'\, n(\boldsymbol{r}'t)/4\pi|\boldsymbol{r}' - \boldsymbol{r}|$. Thus while $\psi(\boldsymbol{r}t)$ commutes with the components of the vector field, $A_i(\boldsymbol{r}t)$, it obeys the commutation rule

$$[A_0(\boldsymbol{r}t), \psi(\boldsymbol{r}'t)] = -\frac{e}{4\pi|\boldsymbol{r} - \boldsymbol{r}'|}\psi(\boldsymbol{r}'t). \tag{52b}$$

with the scalar potential, $A_0(\boldsymbol{r}t)$.

Similar procedures using the above establish the other mappings stated in the theorem. We thus have the following important consequences:

$$\begin{aligned} &D(t) = D[J^\mu, A^\mu, \boldsymbol{R}](t), \quad \text{and} \\ &\langle A(t)\rangle = \text{Tr}D(t)A(t) = A[J^\mu, A^\mu, \boldsymbol{R}](t). \end{aligned} \tag{53}$$

This also holds at the initial time, $t = t_0$.

Theorem II′ (stationary action principle): The physical action

$$W(t_0, t_1) = \frac{1}{2} \int_{t_0}^{t_1} dt \langle\langle 1 | (i\hbar\partial_t - \bar{H}_t) | D(t) \rangle\rangle \tag{54}$$

is stationary with respect to the following variations given the initial specified conditions at $t = t_0$:

$$\frac{\delta W}{\delta n(\boldsymbol{r}t)} = 0; \quad \frac{\delta W}{\delta \boldsymbol{J}(\boldsymbol{r}t)} = 0;$$
$$\frac{\delta W}{\delta \boldsymbol{R}(\ell\kappa t)} = 0; \quad \frac{\delta W}{\delta A_\mu(\boldsymbol{r}t)} = 0. \tag{55}$$

The top two equations lead to the effective Dirac (Schrödinger) equation, the third leads to the effective Newton equation, and the last leads to the effective Maxwell equation, as exhibited in Theorem III′.

Proof: We first note that this theorem is simply a restatement of the stationarity of the action with respect to the density matrix expressed now in terms of the variables of our interest upon using the relation Eq. (53). By separating the operator $\bar{H}_t$ into parts that contain external field terms and those that do not, by referring to Eq. (49) and using the definitions given in Eq. (41), we reexpress the action as a functional of the variables introduced above.

$$\begin{aligned} W(t_0, t_1) &\equiv W[J^\mu, A^\mu, \boldsymbol{R}] \\ &= B[J^\mu, A^\mu, \boldsymbol{R}] + \frac{e}{c} \int_{t_0}^{t_1} dt \int d^3r (J^\mu(\boldsymbol{r}t) a_\mu^{\text{ext}}(\boldsymbol{r}t) + j_\mu^{\text{ext}}(\boldsymbol{r}t) A^\mu(\boldsymbol{r}t)) \\ &\quad - \int_{t_0}^{t_1} dt \sum_{\ell\kappa} \boldsymbol{F}^{\text{ext}}(l\kappa t) \cdot \boldsymbol{R}(l\kappa t). \end{aligned} \tag{56}$$

In the above, the functional $B[J^\mu, A^\mu, \boldsymbol{R}]$ is given by the following expression:

$$\begin{aligned} &B[J^\mu, A^\mu, \boldsymbol{R}] \\ &= \frac{1}{2} \int_{t_0}^{t_1} dt \langle\langle 1 | (i\hbar\partial_t - \bar{H}_e - \bar{H}_i - \bar{H}_f - \bar{H}_{\text{int}(e-f)} - \bar{H}_{\text{int}(e-i)}) | D(t) \rangle\rangle, \end{aligned} \tag{57}$$

where the various parts of the Hamiltonian in the L-space are obtained from Eqs. (50a–e). A similar decomposition holds for the initial time where W is replaced by the free energy functional. The stationary action principle can now be used to obtain the basic coupled equations of motion for the various quantities and the corresponding initial conditions are determined from the free energy functional, as stated in the next theorem.

Theorem III′ (coupled self-consistent equations for the fields): The effective Dirac, Maxwell, and Newton equations, given the associated initial quantities in thermodynamic equilibrium, are

$$i\hbar\frac{\partial\phi_j(\boldsymbol{r}t)}{\partial t}=\left\{\boldsymbol{\alpha}\cdot\left(-i\hbar\boldsymbol{\nabla}-\frac{e}{c}\boldsymbol{A}_{\text{eff}}(\boldsymbol{r}t)\right)+m_0(1-\beta)+A_{\text{eff}_0}(\boldsymbol{r}t)\right\}\phi_j(\boldsymbol{r}t) \tag{58}$$

$$J_\mu(\boldsymbol{r}t)=\sum_i f_i sp(\phi_i^*(\boldsymbol{r}t)\alpha_\mu\phi_i(\boldsymbol{r}t)), \tag{59}$$

$$A_{\text{eff}\mu}(\boldsymbol{r}t)=a_\mu^{\text{ext}}(\boldsymbol{r}t)+A_\mu(\boldsymbol{r}t)+\frac{\delta B_1[J^\mu,A^\mu,\boldsymbol{R}]}{\delta J^\mu(\boldsymbol{r}t)}. \tag{60}$$

Here sp stands for trace over the spinor indices

$$B_1[J^\mu,A^\mu,\boldsymbol{R}]=B_M^s[J^\mu]+B_E^s[A^\mu]+B_i^s[\boldsymbol{R}]+\int_{t_0}^{t_1}dt\,J^\mu(\boldsymbol{r}t)A_\mu(\boldsymbol{r}t)-B_2[J^\mu,A^\mu,\boldsymbol{R}]. \tag{61}$$

$B_M^s[J^\mu]$, $B_E^s[A^\mu]$, and $B_i^s[\boldsymbol{R}]$ are the corresponding "non-interacting" actions of matter, electromagnetic field, and ions:

$$B_M^s[J^\mu]=\frac{1}{2}\int_{t_0}^{t_1}dt\langle\langle 1|i\hbar\partial_t-\bar{T}_e)|D_s(t)\rangle\rangle,$$

$$B_E^s[A^\mu]=\frac{1}{2}\int_{t_0}^{t_1}dt\langle\langle 1|(i\hbar\partial_t-\bar{H}_f)|D_s(t)\rangle\rangle, \tag{62}$$

$$B_i^s[\boldsymbol{R}]=\frac{1}{2}\int_{t_0}^{t_1}dt\langle\langle 1|(i\hbar\partial_t-\bar{T}_i)|D_s(t)\rangle\rangle,$$

and

$$B_2[J^\mu,A^\mu,\boldsymbol{R}]=B[J^\mu,A^\mu,\boldsymbol{R}]-\int_{t_0}^{t_1}dt\int d^3r J^\mu(\boldsymbol{r}t)A_\mu(\boldsymbol{r}t). \tag{63}$$

$$\partial_\mu F^{\mu\nu}(\boldsymbol{r}t)=J_{\text{eff}}^\nu(\boldsymbol{r}t), \tag{64}$$

$$J_{\text{eff}}^\nu(\boldsymbol{r}t)=j_{\text{ext}}^\nu(\boldsymbol{r}t)+J^\nu(\boldsymbol{r}t)+\frac{\delta B_1}{\delta A_\nu(\boldsymbol{r}t)}. \tag{65}$$

And, $M_\kappa\ddot{\boldsymbol{R}}(l\kappa t)=-\boldsymbol{F}_{\text{eff}}(\ell\kappa t)$, (66)

$$\boldsymbol{F}_{\text{eff}}(l\kappa t)=\boldsymbol{F}^{\text{ext}}(l\kappa t)-\frac{\delta B_1}{\delta\boldsymbol{R}(l\kappa t)}. \tag{67}$$

The equilibrium initial conditions are determined by another variational principle on the free energy functional, and lead to corresponding appropriate equations which are not presented here. They serve as the given initial conditions for the time-dependent equations given above. The various equations have the same form as above except that all the B's are replaced by their free energy counterparts Ω, and the time-variable runs along a complex time line $(t_0, t_0 - i\beta)$.

2.5 Effective Action and Φ-Derivable Functional Theory

We now turn our attention to another aspect of the functional theory by first giving a brief account of Baym's Φ-derivable scheme [24]. This is because of our desire to develop transport theory on the same type of self-consistent framework as the TD-functional theory developed thus far. We follow Baym's observation that the quantum theory of transport can be cast in terms of the two particle correlation function

$$L(12;1'2') = G_2(12;1'2') - G(11')G(22') \tag{68}$$

Baym showed that the macroscopic conservation laws for particle number, momentum, and energy will be satisfied if L in (68) is of the form

$$L(12;1'2') = \pm \delta G(11')/\delta U(22')|_{U=0}. \tag{69}$$

The U-dependence on G is given by the usual equation of motion, namely

$$\begin{aligned}&\int d\bar{1}[G_0^{-1}(1\bar{1}) - U(1\bar{1})]G(\bar{1}1';U) = \delta(11') \pm \\ &i\int d\bar{3}V(1\bar{3})G_2(1\bar{3};1'\bar{3}^+;U).\end{aligned} \tag{70}$$

where

$$V(12) = v(\mathbf{r}_1 - \mathbf{r}_2)\delta(t_1 - t_2). \tag{71}$$

$$\text{Write } \pm i\int d\bar{3}V(1\bar{3})G_2(1\bar{3};1'\bar{3}^+;U) \equiv \int d\bar{2}\Sigma(1\bar{2})G(\bar{2}1';U) \tag{72}$$

Thus the "self-energy", containing the exchange-correlation effects of the interaction,

$$\Sigma \textit{ is a functional of } G(U) \textit{ and } V. \tag{73}$$

Then

$$\int d\bar{1}[G_0^{-1}(1\bar{1}) - U(1\bar{1}) - \Sigma(1\bar{1})]G(\bar{1}1') = \delta(11') \tag{74}$$

Invoking gauge transformation (number conservation), translation transformation (linear momentum conservation), and time transformation ø(t) (energy conservation), Eq. (69) is established along with the requirement

$$\delta G(11')/\delta U(2'2) = \delta G(22')/\delta U(1'1) \tag{75}$$

This in turn implies the existence of an effective action functional W such that

$$G(11') = \delta W/\delta U(1'1). \tag{76}$$

$$\begin{aligned}L &= \pm\,\delta G/\delta U = \mp\, G(\delta G^{-1}/\delta U)G \\ &= \mp\, GG \mp G(\delta\Sigma/\delta U)G,\end{aligned} \tag{77}$$

$$\text{and } (\delta\Sigma/\delta U) = \int \delta\Sigma/\delta G \cdot \delta G/\delta U,$$

the above condition (75) may be rewritten as

$$\delta\Sigma(11')/\delta G(2'2) = \delta\Sigma(22')/\delta G(1'1). \tag{78}$$

This in turn implies that another functional exists such that

$$\Sigma(11') = \delta\Phi/\delta G(1'1). \tag{79}$$

Eqs. (76) and (79) together imply that a functional $W_U[G]$ can be constructed such that Eq. (74) for G is a stationary solution:

$$\delta W_U[G]/\delta G = 0 \quad \text{for a fixed value of } U. \tag{80}$$

$$\text{i.e., } W_U[G] = \Phi[G] - \mathrm{tr}(G_0^{-1} - U)G - \mathrm{tr}\ln(-G), \tag{81}$$

$$\text{provided } \delta\Phi/\delta G = \Sigma.$$

This is the central result of Baym. When G is fixed and U varied, we obtain $\delta W_U[G]/\delta U = G$ as required by Eq. (76). Baym also showed that in equilibrium, W is the logarithm of the partition function – which is a variant of Luttinger's variational principle! We may remark here that a similar theory was developed by Klein [25] for the ground state of the many particle system corresponding to the zero-temperature limit of the above theory and has been used by Casida [26] in the DFT context, which we will discuss later. In Appendix A, we show that this stationary principle is the same as the one given in Sect. 2.4 and the two action functionals are different representations of the same quantum action. Consequences of this important result are also given. We will now express the Green function in LQD and reformulate the Baym theory in L-space. For the purposes of mathematical manipulation in what follows, it is convenient to recall Eqs. (32) and (33). We define the Green function in L-space as follows:

$$\mathscr{G}_{t_0}(\mathbf{r}t, \mathbf{r}'t') = \frac{1}{i\hbar}\langle\langle 1 | T(\Psi(\mathbf{r}tt_0)\Psi^\dagger(\mathbf{r}'t't_0)) | D(t_0)\rangle\rangle, \tag{82}$$

$$\Psi(\mathbf{r}tt_0) = \begin{pmatrix} \hat{\psi}(\mathbf{r}tt_0) \\ \tilde{\psi}^\dagger(\mathbf{r}tt_0) \end{pmatrix},$$

where

$$\Psi^\dagger(\mathbf{r}'t't_0) = (\hat{\psi}^\dagger(\mathbf{r}'t't_0)\,\tilde{\psi}(\mathbf{r}'t't_0)). \tag{83}$$

The operators Ψ and $\Psi^\dagger$ obey the usual equal time anticommutation relations. The time-dependence of the field operators appearing here is due to the Heisenberg representation in the L-space. In view of the foregoing development which parallels the traditional Schrödinger quantum theory we may recast the above Green function in terms of the interaction representation in L-space. This leads to the appearance of the S-matrix defined only for real times. We will now indicate the connection of the above to the closed-time path formulation of Schwinger [27] and Keldysh [28] in H-space. Equation (82) can be explicitly written out in the form of a matrix

$$\mathcal{G}_{t_0}(rt, r't') = \frac{1}{i\hbar}\begin{pmatrix} \langle T\hat{\psi}(rtt_0)\hat{\psi}^\dagger(r't't_0)\rangle & \langle T\hat{\psi}(rtt_0)\tilde{\psi}(r't't_0)\rangle \\ \langle T\tilde{\psi}^\dagger(rtt_0)\hat{\psi}^\dagger(r't't_0)\rangle & \langle T\tilde{\psi}^\dagger(rtt_0)\tilde{\psi}(r't't_0)\rangle \end{pmatrix}. \quad (84)$$

Here $\langle \cdots \rangle$ has the same meaning as in Eq. (36). This can be expressed in terms of the corresponding well-known correlation functions, $G^>$ and $G^<$ defined in the H-space:

$$\mathcal{G}_{t_0}(rt, r't') = \begin{pmatrix} G^c(rtt_0;r't't_0) & -G^<(rtt_0;r't't_0) \\ G^>(rtt_0;r't't_0) & -G^{ac}(rtt_0;r't't_0) \end{pmatrix},$$

$$G^c(rtt_0;r't't_0) = \theta(t-t')G^>(rtt_0;r't't_0) + \theta(t'-t)G^<(rtt_0;r't't_0), \quad (85)$$

$$G^{ac}(rtt_0;r't't_0) = \theta(t-t')G^<(rtt_0;r't't_0) + \theta(t'-t)^>(rtt_0;r't't_0).$$

This is precisely the same as the non-equilibrium Green functions defined by Schwinger and Keldysh. It is worth pointing out at this stage that the main advantages of using the LQD approach advocated here are: (1) it allows us to formulate the non-equilibrium time-dependent functional theory and (2) the use of real times in place of closed-time paths is more convenient for calculating self-energy etc.

In order to reformulate the Φ-derivable formalism of Baym in the L-space, we need to introduce an external potential which is nonlocal in both space and time in the Hamiltonian

$$\mathrm{H}^{\mathrm{ext}} = \iint \Psi^\dagger(r't')\mathcal{U}(r't';rt)\Psi(rt)\,dr\,dt\,dr'\,dt' \quad (86)$$

in the notation of Eq. (83). Here, corresponding to the components of Ψ and $\Psi^\dagger$, $\mathcal{U}$ is a 2 by 2 matrix. The equation governing the time evolution of the Green function defined in Eq. (84) in the presence of the above TD-potential is obtained by the usual procedure. This can be recast in the form

$$\mathcal{G}^{-1} = \mathcal{G}_0^{-1} - \mathcal{U} - \Sigma, \quad (87)$$

where Σ is the non-equilibrium self-energy matrix. This equation can be deduced by the stationarity with respect to $\mathcal{G}$ of a functional defined by

$$W_U[\mathcal{G}] = \Phi[\mathcal{G}] - \mathrm{tr}(\mathcal{G}_0^{-1} - \mathcal{U})\mathcal{G} + \mathrm{tr}\ln(-\mathcal{G}), \quad (88)$$

where tr is as defined as before, provided the self-energy Σ is given by the functional derivative of Φ. For completeness, we may point out that by treating $\mathcal{U}$ as the functional variable of W in Eq. (88) we obtain the $\mathcal{G}$ as the functional derivative of W with respect to $\mathcal{U}$. In order to establish this, we consider the definition of the Green function given in Eq. (84). This enables us to write another equivalent expression for $W_U[\mathcal{G}]$ in view of Eq. (86) in the form of a generating functional of $\mathcal{G}$ from which the result stated above follows. Thus

$$W_U[\mathcal{G}] = \ln Z(\beta) + \ln\left\langle\!\left\langle 1 \left| T\left(\exp - \frac{i}{\hbar}\int_{t_0}^{t_f} dt'\bar{H}_I^{\mathrm{ext}}(t')\right)\right| D_{t_0}(t_0)\right\rangle\!\right\rangle, \quad (89)$$

where $Z(\beta)$ is the partition function in the absence of the external field. Moreover we have introduced the notation

$$\bar{H}_I^{\text{ext}}(t) = \left(\exp -\frac{i}{\hbar} t\bar{H}_{t_0}\right)\bar{H}^{\text{ext}}\left(\exp\frac{i}{\hbar} t\bar{H}_{t_0}\right). \tag{90}$$

We observe that in the case of an equilibrium situation, the external field is taken to be time-independent and the time integrals go over the usual complex time domain, $(0, -i\beta)$, and Eq. (89) then reduces to the logarithm of the partition function of the system including the external potential, as before.

We are now in a position to reformulate the Baym scheme in LQD theory and develop a new TD-functional approach. In the traditional pure-state TD-functional theory, one often expresses the stationarity condition equivalently as a one-particle TD-Schrödinger-like equation for some newly defined wave functions. This could of course be recast into an effective one-particle Green function equation with the effective potential appearing here as a self-energy functional. In many-body theory, Baym developed the powerful formalism called Φ-derivable theory described above which maintains all the conservation laws. As shown here, this is based on a rigorous stationary principle involving the one-particle Green function of the system. Here we shall show that a general functional method can now be constructed in a manner analogous to the TD-functional approach. This has the merit that the self-energy deduced in this way is in general complex unlike in the earlier versions of the TD-functional theory. In the LQD, the one-particle Green function, $\mathcal{G}$, is a matrix as was explained earlier. With that understanding and the attendant changes in the notation, we now proceed to develop the Baym formalism in an LQD framework.

The Baym functional $W_{\mathcal{U}}[\mathcal{G}]$ in LQD is now defined as

$$W_{\mathcal{U}}[\mathcal{G}] = \Phi[\mathcal{G}] - \text{tr}\{\mathcal{G}_0^{-1}\mathcal{G} - \ln(-\mathcal{G}) - \mathcal{U}\mathcal{G}\} \tag{91}$$

and is stationary with respect to variations of the one-particle Green function $\mathcal{G}$, provided the exact self-energy Σ is given by

$$\frac{\delta\Phi}{\delta\mathcal{G}} = \Sigma. \tag{92}$$

Here tr stands for the operation:

$$\text{tr}\mathcal{A} = \int dt \int d^3r \sum_i \mathcal{A}_{ii}(rt;rt^+) \tag{93}$$

where the time integration is along the real-time line, and the indices i denote the matrix structure of $\mathcal{A}$, when necessary. Here $\mathcal{U}$, a matrix, is a nonlocal external potential. For the initial thermodynamic equilibrium situation, the time integration runs along the line $(t_0, t_0 - i\beta)$, and $\mathcal{G}$ obeys the Martin-Schwinger time periodicity condition. $W_{\mathcal{U}}[\mathcal{G}]$ then corresponds to the free energy Ω of the system. It is interesting to point out that in this theory the stationarity of the

Baym functional is common to both the non-equilibrium and the equilibrium situations; the difference between the two now arises in the Martin-Schwinger periodicity property of $\mathcal{G}$ for determining the equilibrium Green function. This should be contrasted with the LQD approach given above where the stationarity property for the functional W changes to the minimum free energy functional Ω to obtain the initial equilibrium density matrix. We should recall that the self-energy arises from interactions among the electrons, so that Φ is merely another representation of the interaction contributions. This framework has become a basis for developing approximations in the theory of quantum transport in many diverse systems ranging from semiconductors to superconductors and their nanometric counterparts.

The purpose of the present work is to incorporate this framework into another, very successful formalism in computing the system properties of equally wide ranging magnitude, namely the non-equilibrium time-dependent functional method. Let us define a universal functional $F[\mathcal{G}]$ in LQD to be

$$F[\mathcal{G}] = \Phi[\mathcal{G}] - \mathrm{tr}\mathcal{G}_0^{-1}\mathcal{G} + \mathrm{tr}\ln(-\mathcal{G}). \tag{94}$$

Then

$$W_{\mathcal{U}}[\mathcal{G}] = F[\mathcal{G}] + \mathrm{tr}\mathcal{U}\mathcal{G}. \tag{95}$$

The stationarity property leads to the equation for the one-particle Green function:

$$\frac{\delta F[\mathcal{G}]}{\delta \mathcal{G}} + \mathcal{U} = 0, \quad \text{or} \quad \mathcal{G}^{-1} = \mathcal{G}_0^{-1} + \mathcal{U} + \frac{\delta \Phi}{\delta \mathcal{G}}. \tag{96}$$

Following Kohn, introduce a "non-interacting" system such that

$$\frac{\delta T_S[\mathcal{G}]}{\delta \mathcal{G}} = \mathcal{G}_0^{-1} - \mathcal{G}_S^{-1}, \tag{97}$$

and a one-particle potential $\mathcal{U}_S$ so that

$$\mathcal{G}_0^{-1} - \mathcal{G}_S^{-1} + \mathcal{U}_S = 0. \tag{98}$$

We observe that this corresponds to the "non-interacting" system but in the presence of a one-particle local potential so that from $\mathcal{G}_S$ one can obtain certain one-particle properties the same as for the interacting system, for example, the density of the system. Then write

$$W_{\mathcal{U}}[\mathcal{G}] = T_S[\mathcal{G}] + \tilde{F}[\mathcal{G}] + \mathrm{tr}(\mathcal{U}\mathcal{G}), \tag{99}$$

where

$$\tilde{F}[\mathcal{G}] = F[\mathcal{G}] - T_S[\mathcal{G}], T_S[\mathcal{G}] = \mathrm{tr}(\mathcal{G}_0^{-1}\mathcal{G}_S + \ln(-\mathcal{G}_S)). \tag{100}$$

Then we have for the one-particle equation a Kohn-Sham-type equation when the particles are mutually interacting:

$$\frac{\delta T_S[\mathcal{G}]}{\delta \mathcal{G}} + \frac{\delta \tilde{F}[\mathcal{G}]}{\delta \mathcal{G}} + \mathcal{U} = 0. \tag{101}$$

This gives the equation for $\mathscr{G}$ in the form;

$$\begin{aligned}&\mathscr{G}_0^{-1} - \mathscr{G}^{-1} + \mathscr{V}_{\text{eff}}[\mathscr{G}] = 0, \text{ where}\\&\mathscr{V}_{\text{eff}}[\mathscr{G}] = +\frac{\delta \tilde{F}[\mathscr{G}]}{\delta \mathscr{G}}.\end{aligned} \tag{102}$$

$\mathscr{V}_{\text{eff}}$ here is the self-consistent one-electron potential arising from the electron interactions. Models are constructed based on the suitable choice of the functional $\tilde{F}$, for example, Hartree or Hartree-Fock, or other forms depending on the scheme one wishes to adopt. Formally, $\mathscr{V}_{\text{eff}}$ is the same as the self-energy functional of the many-body theory but, as discussed in Baym's paper, approximation schemes are often employed obeying some basic requirements of the formal theory. The method given here based on LQD folds in the full many-body formalism, including the finite temperature theory, into a Kohn-type scheme. Also it incorporates the Schwinger [27] and Keldysh [28] method when the path-ordered contour on the time path is employed.

In the above development, for the sake of simplicity of presentation, we discussed the electron problem alone. Inclusion of ionic motion and electromagnetic fields brings in new features of the interacting fields. In our earlier works we developed the appropriate generalizations of the TD-functional theory to incorporate such interacting fields. We will now present the corresponding results in the Green-function language thus completing the original program stated in the Introduction.

$$\mathscr{G}_0^{-1} - \mathscr{G}^{-1} + \mathscr{V}_{\text{eff}}[\mathscr{G}, \mathscr{D}, \mathscr{D}^{(\text{em})}] = 0,$$

where

$$\mathscr{V}_{\text{eff}}[\mathscr{G}, \mathscr{D}, \mathscr{D}^{(\text{em})}] = \mathscr{U} + \frac{\delta \tilde{F}[\mathscr{G}, \mathscr{D}, \mathscr{D}^{(\text{em})}]}{\delta \mathscr{G}}. \tag{103}$$

$$\mathscr{D}_0^{-1} - \mathscr{D}^{-1} + \mathscr{K}_{\text{eff}}[\mathscr{G}, \mathscr{D}, \mathscr{D}^{(\text{em})}] = 0,$$

where

$$\mathscr{K}_{\text{eff}}[\mathscr{G}, \mathscr{D}, \mathscr{D}^{(\text{em})}] = \frac{\delta \tilde{F}[\mathscr{G}, \mathscr{D}, \mathscr{D}^{(\text{em})}]}{\delta \mathscr{D}}. \tag{104}$$

$$(\mathscr{D}^{(\text{em})-1})_{ij} = (\mathscr{D}_0^{(\text{em})-1})_{ij} + (\Sigma_{\text{eff}})_{ij},$$

where

$$(\Sigma_{\text{eff}})_{ij} = \frac{\delta \tilde{F}[\mathscr{G}, \mathscr{D}, \mathscr{D}^{(\text{em})}]}{\delta \mathscr{D}_{ij}^{(\text{em})}}. \tag{105}$$

$\tilde{F}$ is given by

$$\begin{aligned}\tilde{F}[\mathscr{G}, \mathscr{D}, \mathscr{D}^{(\text{em})}] &= \Phi[\mathscr{G}, \mathscr{D}, \mathscr{D}^{(\text{em})}] - \text{tr}(\mathscr{G}_0^{-1}\mathscr{G} - \ln(-\mathscr{G}))\\&- \text{tr}(\mathscr{D}_0^{-1}\mathscr{D} - \ln(-\mathscr{D}))\\&- \text{tr}(\mathscr{D}_0^{(\text{em})-1}\mathscr{D}^{(\text{em})} - \ln(-\mathscr{D}^{(\text{em})}))\\&- T_S[\mathscr{G}] - T_i[\mathscr{D}] - T_f[\mathscr{D}^{(\text{em})}].\end{aligned}$$

$\mathcal{K}_{\text{eff}}$ here is the effective "force constant" whereas the $\mathcal{V}_{\text{eff}}$ is the effective one-electron potential containing contributions from all the interactions. Here we have introduced the dyadic Green function associated with the electromagnetic field components, which we here denote by the symbol $\mathcal{D}^{(\text{em})}$ with the corresponding self-energy dyadic Σ so that the corresponding equation determining $\mathcal{D}^{(\text{em})}$ in the presence of other fields is given by Eq. (105). Equations (103)–(105) above are all coupled and should be solved self-consistently.

This method may also be used to derive a set of self-consistent mutually coupled equations for the appropriate Green functions associated with the superconducting electrons (counterpart of Eliashberg equations), the ions, and the electromagnetic fields. These equations are not written down here as they follow by an analysis of the same form given above for the electron-ion-electromagnetic field system. The essential change needed is in the electron Green function which now is a larger matrix to incorporate the Cooper pairing and so must incorporate spin space thus accommodating the Gorkov anomalous terms. With that understanding, formally the equations for the electrons, ions, and electromagnetic fields appear to be in the form similar to those given as above. For details of this derivation one may refer to [16].

Table 2. Implications of Φ-derivable theory to the proposed TD-functional method

TD-functional scheme	Φ-derivable scheme
(1) Dirac–Frenkel stationary action principle involving the functional W determines the exact system density matrix.	(1) A stationary principle involving an effective action functional $W_{\mathcal{U}}$ determines the exact one-body Green function of the system $\mathcal{G}$.
(2) Mapping theorem: (Legendre transformation) $a_\mu^{\text{ext}} \to J_\mu$ makes W a functional of J_μ.	(2) Mapping theorem: (Legendre transformation) *nonlocal* $\mathcal{U} \to \mathcal{G}$ makes $W_{\mathcal{U}}$ a functional of $\mathcal{G}$.
(3) The initial equilibrium condition for the non-equilibrium problem in (1) is determined from the minimum of the free energy functional Ω which is also a functional of J_μ. This serves as the initial condition for determining the non-equilibrium TD-averages.	(3) The initial equilibrium condition for the non-equilibrium problem in (1) is determined from the minimum of the free energy functional of the exact one-body equilibrium $\mathcal{G}$. This serves as the initial condition for determining the non-equilibrium $\mathcal{G}$ required in (1).
(4) Stationary condition in view of (2) leads to an effective equation for the Green function $\mathcal{G}^s$ of the system with a local self-energy determined via a functional $B_1[J_\mu]$: $\Sigma^s = \gamma_\mu \dfrac{\delta B_1}{\delta J_\mu}$.	(4) Stationary condition in view of (2) leads to an equation for the Green function $\mathcal{G}$ of the system with the exact self-energy, Σ, related to a functional Φ: $\Sigma = \dfrac{\delta \Phi}{\delta \mathcal{G}}$. This is called the Φ-derivability condition.
(5) The functional $B_1[J_\mu]$ is set up by physically motivated approximations.	(5) The functional Φ is set up by using diagrammatic analysis of a physically motivated choice of processes.

In Table 2 we exhibit the correspondence between the LQD-based TD-functional scheme and the Φ-derivable, effective action functional scheme.

3 Calculation Schemes

In order to see the implications of this theory, we first re-express Theorem III of the TD-functional scheme in the Green-function language and identify the corresponding self-energy terms. Once this is done, we then proceed to outline the required generalization of the TD-functional scheme to incorporate diagrammatic methods into a self-consistent scheme. This is facilitated by the following definitions of the Green functions:

$$\begin{aligned}
&\mathscr{G}^s(\boldsymbol{r}t;\boldsymbol{r}'t') = \text{Green function associated with Eq. (58)},\\
&\mathscr{D}^s(\ell\kappa\alpha t;\ell'\kappa'\alpha't') = -\frac{i}{\hbar}\frac{\delta R(\ell\kappa\alpha t)}{\delta F^{\text{ext}}(\ell'\kappa'\alpha't')},\\
&\mathscr{D}^{(\text{em})s}_{ij}(\boldsymbol{r}t;\boldsymbol{r}'t') = \frac{c}{4\pi}\frac{\delta A_i(\boldsymbol{r}t)}{\delta j^{\text{ext}}_j(\boldsymbol{r}'t')}.
\end{aligned} \tag{106}$$

We have denoted the Green functions here with a superscript s to distinguish them from those defined in the Baym formulation. After some algebra, we find the following identifications:

$$\begin{aligned}
&\mathscr{V}_{\text{eff}} = \gamma_\mu \frac{\delta B_1}{\delta J_\mu}\delta(\boldsymbol{r}-\boldsymbol{r}')\delta(t-t') \rightarrow \mathscr{U} + \frac{\delta\tilde{F}}{\delta\mathscr{G}},\\
&\mathscr{K}_{\text{eff}} = \frac{\delta^2 B_1}{\delta R\delta R} \rightarrow \frac{\delta\tilde{F}}{\delta\mathscr{D}},\\
&(\Sigma_{\text{eff}})_{ij} = \Pi^{(\text{em})}_{ij} + \frac{\delta^2 B_1}{\delta A_i \delta A_j} \rightarrow \frac{\delta\tilde{F}}{\delta\mathscr{D}^{(\text{em})}_{ij}}.
\end{aligned} \tag{107}$$

Here the electronic polarization tensor is the electromagnetic self-energy due to electron-electromagnetic interaction and is defined by

$$\Pi^{(\text{em})}_{ij} = \frac{\delta J_i}{\delta A_j}. \tag{108}$$

This tensor, in a lowest order calculation, is the familiar relativistic Lindhard-type function. The other self-energies appearing in the above equations expressed in terms of the functional B_1 are in the functional scheme, whereas those expressed in terms of $\tilde{F}$ have their origins in the Φ-derivable scheme. In both formulations, the contributions due to electronic, ionic, and electromagnetic interactions are incorporated into the respective functionals mentioned above. A direct comparison of the two procedures should be illuminating to see how one could employ the Baym scheme in a TD-functional form in order to develop a self-consistent approach to interacting, non-equilibrium problems. First, we enumerate the similarities and then we exhibit a possible connection and indicate application of the method. We choose for the purposes of illustration, to compare the case of an electron system alone. Similar statements hold in general.

4 Numerical Simulation Techniques

We believe that the formal methods given here will be useful in developing foundations for future numerical simulations such as molecular dynamics and Monte Carlo techniques in addressing the problems of non-equilibrium interacting systems, which may also be nanometric. Recently the case of bulk sodium through its melting point has been studied by such methods, employing heuristic physically motivated schemes. Schemes for the "dynamical" simulation of matter in a pure state from 'first principles' are at present based on the Car-Parrinello (CP) optimization method which is founded on a fictitious Lagrangian molecular dynamics framework coupled with density-functional theory. At first, the zero temperature Kohn-Sham procedure was used by introducing a fictitious Lagrangian to generate the dynamics, and, more recently, the finite temperature density functional within a similar fictitious Lagrangian was employed.

Our formalism makes contact with and provides a firm physical and mathematical basis for time-dependent dynamical simulation procedures envisaged by the Car-Parrinello approaches by replacing the fictitious Lagrangian by a precise action principle, as we will show presently. The starting point for making this important connection is our Theorem III. For purposes of illustration here, we focus on the electron-ion system. We thus consider the equations for describing the motion of the electrons as in a Kohn-Sham-type description which can be recast in the usual non-relativistic form and another describing the ionic motion, and the input self-consistent entities entering the formalism. A central feature of these coupled electron-ion equations is that the source of the effective potential for the electrons and the effective force on the ions arise from a single functional. These should be contrasted with the equations studied by Theilhaber [18] who in an attempt to replace the fictitious Lagrangian in the CP method, obtained equations for electrons and ions similar to ours but limited to equilibrium. Apparently he did not realize the presence of a single self-consistent source functional of electron density and ion positions as obtained above, but used a mean-field form for the ion part of the functional. His aim also was to produce a fully self-consistent description of electron-ion motion within a time-dependent density-functional theory which goes beyond adiabatic approximation. Our framework thus leads to the most general form of the equations appearing in such a formulation which meets these requirements. To connect to the CP method, we observe that the CP fictitious kinetic energy functional corresponds to a "penalty" functional to account for its deviation from the true ground state. The corresponding fictitious Lagrangian is a sum of those describing the electrons and the ions which could be taken to be an approximation for the "non-interacting" actions of electrons and ions, which in the CP framework is to be identified with the Kohn-Sham energy functional for fixed ion positions. Moreover, the electron "deviation" field in the resulting time-dependent set-up is treated in a classical sense, i.e. it contains a second-order differential in time for describing the electron field with an arbitrary coefficient which is identified to be a mass parameter for this electron field. The aim of the CP annealing algorithm is to provide an adiabatic evolution of the electronic ground state in the molecular dynamic simulation of the zero temperature theory and the equilibrium ensemble of the finite temperature theory. We observe that our formalism can be adopted to obtain CP as a first approximation for describing the adiabatic evolution of the electronic ground or thermodynamic equilibrium state in a general way as well as pointing a way to more general treatment of real dynamical situations by exhibiting the places where approximations are made to reduce to the CP algorithm.

To illustrate our self-consistent theory involving the electrons and electromagnetic fields, we may mention the new experimental findings in the field of ultra-short laser pulses of very high intensity impinging on atoms. The electric field produced in this case may be of the same order as that caused by the atomic nucleus, and thus traditional perturbation theory is inapplicable to understand this type of phenomena. Ulrich and Gross have recently used the time-dependent density functional theory of Runge-Gross to construct a time-dependent optimized potential for describing the electrons in this case but do not examine the back action on the electric field of the laser. Our framework is admirably suited to deal with this situation as it goes beyond the Runge-Gross method as is clear from our development given here. Thus it should be clear that the construction of the B_1-functional is of paramount importance in the scheme proposed here and the Baym-scheme should be used in this context. With the description of the above two important cases as a subset of several other problems of importance requiring a framework such as ours, we believe that we have established a formal basis for attacking such problems. Much work needs to be done to make this into a computationally viable theory which would take us beyond the present CP-type procedures, with the CP-approaches may be considered as a first step in this direction.

Our aim now is to develop phenomenology to suit the physical situations at hand so as to facilitate the setting up of the required functional of the new TD-theory. (In the stationary state functional theory, the uniform electron gas phenomenology was of importance as a vehicle for theoretical investigation.) In order to gain some insight into the central functional mentioned above, we have developed another theory in parallel to the functional theory described above addressing the same problem within the LQD framework. This is the corresponding generalization of the Baym Φ-derivable method in the many-body theory. It involves a functional of the one-particle electron Green function, the Green function associated with the ion motion, and the dyadic Green function associated with the electromagnetic fields, now expressed in LQD. This theory is also based on a stationary principle and when our TD-functional theory is recast in the Green function language, we observe the similarities between the two methods. This suggests using the diagrammatic methods in Green function theory in setting up suitable functionals needed in the theory.

5 Liouvillean Quantum Transport Theory and Applications

We will now develop the transport equations in L-space from the above Green functions. Following the Keldysh approach in H-space, the transport equations for non-equilibrium plasmas and radiation have been given by DuBois [29]. A similar transport equation for a system of ions may be found in Kwok [30], which is based on the Green function associated with ion positions. In a separate paper [31], we will derive the appropriate transport equations for the coupled system of electrons, ions, and electromagnetic fields.

In terms of the familiar correlation functions, $G^{>}$, and $G^{<}$, the matrix equation for the "super"-Green function, $\mathscr{G}$, is exactly the same as the following expression

$$\mathscr{G} = \begin{pmatrix} G^{c} & -G^{<} \\ G^{>} & -G^{ac} \end{pmatrix}, \tag{109}$$

where G^{c} and G^{ac}, which can be expressed in terms of $G^{>}$ and $G^{<}$, are the chronological and anti-chronological Green functions respectively. Equation (109) is exactly the same as the non-equilibrium matrix Green function expression obtained by other authors [32], using the time contour formulation of Schwinger and Keldysh. Integro-differential transport equations for the matrix elements of $\mathscr{G}$, can be readily obtained from $\mathscr{G}^{-1}\mathscr{G} = \delta$ and its adjoint. We make use of the relations: $F^{<,>\dagger} = -F^{<,>}$, $F^{c\dagger} = -F^{ac}$, to obtain the transport equations for all the matrix elements of $\mathscr{G}$:

$$\begin{aligned} i\hbar(\partial/\partial t + \partial/\partial t')G^{>,<} = [-\hbar^2\nabla^2/2m + \varphi_{\text{eff}} + \text{Re}\Sigma^{r}, G^{>,<}] \\ + [\Sigma^{>,<}, \text{Re}G^{r}] + i\{A, \Sigma^{>,<}\}/2 \\ - i\{\Gamma, G^{>,<}\}/2, \end{aligned} \tag{110}$$

$$i\hbar(\partial/\partial t + \partial/\partial t')G^{c} = [-\hbar^2\nabla^2/2m + \varphi_{\text{eff}} + \Sigma^{c}, G^{c}] + G^{<}\Sigma^{>} - \Sigma^{<}G^{>}, \tag{111}$$

$$i\hbar(\partial/\partial t + \partial/\partial t')G^{ac} = [-\hbar^2\nabla^2/2m + \varphi_{\text{eff}} - \Sigma^{ac}, G^{ac}] + \Sigma^{>}G^{<} - G^{>}\Sigma^{<}, \tag{112}$$

where G^r and Σ^r represent the retarded Green function and its associated self-energy, and φ_{eff} is the effective potential. It is clear from the last two equations that the term $G^{<}\Sigma^{>} - \Sigma^{<}G^{>}$ and its counterpart describe effects beyond the finite-lifetime quasi-particle concept, and represent bonafide non-equilibrium scattering effects. These are similar to those occurring in the last two terms of Eq. (110) for $G^{>,<}$. The equation for $G^{>,<}$ is exactly identical to the Keldysh results [32], while the equations for G^c and G^{ac} also contain collision terms.

The "super" self-energy has formally the same functional form as that of the "zero-temperature" self-energy. In the L-space approach, each of the self-energy matrix elements is calculated using the equation of motion of the "hat" and "tilde" super operators, which is a straightforward application of quantum-field theoretical techniques. Similar transport equations are deduced [31] for the ion and electromagnetic fields from their respective Green functions and self-energies. The self-energies depend on all the field variables exhibiting the mutual interactions among the fields. Thus all the Green functions become mutually coupled, requiring thereby a self-consistent analysis.

For high speed applications, the methods developed here are suited for analyzing very-short-time quantum dynamics of quantized fields, in particular, for interacting quantized fields of electrons, ions (phonons), and electromagnetic fields (photons) in nanoelectronic systems, such as resonant tunneling devices, under varied conditions [33], exciton-polariton effects in multiple-quantum-well systems which may lead to novel coherent light sources and serve as an example of the coupled-field problem (phonon, electron, and photons) of interest in optoelectronics [34], electronic structure, quantum dot saturation spectroscopy as well as transport through quantum dots which are artificially structured semiconducting materials when subjected to intense terahertz laser beams (electron and electromagnetic field system) [35], etc. In all of these systems, self-consistency of the calculation is of great importance for understanding short-time dynamics and nonlinear behavior, and our formalism fits this requirement. In each case, we begin with an approximate but self-consistent self-energy expression which captures the basic physics of the problem at hand and proceeds by iterating the solution until convergence is achieved.

The very-short-time behavior of micro-cavity quantum electrodynamics for characterizing micro-laser arrays for high speed applications, where highly nonlinear coupling of the vacuum fluctuations and the atomic polarizations exist, requires the full time-dependent quantum treatment discussed here.

6 Summary and Conclusions

In this work we have developed a time-dependent functional theory of coupled interacting fields for non-equilibrium situations, employing Liouvillean quantum field theory. The fields considered here are those associated with electrons, ions, and electromagnetic fields, relevant to condensed matter physics. The relevant field operators of physical interest are the electron density, the ion coordinate, and the vector field of the electromagnetic field, whose average values are the functional variables. This then leads us to a multi-variable coupled nonlinear self-consistent equation for these quantities following from an action principle. A single functional of the variables plays a central role, providing the source for effective interaction and correlation among the fields. Setting up this functional for actual applications is crucial in future development of this theory into a calculable scheme. The spirit of this theory is thus similar to the traditional pure-state TD-density functional theory, except that the initial condition of the system being in thermodynamic equilibrium is taken into account in the present theory. To gain some insight into the central functional mentioned above, we have developed another theory in parallel to the functional theory described above, addressing the same problem within the LQD framework. This is a generalization of the Baym Φ-derivable, effective action functional method in many-body theory. It involves a functional of the one-particle electron Green function, the Green function associated with the ion motion, and the dyadic Green function associated with the electromagnetic fields. This theory also is based on a stationary principle and when our TD-functional theory is recast in the Green function language, we observe the similarities between the two methods. This suggests using the diagrammatic methods in Green function theory in setting up suitable functionals needed in the theory. Some numerical calculations comparing the two methods for the electronic structure problem have recently appeared in the literature in this connection [36, 37]. We are presently developing this procedure to address the problem of transport in nanometric systems where the interactions and correlations and the nature of the system require careful accounting of the various contributions. Such a system also provides a testing ground for studying non-equilibrium problems since they can now be fabricated and experimentally characterized.

It may be worthwhile to remark that the use of the ion-coordinate as a variable in the problem instead of ion density will be of great utility in examining the phonon spectrum, electron-phonon interaction, strong coupling superconductivity, etc. when the electron-phonon coupling is strong and must be treated on the same footing in a self-consistent manner. Up to the present time, the traditional density functional method used in determining the phonon properties is to perform several electronic structure and total energy calculations for various different ionic positions and deduce from them the phonon frequencies, etc.

A recent preprint [38] dealing with the chain buckling distortion in the high temperature superconductor, YBCO, is an example of such an analysis and which urgently calls for a new approach to the problem. Our method should be a viable alternate in the near future in addressing such problems. For this we need to provide the functional of two variables, the electron density and the ion coordinate.

We have also pointed out here the formal connection between our formalism and the existing numerical algorithms in special cases (CP-algorithm and time-dependent optimized potential) as well as avenues to go beyond these to include non-adiabatic processes. It should be stressed that unlike the existing theories, our framework is based on a stationary action principle, which facilitates incorporation of the initial constraint of thermodynamic equilibrium. This development is made feasible by working in a superspace formalism. This work thus provides a practical theoretical framework for studying the non-equilibrium statistical mechanics of systems initially in thermodynamic equilibrium.

Theoretical aspects of the time evolution of highly nonlinear interacting quantized field systems are outlined in this paper. Besides the aspect of self-consistency of solutions, specific techniques for solving the time-dependent equations such as the Floquet method for periodic time-dependent problems and transformation to an accelerated reference frame method must be incorporated. The Floquet method has recently been used [39] to study localization, low-frequency, and harmonic generation on a quantum double well driven by a strong laser. The study of wave functions in the presence of a time-dependent field [40], and transient phenomena, as well as the characterization of high speed response in resonant tunneling devices [41] have been reported using the accelerated reference frame method. Only recently, general important consequences of the use of the accelerated reference frame in formulating the time-dependent functional method [42] were discussed.

An important result in Appendix A establishes the equivalence of the two methods given in Sects. (2.4) and (2.5) by showing that the quantum action functional is expressed in two equivalent ways in terms of different representations. This demonstration has important consequences for future computer algorithms to be developed for the TD description of the many-body systems of current interest.

There is a large class of problems involving dynamically coupled fields of electrons (e), ions (i), and electromagnetic fields (f) with the thermodynamic equilibrium of the system as an initial condition, which may be subjected to TD external perturbations. Our formalism is essential to deal with these problems. It may be of interest to give a small set of examples of such problems: (A) change in phonons due to superconducting transition in high TC systems [43] (e–i); (B) self-consistent motion of H, D, or T in PdH, PdD, or PdT systems respectively [44] (e–i); (C) laser interactions with matter [45, 46] (e–f); (D) resonant tunneling device subject to TD bias [47] (e–f); and (E) dynamics of charge transport in both high- and low-TC superconductors [48] (e–i–f). The approaches proposed here also offers practical methods for attacking challenging problems in nanoelectronics and optoelectronics. It may be pointed out that there is a rapidly growing literature in the TD quantum chemistry where a non-adiabatic approach to electron–ion dynamics is essential [45, 50, 51]. For problems involving purely ionic motions as in ferroelectric systems [30, 52], an LQD approach is relevant and is given in [53].

Acknowledgements. This work was supported in part by the Office of Naval Research. Most of the material given here arose out of a series of three lectures on functional theory given by AKR at the Winter School on Some Recent Developments in Quantum Many-Body Problems, Indian Institute of Science, Bangalore, India and lectures on Quantum Transport by FAB given at the Research Center for Theoretical Physics, Jagna, Philippines, and at the University of San Carlos, Cebu, Philippines.

7 Appendix A: Equivalence of the Action Functional [Eq. (54)] and the Baym Functional [Eq. (89)] and its Consequences

In this appendix, we establish the important result that the TD stationary action principle used in the derivation of the TD-functional theory [Eq. (54)] is the same as the stationary principle for the effective action functional of Baym [Eq. (89)]. Thus the action functionals appearing in the two formulations are different representations of the same quantum action leading to two different optimization strategies. We give a version of the work of Jackiw and Kerman [54] (hereafter referred to JK) on this subject, adapted here to LQD. Consequences of this important result in the development of the functional theory presented here are given.

For simplicity of presentation of this important result, we consider, as in JK, the quantum theory of a field Φ in the Hilbert space generated by an external source function, $J(rt)$, governed by a Hamiltonian operator, H. Similar procedures are required to establish this for the fields of interest in this paper. As in Sect. 2.4, the total Hamiltonian is $H_t = H + H_t^{\mathrm{ext}}$, $H_t^{\mathrm{ext}} = \int d^3r J(rt)\Phi(r)$. Consider the stationary solutions of the "action" functional in LQD for this simplified case:

$$W(t_0, t_1) = 1/2 \int_{t_0}^{t_1} dt \langle\langle \Psi(t) | (i\hbar\partial_t - \bar{H}_t) | D_v(t) \rangle\rangle \tag{A.1}$$

with the normalization condition

$$\langle\langle \Psi(t) | D_v(t) \rangle\rangle = 1. \tag{A.2}$$

Introducing a Lagrange multiplier $w(t)$ to incorporate this constraint, we obtain the equations

$$\begin{aligned}(i\hbar\partial_t - \bar{H}_t)|D_v(t)\rangle\rangle &= w(t)|D_v(t)\rangle\rangle,\\ (i\hbar\partial_t - \bar{H}_t)|\Psi(t)\rangle\rangle &= w^*(t)|\Psi(t)\rangle\rangle.\end{aligned} \tag{A.3a,b}$$

Here it should be noted that $w(t)$ is the Lagrange multiplier which incorporates the special property of time in LQD. The physical action of our interest is obtained by taking

$$\langle\langle \Psi(t)| = \langle\langle 1|, \text{ and } |D_v(t)\rangle\rangle = \exp\left(\frac{i}{\hbar}\int dt' w(t')\right)|D(t)\rangle\rangle, \tag{A.4}$$

The solution to Eq. (3a) is contained in Eq. (A.4). Then Eqs. (A.2) and (A.3) become for $|D(t)\rangle\rangle$:

$$\langle\langle 1|D(t)\rangle\rangle = \exp -\left(\frac{i}{\hbar}\int dt' w(t')\right), \text{ and}$$
$$(i\hbar\partial_t - \bar{H}_t)|D(t)\rangle\rangle = 0. \tag{A.5}$$

Moreover, Eq. (A.1), for the physical action becomes

$$W = \int dt' w(t'). \tag{A.6}$$

Using the thermodynamic equilibrium as the initial condition, the second equation in Eq. (A.5) may be solved in the interaction representation and the first equation of Eq. (A.5) may then be expressed in the form

$$\langle\langle 1|D(t)\rangle\rangle = \left\langle\left\langle 1\left|T\exp -\left(\frac{i}{\hbar}\int dt' \bar{H}_I^{\text{ext}}(t')\right)\right|D_{\text{eq}}\right\rangle\right\rangle Z(\beta). \tag{A.7}$$

Thus we have established the equivalence of Eq. (54) with Eq. (89).

There are several important implications of this result: the TD functional theory and the Baym functional theory are versions of the many-body theory in that they invoke different trial functionals in obtaining the stationary solution of the same variational functional, namely, the action functional. The two theories explore the optimization of the action functional in terms of different functional variables: the TD functional employs density, current, ion-coordinates, etc. while the Φ-derivable theory employs the two-point functions such as the Green functions associated with electrons, ions, etc. The above demonstration shows that different optimization strategies may be employed to explore the stationarity of the action functional.

The Berry phase is not explicitly found at first sight in a time-dependent density functional theory. From Eq. (A.5) we find it quite generally resides in the action functional of the theory. It lies buried as the sum over the occupied states of the individual phases associated with the time-dependent Kohn-Sham orbitals. From Eqs. (A.1) and (A.5), the Berry phase may be identified by writing the physical action functional as a sum of two quantities:

$$W \equiv W_B + W_D, \text{ where}$$
$$W_B = \frac{1}{2}\int dt \langle\langle 1|i\hbar\partial_t|D(t)\rangle\rangle; \tag{A.8}$$
$$W_D = -\frac{1}{2}\int dt \langle\langle 1|\bar{H}_t|D(t)\rangle\rangle.$$

In Eq. (A.8), W_B is the Berry phase and W_D is the dynamical phase. Details of this aspect of the action functional are given elsewhere [55].

We may thus conclude that the quantum action functional contains in it many important features of the physics of the system. This opens up the possibilities of computing various dynamical quantities of physical interest.

8 References

1. Hohenberg P, Kohn W (1964) Phys Rev 136: B864
2. Kohn W, Sham LJ (1965) Phys Rev A140: 1133
3. Rajagopal AK (1980) Adv Chem Phys 41: 59
4. Callaway J, March NH (1984) Adv Solid State Phys 38: 135
5. Dreizler RM, Gross EKU (1990) Density Functional Theory. Springer Berlin Heidelberg, New York
6. Lang ND (1973) Adv Solid State Phys 28: 225
7. Parr RG, Yang W (1989) Density-Functional Theory of Atoms and Molecules. Oxford University Press, Oxford
8. Ramana MV, Rajagopal AK (1983) Adv Chem Phys 54: 231
9. Engel E, Müller H, Speicher C, Dreizler RM (1995) in NATO-ASI
10. Gupta U, Rajagopal AK (1982) Physics Reports 87: 259
11. Dandrea RG, Ashcroft NW, Carlsson AE (1986) Phys Rev B34: 2097
12. Rajagopal AK, Buot FA (1995) Phys Rev A51: 1770
13. Rajagopal AK, Buot FA (1995) Phys Rev B15: 6769
14. Runge E, Gross EKU (1984) Phys Rev Letts 52: 997
15. Gross EKU, Kohn W (1990) Adv Quantum Chemistry, 27: 225
16. Rajagopal AK and Buot FA (1994) Phys Lett 195: 312 and (1995) Phys Rev A51: 1883; see also (1995) Int J Qu Chem 56: 389
17. Buot FA, Rajagopal AK (1993) Phys Rev Lett B48: 17217
18. Car R, Parrinello M (1985) Phys Rev Lett 55: 2471
19. Theilhaber J (1992) Phys Rev B46: 12990
20. Grumbach MP, Hohl D, Martin RM, Car R (1994) J Phys: Cond Matt 6: 1999
21. Umezawa H (1993) Advanced Field Theory, American Institute of Physics, New York
22. Schmutz M (1978) Z Phys B30: 97
23. Kaku M (1993) Quantum Field Theory. Oxford University Press, New York, Ch. 4
24. Baym G (1962) Phys Rev 127: 1391
25. Klein A (1960) Phys Rev 121: 950
26. Casida ME (1995) Phys Rev A51: 2005
27. Schwinger J (1961) J Math Phys 2: 407
28. Keldysh LV (1965) Soviet Phys JETP 20: 1018
29. Dubois DF (1967) Lectures in Theoretical Physics. Gordon and Breach, New York, p 469
30. Kwok PCK (1967) Adv Solid State Phys 20: 213
31. Buot FA, Rajagopal AK (1995) (in preparation)
32. Buot FA (1993) Phys Rep 234: 73
33. Buot FA, Rajagopal AK (1995) Material Sc. and Eng. B: Solid State Materials for Advanced Technology 35: 303
34. Jacobson J, Pau S, Cao H, Björk G, Yamamoto Y (1995) Phys Rev A51: 2542
35. Hawrylak P, Schulz PA, Palacios JJ (1995) Solid St Comm 93: 909
36. Serene JW, Hess DW (1992) in: Recent Progress in Many-Body Theories 3: 469
37. Deisz JJ, Eguiluz AG, Hanke W (1993) Phys Rev Lett 71: 2793
38. Pickett WE, Cohen RE, Krakauer H (1994) (preprint)
39. Bavli R, Metiu H (1993) Phys Rev A47: 3299
40. Truscott WS (1993) Phys Rev Lett 70: 1900
41. Buot FA, Rajagopal AK (1993) Phys Rev B48: 17217
42. Dobson JF (1994) Phys Rev Lett 73: 2244; Vignale G (1995) ibid 74: 3233; Rajagopal AK, Vignale G (1995) unpublished
43 Zeyher R (1991) Phys Rev B44: 9596
44. Klein BM, Cohen RE (1992) Phys Rev B45: 12405
45. Faisal FHM (1987) Theory of Multiphoton Processes. Plenum Press, New York; see also Cohen-Tannoudji C, Dupont-Roc J, Grynberg G (1989) Photons and Atoms. Wiley, New York; various articles on strong-field laser interaction especially by Faisal FHM. p. 27 and by Shih-I Chu p. 57 (1991) in: Radiation Effects and Defects in Solids. Part 1, 122–123
46. Ullrich CA, Grossman UJ, Gross EKU (1994) Phys Rev Lett 74: 872
47. Buot FA, Jensen KL (1990) Phys Rev B42: 9429

48. Wacker O-J, Kummel R, Gross EKU (1994) Phys Rev Lett 73: 2915
49. Deumens E, Diz A, Longo R, Öhrn Y (1994) Rev Mod Phys 66: 917
50. Broeckhove J, Lathouwers L (eds) (1992) Time-Dependent Quantum Molecular Dynamics. Plenum Press, New York
51. Allen RE (1994) Phys Rev B50: 18629
52. Singh Y (1991) Phys Reports 207: 351
53. Rajagopal AK, Buot FA (1994) Phys Rev E50: 721
54. Jackiw R, Kerman A (1979) Phys Letts A71: 158
55. Rajagopal AK (1995) (in preparation)

Author Index Volumes 151–181

Author Index Vols. 26–50 see Vol. 50
Author Index Vols. 51–100 see Vol. 100
Author Index Vols. 101–150 see Vol. 150

The volume numbers are printed in italics

Springer-Verlag and the Environment

We at Springer-Verlag firmly believe that an international science publisher has a special obligation to the environment, and our corporate policies consistently reflect this conviction.

We also expect our business partners – paper mills, printers, packaging manufacturers, etc. – to commit themselves to using environmentally friendly materials and production processes.

The paper in this book is made from low- or no-chlorine pulp and is acid free, in conformance with international standards for paper permanency.

SPRINGER NATURE

GPSR Compliance

The European Union's (EU) General Product Safety Regulation (GPSR) is a set of rules that requires consumer products to be safe and our obligations to ensure this.

If you have any concerns about our products, you can contact us on ProductSafety@springernature.com

In case Publisher is established outside the EU, the EU authorized representative is:

Springer Nature Customer Service Center GmbH
Europaplatz 3
69115 Heidelberg, Germany

The manufacturer's authorised representative in the EU is Springer Nature Customer Service Centre GmbH, Europaplatz 3, 69115 Heidelberg, Germany. If you have any concerns regarding our products, please contact ProductSafety@springernature.com

Printed and bound by CPI Group (UK) Ltd, Croydon, CR0 4YY
22/09/2026
02221064-0008